QUANTUM THEORY

Density, Condensation, and Bonding

QUANTUM THEORY

Density, Condensation, and Bonding

Mihai V. Putz, PhD
Associate Professor of Theoretical Physical Chemistry,
Laboratory of Structural and Computational Physical Chemistry;
Biology-Chemistry Department, West University of Timisoara, Romania

Apple Academic Press
TORONTO NEW JERSEY

Apple Academic Press Inc.
3333 Mistwell Crescent
Oakville, ON L6L 0A2
Canada

Apple Academic Press Inc.
1613 Beaver Dam Road, Suite # 104
Point Pleasant, NJ 08742
USA

First issued in paperback 2021

Exclusive worldwide distribution by CRC Press, a Taylor & Francis Group

ISBN 13: 978-1-77463-203-1 (pbk)
ISBN 13: 978-1-926895-14-7 (hbk)

Library of Congress Control Number: 2012935653

Library and Archives Canada Cataloguing in Publication

Quantum theory: density, condensation, and bonding/Mihai V. Putz.

Includes bibliographical references and index.
ISBN 978-1-926895-14-7
1. Quantum theory–Textbooks. I. Putz, Mihai V

QC174.12.Q37 2012 530.12 C2011-908704-9

Contents

List of Abbreviations

AIM	Atoms in molecule
BEC	Bose–Einstein Condensation
BH	Bose–Hubbard
COS	Chemical orthogonal space
DFT	Density functional theory
BEC–DFT	Density functional theory–Bose–Einstein condensation
EE	Electronegativity equalization
ELF	Electronic localization function
XC	Exchange-correlation
FH	Fermi–Hubbard
FB	Fermionic-bondonic
GGA	Generalized gradient approximation
GL	Ginzburg–Landau
GP	Gross–Pitaevsky
GS	Ground state
HSAB	Hard and soft acids and bases
HF	Hartree–Fock
HOMO	Highest occupied molecular orbital
HK	Hohenberg–Kohn
KS	Kohn–Sham
LYP	Lee, Yang, and Parr
LDA	Local density approximation
LUMO	Lowest unoccupied molecular orbital
MaxHD	Maximum hardness
MHP	Maximum hardness principle
MinEL	Minimum electronegativity
MO	Molecular orbital
NLSE	Non-linear Schrödinger equation
SC	Semi-classical
SE	Semiempirical
TF	Thomas–Fermi
2D	Two dimensions
2C	Two-point-central
VSEPR	Valence-shell electron-pair repulsion
VWN	Vosko, Wilk and Nusair

Preface

The book presents in a unitary manner the main actual theories of matter, namely the Density Functional Theory (DFT) for fermions, the Bose-Einstein Condensation (BEC) for bosons, and the Chemical Bonding as a special realization of the first two in the so called mixed fermionic-bosonic states. While the DFT became the main tool for quantum chemistry, being actually the major framework in which the structure of atoms and molecule are accurately computed based on first principles, the BEC models the condensation of bosons through their quantum fields as applied to perfect and diluted gases, to nowadays atomic bosons trapped in optical lattices. Both theories merged in revealing the quantum nature of chemical bonding as a joint characterization by ordinary wave function for bonding and anti-bonding states along a degeneracy driven by the specific order parameter of bosonic condensation of bondons – the quantum particles of chemical bonding. However, the bondonic picture of chemical bond is in detail exposed since constituting one of the recent quantum models of electronic pairing in molecules with a great potential in designing nano-structures carrying certain chemical information. Additionally, the consecrated approaches of chemical bonding such as the Born-Oppenheimer, Valence and Molecular Orbital ones are enriched by the most recent density kernel representations (in both Schrödinger and Dirac pictures) that combine the main density-functional with chemical reactivity ideas in providing the pairing localization across the bonding length in a geometrical intuitive manner, yet in close correspondence with classical molecular orbital approach. As such, the actual book presents in a didactical and consistent manner the modern and ultimately developed quantum theories rooting into the key concepts of density, condensation and bonding: this since, the density, although initially assessed to fermions on DFT, it turns to be the main vehicle of the BEC as well, while condensation, being specific to BEC, appears to have as a particular realization the chemical bonding phenomenon. As a whole, the book compiles, for the first time, the Density Functional Theory with Bose-Einstein Condensation and Chemical Bonding theories in a fresh introductory and novel perspective of many-body structural physical-chemistry.

Scientific University community is the main audience envisaged, either for undergraduate, graduate, and doctorate teaching and research interest. This especially since the modern curricula in either physical or chemical sciences considers license, master and doctorate as flowing steps in a single educational program. The book is therefore introducing the student in the modern theories of matter structure, for fermions (Part I, concerning with Density Functional Theory), for bosons (Part II, concerning with Bose-Einstein Condensation), and then for chemical bonding (Part III) being the last one viewed through considering both sides of matter (i.e., dealing with mixture of fermions and bosons) in explaining the nature of the chemical bond under the consecrated and ultimate quantum paradigms of molecular structure.

Therefore the book is divided in 3 parts, one for each level of studies, namely

PART I: PRIMER DENSITY FUNCTIONAL THEORY (suitable as the undergraduate-license Introduction to Density Functional Theory course in Physics or Chemistry or Natural Sciences faculties/departments)

PART II: PRIMER DENSITY FUNCTIONAL BOSE-EINSTEIN CONDENSATION THEORY (suitable as the graduate-master Introduction to Bose-Einstein Condensation course in Physics or Natural Sciences faculties/departments)

PART III: MODERN QUANTUM THEORIES OF CHEMICAL BONDING (suitable as the post-graduate, master or doctorate Quantum Structure of Molecules course in Chemistry or Natural Sciences faculties/departments)

The relevant subject areas are: Physical-Chemistry in general with a special focus on Structural Physical Chemistry. In this respect, the present book displays several particular features that make it highly desirable for a complete or modern physical-chemical university instruction:

- The book structure is like a succession of three linked courses, from under-graduate (Part I) to graduate (Part II) and post-graduate (Part III) levels in modern quantum theories of many-body systems;
- The book is concrete and direct to the main points on each presented theory, spanning reasonable equilibrated space, yet being didactically arranged as flowing from 4 chapters in the first part (for license lectures), 2 chapters in the second part (for master lectures), up to 3 chapters in the third part (for doctoral lectures) thus assuring the balance between time-line and pyramidal fundamental-to-advanced presentation as demanded by modern superior educational items;
- The book is coined on three main concepts - density, condensation and bonding, this way encompassing the main ideas of each part, chapters and the work as a whole; the first main concept is an intensive quantity (density) that is then employed as the main tool in treating and even linking the intriguing phenomena of condensation (in physics) and bonding (in chemistry) at various levels of comprehension and approximations (e.g. the main field approximation in BEC, the Thomas-Fermi approximation in DFT).
- All parts of the books contain the most celebrated and challenging theories of matter: Density Functional Theory – initially assigned for fermions, along the Bose-Einstein Condensation – originally modeling the bosons; they are then combined in providing, through employing the traditional quantum ideas of atoms-in-molecule, the ultimate (bondonic and fermionic-bosonic) theories of Chemical Bonding;
- The book is dynamic and projected as a textbook for XXI Century in which the foreground quantum information of fermionic with bosonic theories is essentially (deliberately not exhaustively) presented and employed in describing the complex phenomena of Nature: the many-body structures, many-body condensation and chemical bonding. This way, the XXI Century quantum education that allows for the future research and advancement in complex, novel and inter-disciplinary characterization of the micro- and mesoscopic matter (atoms, molecules, nanosystems) is assured within the united structural laws of fermions and bosons in Physical Chemistry.

However, there appears that the proper ordering of the main and related concepts of electronic density, bosonic condensation, and chemical bonding theories culminating with the ultimate author's reported bonding models within both the density functional and bosonic-bondonic condensation frameworks deserves be considered in a unitary textbook publication. It will certainly provide a fresh perspective on the actual curses on quantum theory of structure of physico-chemical systems and will show the students the way for future elaboration and discoveries towards the unification of the physical and chemical concepts of matter.

This and above items aim to inscribe the actual publishing project as an inter-theories textbook of XXI Century in quantum physical-chemistry!

After all, the present book has the potential of benefiting by forthcoming extended editions marking the progress achieved through this open line of united fermionic-bosonic description of the physical-chemical systems.

Nonetheless, author likes to address his sincere gratitute to prof. Haghi (the senior edior of Apple Academics for appreciating this project), Mr. Kumar (director of Apple Acedemics Canada) for professional cooperation in producing fine publishable volume from my initial manuscript. Romaniana Ministry of Education and Research is thanked for suporting the work on this volume within the project CNCS-UEFISCDI-TE16/2010-2013. International Cooperation Center of Free University of Berlin and German Academic Exchange Service (DAAD) are equally acknowledged for supporting over the years my repeated scientific research visits to Free University of Berlin where much of the inspiration for the main thema of this volume was founded out. Nova Science Publishing Inc. and Elsevier BSGV are given credit for their kind permission in non-exclusive reproduction of the author's originally contributions published by them, namely:

- Putz M.V. Conceptual Density Functional Theory: from Inhomogeneous Electronic Gas to Bose-Einstein Condensates. In: *CHEMICAL INFORMATION AND COMPUTATIONAL CHALLENGES IN 21ST*, Mihai V. Putz (Ed.), NOVA Science Publishers, Inc., New York, USA (2012). © Nova Science Publishers, Inc. Reprinted with permission from Nova Science Publishers, Inc. All Rights Reserved.
- Putz M.V. Hidden Side of Chemical Bond: The Bosonic Condensate. In: *ADVANCES IN CHEMISTRY RESEARCH. VOLUME 10*, James C. Taylor (Ed.), NOVA Science Publishers, Inc., New York, USA (2011), pp. 261-298. © Nova Science Publishers, Inc. Reprinted with permission from Nova Science Publishers, Inc. All Rights Reserved.
- Putz M.V. Fulfilling The Dirac Promises on Quantum Chemical Bond. In: *QUANTUM FRONTIERS OF ATOMS AND MOLECULES*, Mihai V. Putz (Ed.), NOVA Science Publishers, Inc., New York, USA (2011), pp. 1-20. © Nova Science Publishers, Inc. Reprinted with permission from Nova Science Publishers, Inc. All Rights Reserved.
- Putz M.V. "Chemical Action and Chemical Bonding", *Journal of Molecular Structure: THEOCHEM*, 900 (1-3) (2009) 64-70. © Elsevier. Reprinted with permission from Elsevier (http://www.elsevier.com). All Rights Reserved.

Much gratitude I owe to Prof. Dr. Hagen Kleinert and Privat.-Doz. Dr. Axel Pelster from Free University Berlin from whom I learn the fascination for Bose-Einstein condensation phenomenology, for their continuous hospitality in their groups and for already a decade of friendship. At last, but most important, special warm thanks are due to my wonderfull Family for understanding and enhancing the importance and efforts I spent for converting the scientific research achivements into an advanced educational discourse.

— Mihai V. Putz, PhD

PART I

PRIMER DENSITY FUNCTIONAL THEORY

1 Basics of Density Functional Theory (DFT)

CONTENTS

1.1 INTRODUCTION

Why another quantum theory of the matter? Why the Density Functional Theory (DFT)? Its pioneers and promoters, Walter Kohn, Robert G. Parr, and Axel Becke – just to name a few, affirm that modeling electronic density features in 3D real space produces much benefits in both conceptually understanding of bonding as well in electronic localization with a bonding impact for the concerned N-many-electronic systems. The present chapter aims to open a unitarily presentation of the quantum fundaments of DFT as well to prospect directions in electronic bonding, reactivity, bridging with the ultimate approach of bosons-bondons allowing further Bose–Einstein condensation. In DFT, the density and their combinations in the density functionals of the total energy plays a primordial role. It fulfills the N-contingency, assures the total energy minimization, influences the different levels of approximation, that is local density or gradient density frameworks, and controls the bonding through electronic reactivity principles of electronegativity, chemical hardness, and chemical action indices (Putz, 2008a, 2008b, 2008c, 2008d, 2008e; Putz, 2011a, 2011b, 2011c, 2011d). Studying the electronic density properties of electronic systems guaranties the universal treatment of whatever systems, no matter how rich in electrons is, from atoms, to molecules, to solids and aggregates.

1.2 THE DFT NECESSITY OR THE WAVE-FUNCTION WALL OF QUANTUM INFORMATION

In Walter Kohn's lecture, with the occasion of receiving his Nobel Prize in Chemistry (Kohn's Nobel Prize, 1998), back in 1998, for DFT theory (Hohenberg and Kohn,

1964; Kohn and Sham, 1965; Kohn et al., 1996), there was formulated a quite provoking assertion affirming that, *heuristically, the general eigen-wave-function* $\Psi(r_1,...,r_N)$ *associated to a system of N electrons fails to be a legitimate scientific concept when* $N \geq N_0 \approx 10^3$.

Nevertheless, this affirmation may be at any time turned in a theorem, eventually as *Kohn's zero DFT theorem*, with a proof following the van Vleck prescription of the so called "exponential wall", leaving with the applicability limits of the conceptually eigen-wave function of multi-electronics systems. However, before proceeding to demonstration, there must be noted that such reality limitation characterized by eigen-wave-function of the multi electronic-systems is not transferable at the quantum mechanics postulates, but providing an alternative quantum scheme, thus paralleling Schrödinger equation, however in a more generally integrated level.

The demonstration of the non-representability of the eigen-function for systems containing more than $N_0 \approx 10^3$ electrons involves two aspects: *the accuracy of representation* by using the eigen-wave function and the *possibility of measuring it* (Putz, 2008a).

Regarding the accuracy of the $\Psi(r_1,...,r_N)$ representation there is widely known that it associates with the density of probability smoothly approaching unity, written in the "liberal" form (according to Kohn):

$$\left|\Psi^*(r_1,...r_N)\Psi(r_1,...,r_N)\right| = 1 - \varepsilon \ , \ \varepsilon = 10^{-2} \tag{1.1}$$

Now, considering a collection of *N'* molecules the total density of probability of this multi-molecular system (and implicit a multi-electronic one) will consequently be:

$$\left|\left(\Psi^*\Psi\right)_1 ...\left(\Psi^*\Psi\right)_{N'}\right| = \left(1-\varepsilon\right)^{N'} \cong \exp(-\varepsilon N') = \exp\left[-10^{-2} N'\right] \tag{1.2}$$

For $N' = 10^3$ molecules in whatever aggregates, for example solids, clusters, supermolecules, or biological macro-molecules, the total density of probability will result from (1.2) as $\exp[-10] \approx 5\times10^{-5}$, meaning that it is determined with much less accuracy compared with the degree of individual eigen-function localization precision in (1.1). Since each molecular system has at least one electron there follows the threshold limit of $N_0 \approx 10^3$ electrons from where the lost in associated wave-function nature is recorded.

This result, relaying on the exponential form (1.2), justifies the title of "exponential wall" for the wave-function limitation.

Then, going to the measurable issue of such eigen-function, let us ask how many bits are necessary for recording its quantum dimension? Assume, again, the working wave-function $\Psi(r_1,...,r_N)$ for all the N electrons in a concerned system. The N electrons in system have a total of $3N$ space variables (in the configuration space); let us now assume an average of q bits necessary in measuring a single variable from the total of $3N$; there results a total information of bits for recording (storing) the total eigen-wave-function of the system.

$$B = q^{3N} \tag{1.3}$$

However, a simple evaluation of the dimension (1.3) shows that for a minimum of $q = 3$ bits/variable and for the above consecrated minimum limit of $N = N_0 \approx 10^3$ electrons in the system the total yield of necessary bits for recording is about of 10^{1000} order – a truly non-realistic dimension. This can be immediately visualized if one recalls that the *total number of baryons* (i.e., all fermions and elementary particles of protonic and neutronic type but not limited only to these) *estimated in entirely Universe* (summing up all existing atoms and free nuclei in the plasma state, but not only limited to these) *gives a result of about* 10^{80} order.

Definitely, the concept of eigen-wave function must be enlarged or modified in such a manner that the quantum description does not be blocked by the exponential wall: from where we can start? Firstly, as was exposed, the eigen-wave function in the configuration space multiplies in an exponential manner the variables accounting for the number and the position of the electrons; thus, the configuration space must be avoided. Then, the density of probability must be reformulated as such the exponential wall for a poly-electronic system is avoided while preserving the dependency of the total number of electrons N.

Fortunately, the above described conceptual project was unfolded in 1963 when Walter Kohn met in Paris (at École Normale Supérieure), during his sabbatical semester, the mate Pierre Hohenberg who was working at the description of the metallic alloys (specially the Cu_xZn_{1-x} systems) by using quantum traditionally methods of averaging crystalline periodic field. Studies of this type of problems often start from the level of the uniform electronic density referential upon which specific perturbation treatments are applied. From this point Kohn and Hohenberg made two crucial further steps in reformulation of the quantum picture of the matter structure: one referred at the electronic density, and another at the relation between electronic density with the externally applied potential on the electronic system; they were consecrated in the so called Hohenberg–Kohn (HK) theorems (DePristo, 1996; Ernzerhof, 1994; Hohenberg and Kohn, 1964; Keller, 1986). They were the fundaments of new emerging quantum DFT that mostly impacted the reformation of the quantum chemistry itself and its foreground principles of structure and transformation.

1.3 HOHENBERG-KOHN-SHAM THEORY

1.3.1 Hohenberg-Kohn Theorems

The first Hohenberg–Kohn (HK1) theorem gives space to the concept of *electronic density of the system* $\rho(\mathrm{r})$ in terms of the extensive relation with the N electrons from the system that it characterizes (Bamzai, 1981; Capelle, 2010; Filhais et al., 2003; Kohanoff, 2006; Koch and Holthausen, 2002; March, 1991; Parr and Yang, 1989; Putz, 2011d; Richard, 2004; Sholl and Steckel, 2009):

$$\int \rho(\mathrm{r})d\mathrm{r} = N \tag{1.4}$$

This relation as much simple it could appears stands as the decisive passage from the eigen-wave function level to the level of total electronic density:

$$\rho(\mathrm{r}) = N\int \Psi^*(\mathrm{r},\mathrm{r}_2,...,\mathrm{r}_N)\Psi(\mathrm{r},\mathrm{r}_2,...,\mathrm{r}_N)d\mathrm{r}_2...d\mathrm{r}_N \tag{1.5}$$

Firstly, last equation satisfies its previous one; this can be used also as simple immediate proof of the relation (1.4) itself. Then, the dependency from the 3N-dimensions of configuration space was reduced at 3-coordinates in the real space, physically measurable.

However, still remains the question: What represents the electronic density functional? Definitely, it neither represents the electronic density in the configuration space nor the density of a single electron, since the N-electronic dependency as multiplication factor of the multiple integral. What remains is that $\rho(r)$ *is simple the electronic density (of the whole concerned system) in* "r" space point. Such simplified interpretation, apparently classics, preserves its quantum roots through the averaging (integral) over the many electronic eigen-function $\Psi(\mathrm{r}_1,\ldots,\mathrm{r}_N)$. Alternatively, the explicit non-dependency of density on the wave function is also possible within the quantum statistical approach where the relation with partition function of the system (the global measure of the distribution of energetic states of a system) is mainly considered.

The major consequence of this theorem consists in defining of the total energy of a system as a function of the electronic density function in what is known as the density functional from where the name of the theory.

$$E[\rho] = F_{HK}[\rho] + C_A[\rho] \tag{1.6}$$

The terms of energy decomposition are identified as: the Hohenberg–Kohn density functional

$$F_{HK}[\rho] = T[\rho] + W[\rho] \tag{1.7}$$

viewed as the summed electronic kinetic $T[\rho]$ and inter-electronic repulsion (interaction, in general) $W[\rho]$, and the so called *chemical action* term (Putz, 2003, 2007a; Putz et al., 2001) being the only explicit functional of total energy.

$$C_A[\rho] = \int \rho(\mathrm{r})V(\mathrm{r})d\mathrm{r} \tag{1.8}$$

Although, not entirely known the HK functional has a remarkably property: It is universally, in a sense that both the kinetic and inter-electronic repulsion are independent of the concerned system. The consequence of such universal nature offers the possibility that once it is exactly or approximately knew the HK functional *for a given external potential* V(r) remain valuable for any other type of potential V'(r) applied on the concerned many electronic system. Let us note the fact that V(r) *should not be reduced only to the Coulombic type* of potentials but is carrying the role of the generic potential applied, that could be of either an electric, magnetic, nuclear, or even electronic nature as far it is external to the system fixed by the N electrons in the investigated system.

Here, another comment is required as well. For instance, the "total energy functional" nomination seems somehow confusing as no single number can explain the wide energy spectrum. In this respect one has to understand that the electronic density ρ is not unique for a system, varying from one eigen-state to eigen-state of another system. Still, this formulation allows the freedom of considering the ground as well the valence state alone in quantum considerations concerning reactivity since the conservation of charge in that state is assured though first HK theorem.

Once "in game", the external applied potential provides the second Hohenberg–Kohn (HK2) theorem. In short, HK2 theorem says that "the external applied potential is determined up to an additive constant by the electronic density of the N-electronic system ground state".

In mathematical terms, the theorem assures the validity of the variational principle applied to the density functional energy relation, that is for every electronic test density $\bar{\rho}$ around the real density r of the ground state.

$$E[\bar{\rho}] \geq E[\rho] \Leftrightarrow \delta E[\rho] = 0 \tag{1.9}$$

The proof of energy variational principle, or, in other words, the one-to-one correspondence between the applied potential and the ground state electronic density, employs the *reductio ad absurdum* procedure. That is to assume that the ground state electronic density $\rho(r)$ corresponds to two external potentials (V_1, V_2) fixing two associate Hamiltonians (H_1, H_2) to which two eigen-total energy (E_1, E_2) and two eigen-wave functions (Ψ_1, Ψ_2) are allowed. Now, if eigen-function Ψ_1 is considered as the true one for the ground state the energy variational principle will cast as the inequality:

$$\mathrm{E}_1[\rho] = \int \Psi_1^* \hat{H}_1 \Psi_1 \delta\tau < \int \Psi_2^* \hat{H}_1 \Psi_2 \delta\tau = \int \Psi_2^* \left[\hat{H}_2 + \left(\hat{H}_1 - \hat{H}_2\right)\right] \Psi_2 \delta\tau \tag{1.10}$$

which is further reduced, on universality reasons of the HK-functional, to the form:

$$E_1[\rho] < E_2[\rho] + \int \rho(\mathrm{r})\left[V_1(\mathrm{r}) - V_2(\mathrm{r})\right] d\mathrm{r} \tag{1.11}$$

On another way, if the eigen-function Ψ_2 is assumed as being the one true ground state wave-function, the analogue inequality springs out as:

$$E_2[\rho] < E_1[\rho] + \int \rho(\mathrm{r})\left[V_2(\mathrm{r}) - V_1(\mathrm{r})\right] d\mathrm{r} \tag{1.12}$$

Taken together the last two relations generate, by direct summation, the evidence of the contradiction:

$$E_1[\rho] + E_2[\rho] < E_1[\rho] + E_2[\rho] \tag{1.13}$$

The removal of such contradiction could be done in a single way, namely, by abolishing, in a reverse phenomenological order, the fact that two eigen-functions, two Hamiltonians and respectively, two external potential exist for characterizing the same ground state of a given electronic system. With this statement the HK2 theorem is formally proofed.

Yet, there appears the so called *V-representability* problem signaling the impossibility of an *a priori* selection of the external potentials types that are in bi-univocal relation with ground state of an electronic system (Chen and Stott, 1991a, 1991b; Kryachko and Ludena, 1991a, 1991b). The problem was revealed as very difficult at mathematical level due to the equivocal potential intrinsic behavior that is neither of universal nor of referential independent value. Fortunately, such principal limitation does not affect the general validity of the energy variational principle regarding the selection of the energy of ground state level from a collection of states with different associated external potentials.

That because, the problem of V-representability can be circumvented by the so called *N-contingency features of ground state electronic density* assuring that, aside of the *N– integrability condition* above, the candidate ground state densities should fulfill the *positivity condition* (an electronic density could not be negative):

$$\rho(\mathrm{r}) \geq 0 \ , \ \forall |\mathrm{r}| \in \Re \tag{1.14}$$

as well as the *non-divergent integrability condition* on the real domain (in relation with the fact that the kinetic energy of an electronic system could not be infinite–since the light velocity restriction):

$$\int_{\Re} \left| \nabla \rho(\mathrm{r})^{1/2} \right|^2 d\mathrm{r} < \infty \tag{1.15}$$

Both last two conditions are easy accomplished by every reasonable density, allowing the employment of the energy variational principle in two steps, according to the so called *Levy–Lieb double minimization algorithm* (Levy and Perdew, 1985): one regarding the intrinsic minimization procedure of the energetic terms respecting all possible eigen-functions folding a trial electronic density followed by the external minimization over all possible trial electronic densities yielding the correct ground state (GS) energy density functional

$$\begin{aligned} E_{GS} &= \min_{\rho} \left[\min_{\Psi \to \rho} \left(\int \Psi^* (T + W + V) \Psi d\tau \right) \right] \\ &= \min_{\rho} \left[\min_{\Psi \to \rho} \left(\int \Psi^* (T + W) \Psi d\tau \right) + \int \rho(\mathrm{r}) V(\mathrm{r}) d\mathrm{r} \right] \\ &= \min_{\rho} \left(F_{HK}[\rho] + C_A[\rho] \right) = \min_{\rho} \left(E[\rho] \right) \end{aligned} \tag{1.16}$$

One of the most important consequences of the HK2 conveys the rewriting of the energy variational principle in the light of above N-contingency conditions of the trial densities as the working Euler type equation (Geerlings and De Proft, 2001, 2003; Chermette, 1999):

$$\delta \{ E[\rho] - \mu N[\rho] \} = 0 \tag{1.17}$$

from where, there follows the Lagrange multiplication factor with the functional definition (Ayers and Parr, 2001; Iczkowski and Margrave, 1961; Parr et al., 1978):

$$\mu = \left(\frac{\delta E[\rho]}{\delta \rho} \right)_{\rho = \rho(V)} \tag{1.18}$$

this way introducing the *chemical potential* as the fundamental quantity of the theory. At this point, the whole chemistry can spring out since identifying the electronic systems electronegativity with the negative of the density functional chemical potential (Parr et al., 1978):

$$\chi = -\mu \tag{1.19}$$

The link with its DFT working definition is quite straight through the series of identities (Putz, 2003; Putz, 2011a-d):

$$\chi = \chi\left(\frac{\partial N}{\partial N}\right)_V = \int \chi\left(\frac{\partial \rho(\mathrm{r})}{\partial N}\right)_V d\mathrm{r} = -\int\left(\frac{\delta E}{\delta \rho(\mathrm{r})}\right)_V\left(\frac{\partial \rho(\mathrm{r})}{\partial N}\right)_V d\mathrm{r} = -\left(\frac{\partial E}{\partial N}\right)_V \tag{1.20}$$

There is therefore proved that electronegativity finds its proper place in the *DFT*, so casting the promise to deliver the most comprehensive analytical expressions, density functionals, and quantification of the chemical reactivity.

Up to now, the Hohenberg-Kohn theorems give new conceptual quantum tools for physico-chemical characterization of an electronic sample by means of electronic density and its functionals, the total energy and chemical potential (electronegativity). These positive density functional premises are in next analyzed towards elucidating of the quantum nature of the chemical bond and reactivity.

1.3.2 Kohn-Sham Equations

Their basic idea consists in assuming a so called orbital basic set for the N-electronic system by replacing the integration with summation over the virtual uni-electronic orbitals $\varphi_i, i=\overline{1,N}$, in accordance with Pauli principle, assuring therefore that the maximal occupancy with a certain spin/orbital within the HK1 frame it is (Keller, 1986):

$$\rho(\mathrm{r}) = \sum_i^N n_i\left|\varphi_i(\mathrm{r})\right|^2,\ 0 \le n_i \le 1,\ \sum_i n_i = N \tag{1.21}$$

Then, the *trial* total eigen-energy may be rewritten as density functional expanded in the original form (Ernserhof, 1994; Kohn and Sham, 1965; Putz, 2008a):

$$\begin{aligned}
E[\rho] &= F_{HK}[\rho] + C_A[\rho] \\
&= T[\rho] + W[\rho] + C_A[\rho] \\
&= T_{KS}[\rho] + J[\rho] + \{(T[\rho] - T_{KS}[\rho]) + (W[\rho] - J[\rho])\} + C_A[\rho] \\
&= \sum_i^N \int \vartheta_i^*(\mathrm{r})\left[-\frac{1}{2}\nabla^2\right]\vartheta_i(\mathrm{r})dr + \frac{1}{2}\iint \frac{\rho(\mathrm{r}_1)\rho(\mathrm{r}_2)}{r_{12}}dr_1 dr_2 + E_{XC}[\rho] + \int V(\mathrm{r})\rho(\mathrm{r})dr
\end{aligned} \tag{1.22}$$

where the contribution of the referential uniform kinetic energy contribution

$$T_{KS}[\rho] = \sum_i^N \int n_i \varphi_i^*(\mathrm{r})\left[-\frac{1}{2}\nabla^2\right]\varphi_i(\mathrm{r})dr \tag{1.23}$$

with the inferior index "KS" referring to as the Kohn-Sham "spherical" or homogeneous attribute and the classical repulsion inter-electronic Coulombic energy

$$J[\rho] = \frac{1}{2}\iint \frac{\rho(\mathrm{r}_1)\rho(\mathrm{r}_2)}{r_{12}}dr_1 dr_2 \tag{1.24}$$

were used as the analytical vehicles to elegantly produce the exchange-correlation energy E_{XC} containing exchange $(W[\rho]-J[\rho])$ and correlation $(T[\rho]-T_{KS}[\rho])$ heuristic terms as the quantum effects of spin anti-symmetry over the classical inter-electronic potential and of corrected homogeneous electronic movement, respectively.

An elegant and instructive method for deriving such equation assumes the same types of orbitals for the density expansion,

$$\rho(\mathrm{r}) = N\varphi^{*}(\mathrm{r})\varphi(\mathrm{r}) \tag{1.25}$$

that, without diminishing the general validity of the results, since preserving the N-electronic character of the system, highly simplifies the analytical discourse.

Actually, with the trial density replaced throughout the energy expression has to undergo the minimization procedure, with the practical equivalent integral variant:

$$\int \frac{\delta\left(E[\rho] - \mu N[\rho]\right)}{\delta\varphi^{*}} \delta\varphi^{*} d\mathrm{r} = 0 \tag{1.26}$$

Note that, in fact, we chose the variation in the conjugated uni-orbital $\varphi^{*}(\mathrm{r})$ providing the useful differential link:

$$\delta\rho(\mathrm{r}) = N\varphi(\mathrm{r})\delta\varphi^{*}(\mathrm{r}) \Rightarrow \left[\delta\varphi^{*}(\mathrm{r})\right]^{-1} = N\varphi(\mathrm{r})\frac{1}{\delta\rho(\mathrm{r})} \tag{1.27}$$

Now, one firstly gets (Neal, 1998)

$$\frac{\delta}{\delta\varphi^{*}(\mathrm{r})}\left\{\begin{aligned}&-\frac{N}{2}\int\varphi^{*}(\mathrm{r})\nabla^{2}\varphi(\mathrm{r})d\mathrm{r} + J[\rho] + E_{XC}[\rho]\\&+N\int V(\mathrm{r})\varphi^{*}(\mathrm{r})\varphi(\mathrm{r})d\mathrm{r} - \mu N\int\varphi^{*}(\mathrm{r})\varphi(\mathrm{r})d\mathrm{r}\end{aligned}\right\} = 0 \tag{1.28}$$

By performing the required partial functional derivations respecting the uni-orbital $\varphi^{*}(\mathrm{r})$ and by taking account of the equivalence in derivation of $J[\rho]$ and $E_{XC}[\rho]$ terms, we have:

$$-\frac{N}{2}\nabla^{2}\varphi(\mathrm{r}) + N\varphi(\mathrm{r})\frac{\delta J[\rho]}{\delta\rho} + N\varphi(\mathrm{r})\frac{\delta E_{XC}}{\delta\rho} + NV(\mathrm{r})\varphi(\mathrm{r}) - \mu N\varphi(\mathrm{r}) = 0 \tag{1.29}$$

After immediate suppressing of the N factor in all the terms and by considering the exchange-correlation potential with the formal definition (Gaspar and Nagy, 1987; Guo and Whitehead, 1991; Lam et al., 1998; Levy, 1991; Zhao and Parr, 1992):

$$V_{XC}(\mathrm{r}) = \left(\frac{\delta E_{XC}[\rho]}{\delta\rho(r)}\right)_{V(\mathrm{r})} \tag{1.30}$$

we get the simplified equation (Putz and Chiriac, 2008):

$$\left[-\frac{1}{2}\nabla^{2} + \left(V(\mathrm{r}) + \int\frac{\rho(\mathrm{r}_2)}{|\mathrm{r} - \mathrm{r}_2|}d\mathrm{r}_2 + V_{XC}(\mathrm{r})\right)\right]\varphi(\mathrm{r}) = \mu\varphi(\mathrm{r}) \tag{1.31}$$

Moreover, once introducing the so called KS-effective potential:

$$V_{KS}(\mathrm{r}) = V(\mathrm{r}) + \int \frac{\rho(\mathrm{r}_2)}{|\mathrm{r} - \mathrm{r}_2|} d\mathrm{r}_2 + V_{XC}(\mathrm{r}) \tag{1.32}$$

the resulted equation recovers the traditional Schrödinger shape:

$$\left[-\frac{1}{2}\nabla^2 + V_{KS}\right]\varphi(\mathrm{r}) = \mu\varphi(\mathrm{r}) \tag{1.33}$$

This is why, the functions $\varphi(\mathrm{r})$ are used to be called as Kohn-Sham (KS) orbitals; they provide the orbital set solutions of the associate KS equations (Hohenberg and Kohn, 1964; Kohn and Sham, 1965; Kohn et al., 1996):

$$\left[-\frac{1}{2}\nabla^2 + V_{KS}\right]\varphi_i(\mathrm{r}) = \mu_i\varphi_i(\mathrm{r}),\ i = \overline{1, N} \tag{1.34}$$

once one reconsiders electronic density back with general case.

Yet, KS equations, a part of delivering the KS wave-functions $\varphi_i(\mathrm{r})$, associate with another famous physico-chemical figure, the orbital chemical potential μ_i, which in any moment can be seen as the negative of the orbital electronegativities on the base of the chemical potential-electronegativity relationship.

Going now to a summative characterization of the above optimization procedure worth observing that the *N*-electronic in an arbitrary external *V*-potential problem is conceptual-computationally solved by means of the following self-consistent algorithm (Putz, 2008a):

(i) It starts with a trial electronic density satisfying the *N*-contingency conditions;
(ii) With trial density the effective potential containing exchange and correlation is calculated;
(iii) With computed V_{KS} the KS equations are solved for $\varphi_i(\mathrm{r}),\ i = \overline{1, N}$;
(iv) With the set of functions $\{\varphi_i(\mathrm{r})\}_{i=\overline{1,N}}$ the new density is recalculated;
(v) The procedure is repeated until the difference between two consecutive densities approaches zero;
(vi) Once the last condition is achieved one retains the last set $\{\varphi_i(\mathrm{r}), \mu_i = -\chi_i\}_{i=\overline{1,N}}$

The electronegativity orbital observed contributions are summed up with the expression:

$$-\sum_i^N \langle\chi_i\rangle = \sum_i^N \int n_i\varphi_i^*(\mathrm{r})\left[-\frac{1}{2}\nabla^2 + V_{KS}(\mathrm{r})\right]\varphi_i(\mathrm{r})d\mathrm{r} = T_{KS}[\rho] + \int V_{KS}\rho(\mathrm{r})d\mathrm{r} \tag{1.35}$$

while the density functional of the total energy for the *N*-electronic system will take the final figure:

$$E[\rho] = -\sum_i^N \langle\chi_i\rangle - \frac{1}{2}\iint \frac{\rho(\mathrm{r}_1)\rho(\mathrm{r}_2)}{|\mathrm{r}_{12}|} d\mathrm{r}_1 d\mathrm{r}_2 + \left\{E_{XC}[\rho] - \int V_{XC}(\mathrm{r})\rho(\mathrm{r})d\mathrm{r}\right\} \tag{1.36}$$

showing that the optimized many-electronic ground state energy is directly related with global or summed over orbital electronegativities. One can observe for this that even in the most optimistic case in which the last two terms in are hopefully canceling each other there still remains a (classical) correction to be added on global electronegativity in total energy. Or, in other terms, electronegativity alone is not enough to better describe the total energy of a many-electronic system, and still its correction can be modeled in a global (almost classical) way. Such considerations stressed upon the accepted semi classical behavior of the chemical systems, at the edge between the full quantum and classical treatments.

1.4 QUANTUM FORMALIZATION OF DENSITY FUNCTIONAL THEORY

1.4.1 Electronic Density Theorems

In what follows further formalization of DFT will be presented (Putz, 2011d).

DFT1 Definition: For every external potential $V \in \{V\}$ a subspace of wave functions is formed:

$$\Psi_V = \left\{ |\Psi\rangle = \sum_{i=1}^{q} c_i |\Psi_i\rangle \right\} \tag{1.37}$$

formed on the set of degenerated wave functions, ortho-normalized, of the fundamental state associated with the potential $V \in \{V\}$. The entire formed space on the multitude of external potentials will be as follows

$$\Psi = \bigcup_{V \in \{V\}} \Psi_V \tag{1.38}$$

and in similar way for the distributed subspace :

$$\rho_V = \left\{ \rho(r) \middle| \rho(r) = \langle \Psi | \hat{\rho}(r) | \Psi \rangle \,,\, |\Psi\rangle \, \hat{I} \, \Psi_V \right\} \tag{1.39}$$

and their assembly

$$\rho = \bigcup_{V \in \{V\}} \rho_V \tag{1.40}$$

Analogous with the previous section the following applications can be formed:

$$C^{-1} : \{\Psi\} \rightarrow \{V\} \tag{1.41}$$

$$D : \{\Psi\} \rightarrow \{\rho\} \tag{1.42}$$

The situation in which the proper subspaces of wave functions to the distinct external potentials remains valid are always associated with the distinct subspaces of distribution, in other words, also in the case of the degenerated fundamental state the application $(CD)^{-1} : \{\rho\} \rightarrow \{V\}$ is a unique application.

However, the case of fundamental degenerated state maintains also the situation when for an external potential given, two non-identical fundamental degenerated states $\Psi_V \neq \Psi'_V$ can correspond to identical densities, $\rho_V = \rho'_V$. This means that for a density does not corresponds only one fundamental state, so D is not invertible.

As a consequence, the HK1 theorem is not valid for fundamental degenerated state, which rises the problem of finding of a unique correlated functional associated with the density of a given external potential. Using the correspondence of the two non-identical fundamental degenerated states $\Psi_V \neq \Psi'_V$ with the unique value of a fundamental state's energy associated with the density of the associated V external potential

$$\left\langle \Psi[\rho] \middle| \hat{T} + \hat{W} + \hat{V} \middle| \Psi[\rho] \right\rangle = \left\langle \Psi'[\rho] \middle| \hat{T} + \hat{W} + \hat{V} \middle| \Psi'[\rho] \right\rangle = E \tag{1.43}$$

results exactly HK functional as being uniquely defined by density even for fundamental degenerated state according to the defined expression.

$$F_{HK}[\rho] := E - \int \rho(\mathrm{r}) V([\rho], \mathrm{r}) d\mathrm{r} \tag{1.44}$$

Also, the variational principle is valid for fundamental degenerated state too.

Two fundamental problems are now in discussion: *are all nonnegative defined density functions, V-representable? And if not, can the HK functional domain be extended to the non V-representable functions?*

This issue is taken into discussion especially for the fact that in the minimization process is considered that a density multitude of which only one determines *uniquely the bare* applied external potential, when the total energy of the system reaches its minimum.

Anyhow, to this terminus point (that cannot be initially accessed, neither guessed) the variational process is initialized with a test density and then with another, etc., as they not representing in a unique way the external (real) potential, but neither can they be excluded from the minimization process. So, the necessity of defining a "collective" non-pure but V-representable density appears.

DFT2 Definition: Considering an independent fundamental q states system (test states) with the help of which the statistic operator (von Neumann) of density is formed

$$\hat{D} = \sum_{i=1}^{q} d_i \left|\Psi_i\right\rangle \left\langle \Psi_i \right| \tag{1.45}$$

with

$$d_i^* = d_i \geq 0 \,,\, \sum_{i=1}^{q} d_i = 1 \tag{1.46}$$

through which the assembly density can be calculated as

$$\rho_D(\mathrm{r}) = Tr\left\{\hat{D}\hat{\rho}(\mathrm{r})\right\} = \sum_{i=1}^{q} d_i \rho_i(\mathrm{r}) \tag{1.47}$$

with

$$\rho_i(\mathrm{r}) = \left\langle \Psi_i \middle| \hat{\rho}(\mathrm{r}) \middle| \Psi_i \right\rangle \tag{1.48}$$

It is demonstrated that ρ_D is *non pure V-representable* but as general effect it is *V-representable*. The following theorem can be formulated (Levy and Gorling, 1996).

DFT1 Theorem (of V – representability): The HK functional domain, $F_{HK}[\rho]$, can be extended so as it can include the V-representable functions by changing the space of wave function with the matrix density for the constant V- external potential:

$$D_V = \left\{ \hat{D} = \sum_{i=1}^{q} d_i |\Psi_i\rangle\langle\Psi_i| \,,\, d_i^* = d_i \geq 0 \,,\, \sum_{i=1}^{q} d_i = 1 \right\} \tag{1.49}$$

and considering as proper, the new ρ_V space of the V-representable densities ensemble $\rho_D(r)$. On these terms the extension of the HK functional to the V-representable densities ensemble will be defined

$$F_{HK}[\rho] := \mathrm{Tr}\left\{ \hat{D}[\rho]\left(\hat{T} + \hat{W}\right) \right\} \tag{1.50}$$

in which $\hat{D}[\rho]$ can be any of the matrix density associated with an external unique potential $\hat{V}[\rho]$ that generates a ρ_D density in the V-representable ensemble (Chen and Stott, 1991a, 1991b).

A version of this extension was also proposed by Levy and Lieb under the form (Lieb, 1976; Lieb, 1981; Levy, 1982)

$$F_{HKL(I)}[\rho] := \inf_{\Psi \to \rho} \langle \Psi | \hat{T} + \hat{W} | \Psi \rangle \tag{1.51}$$

in which $\Psi \to \rho$ indicated the fact that the minimum is searched among every anti-symmetrical, normalized, N particles of wave functions, prescribed by the densities of the *V-representable* ensemble.

More so, a general version recommends the functional form:

$$F_{KSL(II)}[\rho] := \inf_{\hat{D} \to \rho} \mathrm{Tr}\left\{ \hat{D}[\rho]\left(\hat{T} + \hat{W}\right) \right\} \tag{1.52}$$

$$\hat{D} = \sum_{i=1}^{\infty} d_i |\Psi_i\rangle\langle\Psi_i| ,\, d_i^* = d_i \geq 0 \,,\, \sum_{i=1}^{\infty} d_i = 1 \tag{1.53}$$

For every version of functional density the total energy of fundamental (ground) state (GS) satisfies the variational principle

$$E_{GS} = \inf_{\rho(\mathrm{r})} E_V[\rho] = \inf_{\rho(\mathrm{r})} \left[F_{KSL(I/II)}[\rho] + \int \rho(\mathrm{r}) V(\mathrm{r}) d\mathrm{r} \right] \tag{1.54}$$

These functionals $F_{KSL(I/II)}[\rho]$ are defined for every functions $\rho(\mathrm{r})$ that are represented as N particles of anti-symmetrical functions $\Psi(\mathrm{r}_1, \mathrm{r}_2, ..., \mathrm{r}_N)$ not necessary as fundamental state. This observation implicates the following theorem.

DFT2 Theorem (of N – representability): All nonnegative and integrable functions are N-representable (Kryachko and Ludena, 1991a, 1991b).

The demonstration is shown by using the proper determinant of N particles starting from a nonnegative function $\rho(\mathrm{r})$ that satisfies the fundamental relation (1.4). Based on Harriman's works (Harriman, 1978a, 1978b, 1983, 1984, 1986) the following construction is proposed.

DFT3 Definition: An auxiliary function which satisfies the differential equation $f(\mathrm{r})$ is introduced

$$df(\mathrm{r}) = \frac{2\pi}{N}\rho(\mathrm{r})d\mathrm{r} \tag{1.55}$$

of which the direct integration reproduces the initial function under focus

$$f(\mathrm{r}) = \frac{2\pi}{N}\int_{-\infty}^{\mathrm{r}} \rho(\mathrm{r}')\, d\mathrm{r}' \tag{1.56}$$

with the help of which we can define the set of uni-particle orbitals:

$$\varphi_k(\mathrm{r}) := \sqrt{\frac{\rho(\mathrm{r})}{N}} \exp\{i[kf(\mathrm{r}) + \varphi(\mathrm{r})]\}, k \in Z\,, k \in Z \tag{1.57}$$

in which the phase function $\varphi(\mathrm{r})$, is real.

With this definition it can be shown that the introduced orbitals are ortho-normalized

$$\begin{aligned}\int_{-\infty}^{+\infty} \varphi_{k'}^*(\mathrm{r})\varphi_k(\mathrm{r})d\mathrm{r} &= \int_{-\infty}^{+\infty} \frac{\rho(\mathrm{r})}{N}\exp[i(k-k')f(\mathrm{r})]d\mathrm{r} \\ &= \frac{1}{2\pi}\int_{-\infty}^{+\infty} \exp[i(k-k')f(\mathrm{r})]df = \delta_{kk'}\end{aligned} \tag{1.58}$$

and completely defined

$$\begin{aligned}\sum_{k\in Z}\varphi_k(\mathrm{r})\varphi_k^*(\mathrm{r}') &= \exp\{i[\varphi(\mathrm{r}) - \varphi(\mathrm{r}')]\}\frac{\sqrt{\rho(\mathrm{r})\rho(\mathrm{r}')}}{N}\sum_{k\in Z}\exp\{ik[f(\mathrm{r}) - f(\mathrm{r}')]\} \\ &= \exp\{i[\varphi(\mathrm{r}) - \varphi(\mathrm{r}')]\}\frac{\sqrt{\rho(\mathrm{r})\rho(\mathrm{r}')}}{N}2\pi\delta(f(\mathrm{r}) - f(\mathrm{r}')) \\ &= \frac{\delta(f(\mathrm{r}) - f(\mathrm{r}'))}{\left|\dfrac{d\mathrm{r}}{df(\mathrm{r})}\right|} = \delta(\mathrm{r} - \mathrm{r}')\end{aligned} \tag{1.59}$$

Thus, the associated Slater determinant can be built

$$\chi_{k_1,k_2,\ldots,k_N} = \frac{1}{\sqrt{N!}}\det[\varphi_{k_1},\ldots,\varphi_{k_N}]\ ,\ k_1,\ldots,k_N \in Z\ ,\ k_i \neq k_j\ \ for\ i \neq j \tag{1.60}$$

through which is found the integrable nonnegative density function that satisfies the equality

$$\left\langle \chi_{k_1,\ldots,k_N} \middle| \hat{\rho}(\mathrm{r}) \middle| \chi_{k_1,\ldots,k_N} \right\rangle = \sum_{i=1}^{N} \left|\varphi_{k_i}(\mathrm{r})\right|^2 = \rho(\mathrm{r}) \tag{1.61}$$

The orbitals above are named *equi-density orbitals* and are used for the rigorous construction of functionals which are correlated with the exact energy of the fundamental state.

As an example, the kinetic energy contribution to the fundamental state energy of a system in interaction will be given by construction:

$$T = \int dr \frac{\hbar^2}{2m} [\nabla_r \cdot \nabla_{r'} \gamma(r, r')]_{r' \to r} \tag{1.62}$$

in which $\gamma(r, r')$ stays for the uni-particle density matrix that may be exactly written with the help of the above orbitals:

$$\gamma(r, r') = \sum_{k,k' \in Z} \gamma_{kk'} \varphi_k(r) \varphi^*_{k'}(r') = \sqrt{\rho(r)\rho(r')}\Gamma(r, r') \tag{1.63}$$

$$\Gamma(r, r') = \frac{1}{N} \exp\{i[\varphi(r) - \varphi(r')]\} \sum_{k,k' \in Z} \gamma_{kk'} \exp\{i[kf(r) - k' f(r')]\} \tag{1.64}$$

Within these conditions the expression for the kinetic energy functional results to be

$$T[\rho] = T_W[\rho] + \int drA[\rho(r), r] \tag{1.65}$$

in which the first functional recovers the kinetic term suggested by von Weizsäcker since 1935 (Dufek et al., 1994; Murphy, 1981; Romera and Dehesa, 1994)

$$T_W[\rho] = \frac{\hbar^2}{8m} \int dr \frac{(\nabla \rho(r))^2}{\rho(r)} \tag{1.66}$$

and the other functional term looks like:

$$A[\rho(r)] = \frac{\hbar^2}{2m} \rho(r) [\nabla_r \cdot \nabla_{r'} \Gamma(r, r')]_{r' \to r} \tag{1.67}$$

Worth noting that, in approximating a nonlocal density for the electronic gas considered homogeneous, we have the consecrated expression (Parr and Yang, 1989):

$$\Gamma_0(r, r') = \frac{\sin(k_F y) - (k_F y)\cos(k_F y)}{\pi^2 y^3}, \quad y = |r' - r| \tag{1.68}$$

with the help of which we obtain the functional:

$$A[\rho(r)] = \frac{3\hbar^2}{10m} \rho(r) k_F^{\ 2} \tag{1.69}$$

in which, making the replacement $k_F \to (3\pi^2 \rho(r))^{1/3}$ one rediscovers the form of density functional suggested by Thomas and Fermi, in case of the above exposed theory Thomas-Fermi (TF), since 1920 (Lieb, 1991):

$$T_{TF}[\rho]=\int d\mathrm{r}A_{TF}[\rho(\mathrm{r})]=\frac{3\hbar^2}{10m}\left(3\pi^2\right)^{2/3}\int\rho(\mathrm{r})^{5/3}d\mathrm{r} \tag{1.70}$$

In this way, the actual theory of the functional density admits as a particular case functionals that appear in a "sub-theory" at a level of quasi-uniform electronic distributions.

1.4.2 Density Functionals Theorems

DFT4 Definition: The exchange-correlation density functional is written from the universal Hohenberg - Kohn functional as

$$E_{XC}[\rho]=F_{HK}[\rho]-\frac{1}{2}\iint d\mathrm{r}d\mathrm{r}'\rho(\mathrm{r})W(\mathrm{r},\mathrm{r}')\rho(\mathrm{r}')-T_{KS}[\rho] \tag{1.71}$$

through which the exchange-correlation potential is recognized

$$V_{XC}([\rho_0];\mathrm{r})=\left.\frac{\delta E_{XC}[\rho]}{\delta\rho(\mathrm{r})}\right|_{\rho_0} \tag{1.72}$$

With these, the variational principle of the fundamental/ground state level can be applied

$$\begin{aligned}0=\delta E_V&=E_V[\rho_0+\delta\rho]-E_V[\rho_0]\\&=\delta T_{KS}+\int d\mathrm{r}\delta\rho(\mathrm{r})\left\{V(\mathrm{r})+\int W(\mathrm{r},\mathrm{r}')\rho_0(\mathrm{r}')d\mathrm{r}'+V_{XC}([\rho_0];\mathrm{r})\right\}\end{aligned} \tag{1.73}$$

in which, for writing the unitary principle, we have to express also the differential variation of kinetic energy

$$\begin{aligned}\delta T_{KS}&=\sum_{i=1}^{N}\int d\mathrm{r}\left[\delta\varphi_i^*(\mathrm{r})\left(-\frac{\hbar^2}{2m}\nabla^2\right)\varphi_{i,0}(\mathrm{r})+\varphi_{i,0}^*(\mathrm{r})\left(-\frac{\hbar^2}{2m}\nabla^2\right)\delta\varphi_i(\mathrm{r})\right]\\&=\sum_{i=1}^{N}\int d\mathrm{r}\left[\delta\varphi_i^*(\mathrm{r})\left(-\frac{\hbar^2}{2m}\nabla^2\right)\varphi_{i,0}(\mathrm{r})+\delta\varphi_i(\mathrm{r})\left(-\frac{\hbar^2}{2m}\nabla^2\right)\varphi_{i,0}^*(\mathrm{r})\right]\end{aligned} \tag{1.74}$$

For the last equality, through using the Green theorem there follows that the kinetic operators take action on the uni-particle orbitals only, yielding, by employing also the Schrödinger equation, the expression

$$\delta T_{KS}=\sum_{i=1}^{N}\int d\mathrm{r}\left\{\delta\varphi_i^*(\mathrm{r})[\varepsilon_i-V_{KS}(\mathrm{r})]\varphi_{i,0}(\mathrm{r})+\varphi_{i,0}^*(\mathrm{r})[\varepsilon_i-V_{KS}(\mathrm{r})]\delta\varphi_i(\mathrm{r})\right\} \tag{1.75}$$

Note that taking into consideration the second order terms in the orbital density differential

$$\delta\left|\varphi_i(\mathrm{r})\right|^2=\left|\varphi_{i,0}(\mathrm{r})+\delta\varphi_i(\mathrm{r})\right|^2-\left|\varphi_{i,0}(\mathrm{r})\right|^2=\varphi_{i,0}^*(\mathrm{r})\delta\varphi_i(\mathrm{r})+\varphi_{i,0}(\mathrm{r})\delta\varphi_i^*(\mathrm{r}) \tag{1.76}$$

the last expression resumes to

$$\delta T_{KS} = \sum_{i=1}^{N} \varepsilon_i \int d\mathrm{r}\delta|\varphi_i(\mathrm{r})|^2 - \sum_{i=1}^{N} \int d\mathrm{r} V_{KS}(\mathrm{r})\delta|\varphi_i(\mathrm{r})|^2 = -\int d\mathrm{r} V_{KS}(\mathrm{r})\delta\rho(\mathrm{r}) \tag{1.77}$$

taking into consideration for the last equality the fact that $\varphi_{i,0}(\mathrm{r})$ and also $[\varphi_{i,0}(\mathrm{r}) + \delta\varphi_i(\mathrm{r})]$ are normalized to unity and so the term with the appropriate uni-orbital energies disappears.

Going back to the variation equation of total energy results the auxiliary potential expression (also named as the Kohn-Sham potential and noted like V_{KS})

$$V_{KS}(\mathrm{r}) = V(\mathrm{r}) + \int d\mathrm{r} W(\mathrm{r},\mathrm{r}')\rho_0(\mathrm{r}') + V_{XC}([\rho_0];\mathrm{r}) \tag{1.78}$$

with the contribution of which the classic Kohn-Sham theorem turns formulated and defined completely, with the self-consistent multi-orbital formed system.

We can also formulate the following theorem.

DFT3 Theorem: In the exchange-correlation energy density functional in which the HK functional is interpreted from the DFT4 Definition

$$E_{XC}[\rho] = \left\{W[\rho] - \frac{1}{2}\iint d\mathrm{r}d\mathrm{r}'\rho(\mathrm{r})W(\mathrm{r},\mathrm{r}')\rho(\mathrm{r}')\right\} + \{T[\rho] - T_{KS}[\rho]\} \tag{1.79}$$

the exchange-correlation kinetic contribution is always non-negative

$$T_{XC}[\rho] := T[\rho] - T_{KS}[\rho] \geq 0 \tag{1.80}$$

The demonstration is immediate, and it is based on the Rayleigh-Ritz principle

$$T_{KS}[\rho] = \langle\Phi_{KS}[\rho]|\hat{T}|\Phi_{KS}[\rho]\rangle \leq \langle\Psi[\rho]|\hat{T}|\Psi[\rho]\rangle = T[\rho] \tag{1.81}$$

in which for a given density for a fundamental state, $\rho(\mathrm{r})$, the wave function $\Phi_{KS}[\rho]$, is the one that minimizes the system without the inter-particle interaction, $\langle\psi|\hat{T}|\psi\rangle$, opposite to $\Psi[\rho]$ wave function, that minimizes the system with the included inter-particle interaction, $\langle\psi|\hat{T}+\hat{W}|\psi\rangle$.

In view of the above exchange and correlation functional density formulation, the case of not-degenerated fundamental states, it can be easily reduces to that of the degenerated fundamental states, taking into account the DFT1 Theorem of V-representability and the extended HK functional.

DFT4 Theorem: For a system having interacting fermions, with the degenerated fundamental state and von Neumann type statistic densities, the V-representable ensemble admits for the kinetic energy of the auxiliary system, that is for the non-interacting fermions, the extended functional representation

$$T_{KS-Neumann}[\rho] = \inf_{\substack{\hat{D}\to\rho,\\ \hat{D}=\sum_K d_K|\Psi_K\rangle\langle\Psi_K|,\\ \sum_K d_K=1}} \mathrm{Tr}\left\{\hat{D}\hat{T}\right\} \tag{1.82}$$

with the help of which we can properly write the exchange and correlation energy as the density functional

$$E_{XC}[\rho]=F_{KS-Neumann}[\rho]-\frac{1}{2}\iint drdr'\rho(r)W(r,r')\rho(r')-T_{KS-Neumann}[\rho] \tag{1.83}$$

In these terms we can introduce the following theorem (Dreizler and Gross, 1990).

DFT5 Theorem (extended KOHN – SHAM): The fundamental state density of an arbitrary system (also degenerated) of interacted fermions found in a variational manner for an external applied potential $V(r)$ can be obtained from resolving the following self -consistent set of equations

$$\left[-\frac{\hbar^2}{2m}\nabla^2+V(r)+\int dr'W(r,r')\rho(r')+V_{XC}([\rho];r)\right]\varphi_i(r)=\varepsilon_i\varphi_i(r)\ ,\ \ \varepsilon_1\le\varepsilon_2\le... \tag{1.84}$$

$$V_{XC}([\rho];r)=\frac{\delta E_{XC}[\rho]}{\delta\rho(r)}=\frac{\delta}{\delta\rho(r)}\left\{\begin{matrix}F_{KS-Neumann}[\rho]\\ -\frac{1}{2}\iint drdr'\rho(r)W(r,r')\rho(r')\\ -T_{KS-Neumann}[\rho]\end{matrix}\right\} \tag{1.85}$$

$$\rho(r)=\sum_{i=1}^{\infty}\gamma_i\left|\varphi_i(r)\right|^2 \tag{1.86}$$

$$\sum_{i=1}^{\infty}\gamma_i=N \tag{1.87}$$

with γ_i being the uni-orbital occupation number taking the values:

$$\begin{aligned}\gamma_i=1:\ \varepsilon_i<\mu\\ 0\le\gamma_i\le1:\varepsilon_i=\mu\\ \gamma_i=0:\varepsilon_i>\mu\end{aligned} \tag{1.88}$$

with m as the energy of the highest occupied uni-particle level (i.e., the Fermi energy).

In this context it can be obtained the useful formula for the proper energy of the fundamental state, inserting the kinetic term of the system without inter-particle interaction correlated to the uni-orbital occupation number:

$$T_{KS-Neumann}[\rho]=\sum_{i=1}^{\infty}\gamma_i\int dr\varphi_i^*(r)\left(-\frac{\hbar^2}{2m}\nabla^2\right)\varphi_i(r) \tag{1.89}$$

which can be rewritten using the uni-particle energy expression

$$\varepsilon_i=\int dr\varphi_i^*(r)\left(-\frac{\hbar^2}{2m}\nabla^2\right)\varphi_i(r)+\int dr\rho(r)V_{KS}(r) \tag{1.90}$$

as:

$$T_{KS-Neumann}[\rho] = \sum_{i=1}^{\infty} \gamma_i \varepsilon_i - \int d\mathrm{r}\rho(\mathrm{r})V_{KS}(\mathrm{r}) \tag{1.91}$$

Taking into account the expression of the V_{KS} potential, replaced in the last expression, it can be written for the total energy, with T_{KS} extended to $T_{KS\text{-}Neumann}$, providing the working formula:

$$E_0[\rho] = \sum_{i=1}^{\infty} \gamma_i \varepsilon_i - \frac{1}{2}\iint d\mathrm{r}d\mathrm{r}'\rho(\mathrm{r})W(\mathrm{r},\mathrm{r}')\rho(\mathrm{r}') + E_{XC}[\rho] - \int \rho(\mathrm{r})V_{XC}([\rho];\mathrm{r})d\mathrm{r} \tag{1.92}$$

Next, the modification of the HK universal functional terms will be followed in the conditions of increasing the electronic cloud without expanding the nuclei system (Born – Oppenheimer approximation). For this the scaled energetic terms are introduced by using the following definition (Chan and Handy, 1999).

DFT5 Definition: Considering $\varphi(\mathrm{r}_1, \mathrm{r}_2, ..., \mathrm{r}_N)$ a normalized eigen-state of the Hamiltonian $\hat{H} = \hat{T} + \hat{W} + \hat{V}$ associated with the system of N fermions being in stationary conditions in interaction and in external applied field. Then, the normalized scaled eigen-state will be

$$\varphi_\gamma(\mathrm{r}_1, \mathrm{r}_2, ..., \mathrm{r}_N) \equiv \gamma^{3N/2}\varphi(\gamma\mathrm{r}_1, \gamma\mathrm{r}_2, ..., \gamma\mathrm{r}_N) \tag{1.93}$$

Inducing for the component terms of the Hamiltonian the following transformation relations

$$\langle \varphi_\gamma | \hat{T} | \varphi_\gamma \rangle = \gamma^2 \langle \varphi | \hat{T} | \varphi \rangle \tag{1.94}$$

$$\langle \varphi_\gamma | \hat{W} | \varphi_\gamma \rangle = \gamma \langle \varphi | \hat{W} | \varphi \rangle \tag{1.95}$$

$$\langle \varphi_\gamma | \hat{V} | \varphi_\gamma \rangle = \gamma^2 \langle \varphi | \hat{V} | \varphi \rangle \tag{1.96}$$

and for the fundamental state density a volume-type scaled expression:

$$\rho_\gamma(\mathrm{r}) = \gamma^3 \rho(\gamma\mathrm{r}) \tag{1.97}$$

With these, the next theorem can be stated.

DFT6 Theorem (of Hohenberg-Kohn scaling): For a system of fermions in interaction, being in a fundamental state the next inequalities are true (Kryachko and Ludena, 1991a, 1991b):

$$T[\rho_\gamma] < \gamma^2 T[\rho], \quad \gamma > 1 \tag{1.98}$$

$$T[\rho_\gamma] > \gamma^2 T[\rho], \quad \gamma < 1 \tag{1.99}$$

$$W[\rho_\gamma] < \gamma W[\rho], \quad \gamma < 1 \tag{1.100}$$

$$W[\rho_\gamma] > \gamma W[\rho], \quad \gamma > 1 \tag{1.101}$$

The demonstration is made simply by using the stationary Schrödinger equation for two wave functions associated to the system, $\Psi_{\rho_\gamma} \neq \phi_\gamma$, $\gamma \neq 1$. Then,

(a) When considering:

$$\left\{\hat{T}[\rho_\gamma]+\hat{W}[\rho_\gamma]+\hat{V}[\rho_\gamma]\right\}\psi_{\rho_\gamma} = E_{\rho_\gamma}\psi_{\rho_\gamma} \tag{1.102}$$

then through the *constrained searching* formalism (Levy, 1982) for the HK universal functional, is obtained:

$$\begin{aligned} T[\rho_\gamma]+W[\rho_\gamma] &= F_{HK}[\rho_\gamma] = \inf_{\psi\to\rho_\gamma}\langle\psi|\hat{T}+\hat{W}|\psi\rangle \\ &= \left\langle\psi_{\rho_\gamma}\left|\hat{T}+\hat{W}\right|\psi_{\rho_\gamma}\right\rangle < \langle\varphi_\gamma|\hat{T}+\hat{W}|\varphi_\gamma\rangle,\ \gamma\neq 1 \end{aligned} \tag{1.103}$$

from which taking into account the scaled properties the fundamental inequality is obtained

$$T[\rho_\gamma]+W[\rho_\gamma] < \gamma^2 T[\rho]+\gamma W[\rho]\ ,\ \gamma\neq 1 \tag{1.104}$$

(b) However if we write

$$E\varphi_\gamma(\mathrm{r}_1,\ldots,\mathrm{r}_N) = \hat{H}(\gamma\mathrm{r}_1,\ldots,\gamma\mathrm{r}_N)\left[\gamma^{3N/2}\varphi(\gamma\mathrm{r}_1,\ldots,\gamma\mathrm{r}_N)\right] = \left(\frac{1}{\gamma^2}\hat{T}+\frac{1}{\gamma}\hat{W}+\frac{1}{\gamma}\hat{V}\right)\varphi_\gamma(\mathrm{r}_1,\ldots,\mathrm{r}_N) \tag{1.105}$$

while being multiplied with γ^2 ($\gamma \neq 0$) it reduces to

$$\left(\hat{T}+\gamma\hat{W}+\gamma\hat{V}\right)\varphi_\gamma = \left(\gamma^2 E\right)\varphi_\gamma \tag{1.106}$$

from where, by applying once again the formalism of constrained minimization

$$\begin{aligned} T[\rho_\gamma]+\gamma W[\rho_\gamma] &= \min_{\substack{\varphi_\gamma \\ \psi\to\rho_\gamma}}\langle\psi|\hat{T}+\gamma\hat{W}|\psi\rangle \\ &= \left\langle\varphi_\gamma\left|\hat{T}+\gamma\hat{W}\right|\varphi_\gamma\right\rangle > \left\langle\psi_{\rho_\gamma}\left|\hat{T}+\gamma\hat{W}\right|\psi_{\rho_\gamma}\right\rangle, \gamma\neq 1 \end{aligned} \tag{1.107}$$

leaving with, through using once again the scaled orbital formulas, another inequality

$$T[\rho_\gamma]+\gamma W[\rho_\gamma] > \gamma^2\left(T[\rho]+W[\rho]\right),\ \gamma\neq 1 \tag{1.108}$$

Combining the inequalities from the steps (a) and (b) in a system from which we remove one on a time the kinetic or the Coulombic term, taking into account the sign of the γ parameter, the above energetic scaled inequalities are confirmed.

However, one should observe the fact that for a system of interacting electrons the scaled transforming relations of the previous type cannot be written:

$$T[\rho_\gamma] \neq \gamma^2 T[\rho] \tag{1.109}$$

$$W[\rho_\gamma] \neq \gamma W[\rho] \tag{1.110}$$

but only for the systems without internal interaction, $\hat{W} \equiv 0$, when due to the stated Hohenberg and Kohn theorem for the unique determination of the external potential, $\hat{V}[\rho_\gamma]=\gamma\hat{V}$, it turns out that the two associated wave functions have to be identical $\psi_{\rho_\gamma}=\varphi_\gamma$, $\gamma \neq 1$, which determines the equality:

$$T_{KS}[\rho_\gamma]=\gamma^2 T_{KS}[\rho] \tag{1.111}$$

Remarkably, although the kinetic energies functionals and the Coulombic interaction do not satisfy the above presented scaled equations in general fermionic systems, they satisfy relations of variational type as in the following theorem (Nagy, 1998; Nagy et al., 1999; Liu et al., 1999; Ou-Yang and Levy, 1991).

DFT7 Theorem (of the virial): The kinetic energy functional of system with Coulombic interaction fulfills the equation:

$$2T[\rho]=-W[\rho]-\int V(\mathrm{r})\rho(\mathrm{r})d\mathrm{r} \tag{1.112}$$

The proof is immediate taking into account the virial equation for a Coulombic system:

$$2\langle\varphi|\hat{T}|\varphi\rangle=-\langle\varphi|\hat{V}|\varphi\rangle-\langle\varphi|\hat{W}|\varphi\rangle \tag{1.113}$$

in which we replace the wave function of the ground state according to definition DFT1, $\varphi=\varphi[\rho]$.

We have to underline the fact that this equation is not universal, being true only for the external potential $V([\rho];\mathrm{r})$ prescribed by the density $\rho(\mathrm{r})$

For the rigorous relations containing universal functionals have been calculated one may write (Levy and Perdew, 1985)

$$2T[\rho]+\int d\mathrm{r}\rho(\mathrm{r})\mathrm{r}\cdot\nabla\frac{\delta T[\rho]}{\delta\rho(\mathrm{r})}=-W[\rho]-\int d\mathrm{r}\rho(\mathrm{r})\mathrm{r}\cdot\nabla\frac{\delta W[\rho]}{\delta\rho(\mathrm{r})} \tag{1.114}$$

$$2T_{KS}[\rho]=\int d\mathrm{r}\rho(\mathrm{r})\mathrm{r}\cdot\nabla V_{KS}([\rho];\mathrm{r}) \tag{1.115}$$

respectively for the systems with or without internal Coulombic interaction in which $V_{KS}([\rho];\mathrm{r})$ represents the exact Kohn-Sham potential.

Starting from these two relations one can form the next theorem.

DFT8 Theorem: The correlation and exchange energy density functional satisfies the equation

$$0\geq E_{XC}[\rho]+\int d\mathrm{r}\rho(\mathrm{r})\mathrm{r}\cdot\nabla V_{XC}([\rho];\mathrm{r})=T_{KS}[\rho]-T[\rho] \tag{1.116}$$

The inequality results immediately from the fact that the exchange and correlation kinetic contribution is always positive

$$T_{XC}[\rho]:=T[\rho]-T_{KS}[\rho]\geq 0 \tag{1.117}$$

according with the DFT3 Theorem.

From the last inequalities one may also separately writing the exchange and correlation density functionals as follows:

$$E_X^{KS}[\rho]+\int d\mathrm{r}\rho(\mathrm{r})\mathrm{r}\cdot\nabla V_X^{KS}([\rho];\mathrm{r})=0 \tag{1.118}$$

$$E_C^{KS}[\rho]+\int d\mathrm{r}\rho(\mathrm{r})\mathrm{r}\cdot\nabla V_C^{KS}([\rho];\mathrm{r})=T_{KS}[\rho]-T[\rho] \tag{1.119}$$

taking into account the notations:

$$V_X^{KS}([\rho];\mathrm{r})=\frac{\delta E_X^{KS}[\rho]}{\delta\rho(\mathrm{r})} \tag{1.120}$$

$$V_C^{KS}([\rho];\mathrm{r})=\frac{\delta E_C^{KS}[\rho]}{\delta\rho(\mathrm{r})} \tag{1.121}$$

It is noticed that the exchange and correlation density functional $E_{XC}^{KS}[\rho]$ determines in a single manner the exchange and correlation density functional of the kinetic contribution, $T_{XC}[\rho]=T[\rho]-T_{KS}[\rho]$, without the intervention of Kohn-Sham exchange potential contribution. However, still remains the issue that exchange and correlation density *functionals* energy remain unknown in general analytical formulation.

Nevertheless, one method of calculation for the exchange and correlation kinetic term is via the application of Hellmann-Feynman theorem under the next form.

DFT9 Theorem (of Hellmann-Feynman type): It is considered the Hamiltonian of a fermions' system with the Coulombic interaction modified by introducing l and g parameters under the form:

$$\hat{H}(\lambda,g):=\hat{T}+g\hat{W}+\hat{V}+\lambda\hat{A}+\hat{V}_{eff}([\rho];\lambda,g) \tag{1.122}$$

in which $V_{eff}([\rho];\lambda,g)$ is the actual local potential so chosen that it generates the density of the fundamental state, $\hat{A}$ is an arbitrary operator whom the density functional is searched. In particular, $\hat{H}(\lambda=0,g=0)$ is the exactly Kohn-Sham Hamiltonian. If the Hamiltonian $\hat{H}(\lambda,g)$ is evaluated in the fundamental state, making the integration in comparison with the g coupling constant, and the limit $\lambda\to 0$ considering the Hellmann-Feynman theorem, then for $\hat{A}$ operator density functional will obtain the expression:

$$A[\rho]=\left\langle\hat{A}\right\rangle_{KS}+\frac{d}{d\lambda}E_{XC}([\rho],\lambda)\big|_{\lambda=0} \tag{1.123}$$

in which $\left\langle\hat{A}\right\rangle_{KS}$ shows the expected value of the $\hat{A}$ operator for the Slater determinant built on Kohn-Sham orbitals.

For the application of calculating the correlation and exchange Kinetic term, $\hat{A}=\hat{T}$, the improved Hamiltonian is proposed ("the augmented Hamiltonian"), opposite to $\hat{H}=\hat{T}+\hat{W}+\hat{V}$, in parameterized form:

$$\hat{T}+\hat{W}+\hat{V}+\lambda\hat{T}=e^2\left\{[(1+\lambda)a_0]\sum_{i=1}^{N}\frac{\nabla_i^2}{2}+\frac{1}{2}\sum_{i\neq j}\frac{1}{|\mathrm{r}_i-\mathrm{r}_j|}-\sum_{i=1}^{N}\frac{Z}{|\mathrm{r}_i|}\right\} \tag{1.124}$$

with $a_0=h^2/\left(4\pi^2me^2\right)$ Bohr radius.

Counting the derivative term from the expression of A[ρ] with the last expression will obtain an useful formula of the correlation and exchange density kinetic contribution depending on the whole correlation and exchange energy

$$T_{XC}[\rho]=a_0\frac{dE_{XC}[\rho]}{da_0} \tag{1.125}$$

that for the homogenous electronic gas approximation (LDA: "Local Density Approximation") will result in the form of one virial equation as follows (Bartolotti and Acharya, 1982; Zhao et al., 1994; Dunlap and & Andzelm, 1990, 1992; Liu, 1996)

$$t_{XC}[\rho]=3V_{XC}[\rho]-4\varepsilon_{XC}[\rho] \tag{1.126}$$

in which the included terms represent the exchange-correlation energies per particle $\varepsilon_{XC}[\rho]$ alongside the associate exchange-correlation potential

$$V_{XC}[\rho]=\frac{d}{d\rho}(\rho\varepsilon_{XC}[\rho]) \tag{1.127}$$

Taking into account that the kinetic energy does not have a separate exchange contribution the last equality can be divided in distinct exchange and correlation relationships as follows

$$0=3V_X[\rho]-4\varepsilon_X[\rho] \tag{1.128}$$

$$t_{XC}[\rho]=3V_C[\rho]-4\varepsilon_C[\rho] \tag{1.129}$$

in which only the exchange and correlation kinetic energy was retained as contributing the correlation energetic effects.

The issue of how exchange and correlation contribution in general and the Kohn-Sham system in special will behave at considering bosonic particles and their associate fields is to be in Chapter 6 addressed through introducing the Bose-Einstein phenomenology and the allied features and working concepts.

KEYWORDS

- **Density functional theory**
- **Eigen-wave function**
- **Electronegativity**
- **Ground state**
- **Hohenberg-Kohn**

2 Physical Realizations of DFT

CONTENTS

2.1 INTRODUCTION

Once the general principles of Density Functional Theory (DFT) were presented (in previous chapter), further explorations of its extension will be here undertaken, that is for many electronic coupled spin systems (without discussing the diamagnetic effects that will imply the inclusion of the orbital's currents also), for the excited and partially occupied states (without considering the junction with the spin coupling), for systems with internal symmetry in what regards their atomic or molecular orbital's populations, for the non-Born-Oppenheimer molecular systems, for the non-stationary (or temporally dependent) systems, up to the systems with finite temperature in both fundamental or as the local inhomogeneous electronic fluid thermodynamics in Mermin and Parr pictures, respectively (Putz, 2008e).

2.2 DFT REALIZATION OF SPIN STATES

Will consider the extended Hamiltonian so that it includes also the interaction of the spin-polarized system characterized by the operator of the magnetic moment density $\hat{\vec{m}}(\mathrm{r})$ with an external magnetic field of induction vector operator $\hat{\vec{B}}(\mathrm{r})$:

$$\hat{H} = \hat{T} + \hat{W} + \hat{U} \tag{2.1}$$

$$\hat{U} = \int d\mathrm{r} \left[V(\mathrm{r})\hat{\rho}(\mathrm{r}) - \vec{B}(\mathrm{r}) \cdot \hat{\vec{m}}(\mathrm{r}) \right] \tag{2.2}$$

$$\hat{\rho}(\mathrm{r}) = \sum_a \hat{\psi}_a^+(\mathrm{r})\hat{\psi}_a(\mathrm{r}) = \hat{\rho}_+(\mathrm{r}) + \hat{\rho}_-(\mathrm{r}) \tag{2.3}$$

$$\hat{\vec{m}}(\mathrm{r}) = -\mu_0 \sum_{a,b} \hat{\psi}_a^+(\mathrm{r})\sigma_{ab}\hat{\psi}_b(\mathrm{r}) \tag{2.4}$$

where σ_{ab} represent the Pauli matrices, and $\mu_0 = e\hbar/(2mc)$ corresponds to Bohr magneton. Taking these relations into consideration, the mono-particle operator can be rewritten as follows:

$$\hat{U} = \sum_{a,b} \hat{\psi}_a^+(\mathrm{r})U_{ab}\hat{\psi}_b(\mathrm{r}) \tag{2.5}$$

$$U_{ab}(\mathrm{r}) = V(\mathrm{r})\delta_{ab} - \mu_0\vec{B}(\mathrm{r})\cdot\vec{\sigma}_{ab} \tag{2.6}$$

Now it can be formulated the Hohenberg – Kohn type theorem for the spin polarized systems.

ExtDFT1 Theorem: For two different fundamental states $|\Psi\rangle \neq |\Psi'\rangle$ of a system also characterized by spin, they always correspond to two different spin matrix $\rho_{ab}(\mathrm{r}) \neq \rho_{ab}'(\mathrm{r})$, or equivalent, to two different 4(quadri)-vectors, $(\rho(\mathrm{r}), \vec{m}(\mathrm{r})) \neq (\rho'(\mathrm{r}), \vec{m}'(\mathrm{r}))$, where:

$$\rho_{ab}(\mathrm{r}) = \langle\Psi|\hat{\psi}_a^+(\mathrm{r})\hat{\psi}_b(\mathrm{r})|\Psi\rangle \tag{2.7}$$

$$\rho(\mathrm{r}) = \langle\Psi|\hat{\rho}(\mathrm{r})|\Psi\rangle \tag{2.8}$$

$$\vec{m}(\mathrm{r}) = \langle\Psi|\hat{\vec{m}}(\mathrm{r})|\Psi\rangle \tag{2.9}$$

In these conditions, the D application: $|\Psi\rangle \to (\rho(\mathrm{r}), \vec{m}(\mathrm{r}))$ is irreversible.

The demonstration follows the course of classical HK theorem (of Chapter 1) with the distinction that the energetic inequality will add one more term, that is:

$$E < E' + \int d\mathrm{r}\,\rho'(\mathrm{r})[V(\mathrm{r}) - V'(\mathrm{r})] - \int d\mathrm{r}\vec{m}'(\mathrm{r})\cdot\left[\vec{B}(\mathrm{r}) - \vec{B}'(\mathrm{r})\right] \tag{2.10}$$

The *D*-application irreversibility as taken from the DFT1 definition, see equation (1.42) of Chapter 1, allows us to introduce the density functional and the magnetic moment density of the total energy as follows:

$$E_{V_0,\vec{B}_0}[\rho,\vec{m}] = F[\rho,\vec{m}] + \int d\mathrm{r}\left[V_0(\mathrm{r})\rho(\mathrm{r}) - \vec{B}_0(\mathrm{r})\cdot\vec{m}(\mathrm{r})\right] \tag{2.11}$$

where:

$$F[\rho,\vec{m}] = \langle\Psi[\rho,\vec{m}]|(\hat{T}+\hat{W})|\Psi[\rho,\vec{m}]\rangle \tag{2.12}$$

is a universal functional of quadri-vector density $(\rho(\mathrm{r}), \vec{m}(\mathrm{r}))$.

From here, immediately will result the variational property of the quadri-density functional of the total energy:

$$E_0 = E_{V_0,\vec{B}_0}\left[\rho_0, \vec{m}_0\right] \tag{2.13}$$

$$E_0 < E_{V_0,\vec{B}_0}\left[\rho_0, \vec{m}_0\right], \text{for } \left(\rho(\mathrm{r}), \vec{m}(\mathrm{r})\right) \neq \left(\rho_0(\mathrm{r}), \vec{m}_0(\mathrm{r})\right) \tag{2.14}$$

in which E_0 is the total energy of the fundamental state, $\left(\rho_0(\mathrm{r}), \vec{m}_0(\mathrm{r})\right)$ is the quadri-vector density of the fundamental state, for a system located in the external quadri-potential $\left(V_0(\mathrm{r}), \vec{B}_0(\mathrm{r})\right)$.

Consequently one can made the following affirmation.

ExtDFT1 Proposition: taking into consideration the external quadri-vector, with scalar and vectorial components, the C bijection application from the DFT1 Definition, see equation (1.41) of Chapter 1, *is no longer valid:* $|\Psi\rangle \to \left(V(\mathrm{r}), \vec{B}(\mathrm{r})\right)$, *or, more precisely, for the fundamental state of the system the density quadri-vector does not always correspond to a unique value.*

This can be easily observed writing the state equation for two identical fundamental states under external quadri-vectorial potentials:

$$|\Psi\rangle = |\Psi'\rangle \Rightarrow (\hat{U} - \hat{U}')|\Psi\rangle = (E_0 - E'_0)|\Psi\rangle \tag{2.15}$$

from where, the resulting condition:

$$\det\left\{(\Delta U)_{ab} - (\Delta E_0)\delta_{ab}\right\} = 0 \tag{2.16}$$

is not in contradiction with the situation in which the two external quadri-potentials are differing one from each other through more than one constant, $U \neq U'+const.$. This was demonstrated (von Barth and Hedin, 1972) through the construction of an infinite set of external fields which for one Hamiltonian without Coulomb interaction satisfies a common fundamental state.

It is noticed the situation in which $\vec{B} \to 0$, when the formalism is reduced in approaching the density functionals for a separate density on "upward", respective "downward" spin polarities.

Generalizing the Kohn – Sham scheme from the DFT5 Theorem (extended Kohn–Sham of Chapter 1) results naturally for the spin characterized systems, in which, to simplify, one will consider the external magnetic field and the system's magnetic moment density having the non-zero only direction component 0z, respecting an arbitrary reference system: $\vec{B}(\mathrm{r}) = (0, 0, B(\mathrm{r}))$, $\vec{m}(\mathrm{r}) = (0, 0, m(\mathrm{r}))$.

ExtDFT2 (Kohn–Sham with spin) Theorem: The exact spin densities of the fundamental states of an arbitrarily fermionic system in internal Coulomb interaction and in a scalar external field and one magnetic field satisfy the self-consistent set of equations, with a (taking + or – values) meaning the "upward" and "downward" spin projection on 0z direction:

$$\left[\begin{array}{l} -\dfrac{\hbar^2}{2m}\nabla^2 + V(\mathrm{r}) - a\mu_0 B(\mathrm{r}) \\ +\int d\mathrm{r}' W(\mathrm{r},\mathrm{r}')\rho(\mathrm{r}') + V_{XC}\left([\rho_+;\rho_-];\mathrm{r}\right) \end{array}\right]\varphi_i^{(a)}(\mathrm{r}) = \varepsilon_i^{(a)}\varphi_i^{(a)}(\mathrm{r}),\ \varepsilon_i^{(a)} \le \varepsilon_j^{(a)} \le \ldots \tag{2.17}$$

$$\begin{aligned}V_{XC}^{(a)}([\rho_+,\rho_-];\mathrm{r}) &= \frac{\delta E_{XC}[\rho_+,\rho_-]}{\delta\rho_a(\mathrm{r})}\\ &= \frac{\delta}{\delta\rho_a(\mathrm{r})}\left\{F_L[\rho_+,\rho_-]-\frac{1}{2}\iint d\mathrm{r}d\mathrm{r}'\rho(\mathrm{r})W(\mathrm{r},\mathrm{r}')\rho(\mathrm{r}')-T_L[\rho_+,\rho_-]\right\}\end{aligned} \tag{2.18}$$

$$F_L[\rho_+,\rho_-] = \inf_{\substack{\hat{D}\to(\rho_+,\rho_-)\\ \hat{D}=\sum_K d_K|\Psi_K\rangle\langle\Psi_K|,\ \sum_K d_K=1}} \mathrm{Tr}\left\{\hat{D}\left(\hat{T}+\hat{W}\right)\right\} \tag{2.19}$$

$$T_L[\rho_+,\rho_-] = \inf_{\substack{\hat{D}\to(\rho_+,\rho_-)\\ \hat{D}=\sum_K d_K|\Psi_K\rangle\langle\Psi_K|,\ \sum_K d_K=1}} \mathrm{Tr}\left(\hat{D}\hat{T}\right) \tag{2.20}$$

$$\rho_a(\mathrm{r}) = \sum_{i=1}^{\infty}\gamma_i^{(a)}\left|\varphi_i^{(a)}(\mathrm{r})\right|^2 \tag{2.21}$$

$$\sum_{i=1}^{\infty}\gamma_i^{(a)} = N_a\ ,\quad N_+ + N_- = N \tag{2.22}$$

$$\begin{aligned}\gamma_i^{(a)} &= 1:\ \varepsilon_i^{(a)} < \mu^{(a)}\\ 0\le\gamma_i^{(a)} &\le 1:\ \varepsilon_i^{(a)} = \mu^{(a)}\\ \gamma_i^{(a)} &= 0:\ \varepsilon_i^{(a)} > \mu^{(a)}\end{aligned} \tag{2.23}$$

$\gamma_i^{(a)}$ being the mono-orbital occupation number and where $\mu^{(a)}$ represents the energy of the highest mono-particle level occupied in spin state "*a*" (identified with Fermi energy).

Now results also the relation in analogy with that one without the spin quantification, but with the separation specificity on the upward/downward spin density variables, as it follows:

$$\begin{aligned}E_0[\rho_+,\rho_-] &= \sum_{i=1}^{\infty}\sum_a\gamma_i^{(a)}\varepsilon_i^{(a)} - \frac{1}{2}\iint d\mathrm{r}d\mathrm{r}'\rho(\mathrm{r})W(\mathrm{r},\mathrm{r}')\rho(\mathrm{r}')\\ &+E_{XC}[\rho_+,\rho_-]-\sum_a\int d\mathrm{r}\,V_{XC}^{(a)}([\rho_+,\rho_-];\mathrm{r})\rho_a(\mathrm{r})\end{aligned} \tag{2.24}$$

This way the specific extension of the Kohn–Sham formalism of the total energy density functional of the fundamental state (degenerated or not) for the systems for which is taken into consideration the spin contribution was performed.

2.3 DFT REALIZATION OF EXCITED STATES

For developing the extension to the excited states we can start from the fundamental state energy of the density functional $E_V[\rho]$ for a system found in external scalar uni-potential field, $V(\mathrm{r})$, which contains specific information about the excited states which can be summarized in the following Proposition (Levy and Perdew, 1993).

ExtDFT2 Proposition: (a) Each density $\rho_i(\mathrm{r})$ *which optimizes the density functional of the fundamental state energy* $E_V[\rho]$ *corresponds to* E_i *energy that is formed as being an exact stationary state of the system (including an excited state). Minimum absolute density corresponds to fundamental sate energy. (b) Not every excited state density* $\rho_i(\mathrm{r})$ *corresponds to an extreme of the density functional* $E_V[\rho]$*. (c) If* $\rho_i(\mathrm{r})$ *is an arbitrary density of an excited state* E_i *then it is low limited by the density functional of the exact energy of the fundamental state,* $E_V[\rho_i] \leq E_i$*, equality that takes place only if* $\rho_i(\mathrm{r})$ *is a function that extremizes the functional* $E_V[\rho]$.

Based on these assertions, one understand the approach (Valone and Capitani, 1981) dealing with arbitrary excited state as an extinction of Rayleigh–Ritz principle for a Hamiltonian developed into MacDonald series (MacDonald, 1934), which shows fulfill the following theorem.

ExtDFT3 Theorem: If E_b *is the energy of an excited state that minimizes the polynomial* $(E-\lambda)^2$*, with l parameter arbitrary fixed as real, then there takes place the inequality:*

$$\langle\psi|\left(\hat{H}-\lambda\right)^2|\Psi\rangle \geq (E_b-\lambda)^2 \tag{2.25}$$

satisfied for every N-particles functions, $|\Psi\rangle$*, anti-symmetrically normalized.*

A different way to approach excited states is achieved on the basis of variational Rayleigh–Ritz as applied on generalized fractional occupied excited states, based on the following theorem (Gross et al., 1988a, 1988b).

For any integers, $m \leq M$*, there is defined* L_m *as the subspace covered by the states* $|j\rangle$ *of the Hamiltonian* $\hat{H}$ *with* $E_j < E_{m'}$ *and by subspace* U_m *covered by eigen-states* $|j\rangle$ *of the Hamiltonian* $\hat{H}$ *with* $E_j \leq E_m$ *such that* U_m *contains the subspace* L_m *and all the multiplet states with energy* $E_{m'}$*. Then, it takes place: (a) If* $\eta_1, \ldots, \eta_M$ *are positive real numbers so that* $\eta_1 \geq \eta_2 \geq \ldots \geq \eta_M > 0$*, then, for any ortho-normalized set of functions* $\{|\Psi_1\rangle, |\Psi_2\rangle, \ldots, |\Psi_M\rangle\}$ *the generalized Rayleigh-Ritz principle occurs:*

$$\eta_1\langle\Psi_1|\hat{H}|\Psi_1\rangle + \ldots + \eta_M\langle\Psi_M|\hat{H}|\Psi_M\rangle \geq \eta_1 E_1 + \ldots + \eta_M E_M \tag{2.26}$$

(b) The previous relationship becomes equality if and only if $m = M$*, while for any other values of m with* $\eta_m \neq \eta_{m+1}$ *the inclusion occurs:*

$$L_m \subset \{|\Psi_1\rangle, |\Psi_2\rangle, \ldots, |\Psi_M\rangle\} \subset U_m \tag{2.27}$$

Actually, to demonstrate the theorem is equivalent to show that:

$$(\Delta E)_M = \sum_{m=1}^{M} \eta_m \left[\langle\Psi_m|\hat{H}|\Psi_m\rangle - E_m\right] \tag{2.28}$$

is nonnegative in the first part of the theorem and cancels in the second part if and only if the relationship (2.27) takes place for $m = M$, while for any other m provides $\eta_m \neq \eta_{m+1}$.

If one assumes the formally setting $\eta_{M+1} = 0$, then we have for (2.28):

$$(\Delta E)_M = \sum_{m=1}^{M}(\eta_m - \eta_{m+1})\sum_{k=1}^{m}\left[\langle\Psi_k|\hat{H}|\Psi_k\rangle - E_k\right] = \sum_{m=1}^{M}(\eta_m - \eta_{m+1})(\Delta E)_m^{EQUI} \tag{2.29}$$

since recognizing the equi-occupied states with

$$\eta_1 = \eta_2 = \ldots = \eta_m = 1, \eta_{j>m} = 0 \tag{2.30}$$

Therefore, taking into account that $\eta_1 \geq \eta_2 \geq \ldots \geq \eta_M > 0$, practically, it must be shown that:

$$(\Delta E)_m^{EQUI} = \sum_{k=1}^{m}\left[\langle\Psi_k|\hat{H}|\Psi_k\rangle - E_k\right]\begin{cases} > 0 \ , & (a) \\ = 0 \ , & (b) \end{cases} \tag{2.31}$$

To this aim one unfolds the occupied states wave functions $|\Psi_k\rangle$, $k = 1,\ldots, M$ in the complete set of functions of the Hamiltonian $\hat{H}$:

$$|\Psi_k\rangle = \sum_{j=1}^{\infty}\alpha_{kj}|j\rangle, \quad \sum_{j=1}^{\infty}|\alpha_{kj}|^2 = 1 \tag{2.32}$$

with which the relation (2.31) is rewritten:

$$(\Delta E)_m^{EQUI} = \sum_{j=1}^{\infty}\gamma_j E_j - \sum_{j=1}^{\infty}E_j \tag{2.33}$$

In equation (2.33) γ_i is the orbital occupation number, respectively the probability that one has its eigen-state $|j\rangle$ as belonging to the subspace $\{|\Psi_1\rangle, |\Psi_2\rangle, \ldots, |\Psi_m\rangle\}$ and having the obvious properties:

$$0 \leq \gamma_j = \sum_{k=1}^{m}|\alpha_{kj}|^2 \leq 1, \sum_{j=1}^{\infty}\gamma_j = m \tag{2.34}$$

The relationship (2.33) can be now rewritten also under the form:

$$(\Delta E)_m^{EQUI} = \sum_{j=1}^{m}(\gamma_j - 1)E_j + \sum_{j=m+1}^{\infty}\gamma_j E_j \tag{2.35}$$

which if added with the null terms:

$$\sum_{j=1}^{m}(1-\gamma_j)E_m = 0 \ , \ \sum_{j=m+1}^{\infty}\gamma_j E_m = 0 \tag{2.36}$$

will look as follows:

$$(\Delta E)_m^{EQUI} = \sum_{j=1}^{m}(1-\gamma_j)(E_m - E_j) + \sum_{j=m+1}^{\infty}\gamma_j(E_j - E_m) \tag{2.37}$$

that is a positive expression for occupation numbers (2.34) and accounting for their relative energy values fixed by the summation limits ranges. Thus, the first part of the theorem was proved.

For the second part of the theorem one will reconsider the writing the energy variation (2.37) such that the first term to stop at the last state r, $r<m$, within the space L_m:

$$(\Delta E)_m^{EQUI} = \sum_{j=1}^{r}(1-\gamma_j)(E_m - E_j) + \sum_{j=s+1}^{\infty}\gamma_j(E_j - E_m) \tag{2.38}$$

From equation (2.38) there is observed that the result is obtained only when $\gamma_j = 1$ for $j \leq r$ that implies $m = M$ and, respectively, when $\gamma_j = 0$ for $j>s$ that is equivalent to assertion (b) of the theorem.

Basically the theorem ExtDFT4 tells us that the excited eigen-states are allowed with fractional occupation numbers.

Next, we aim a methodology for acquiring Kohn–Sham formulation of for excited states.

For a given set of occupation numbers, $\vec{\gamma} \equiv (\gamma_1, \gamma_2, ...)$ the KS orbital energies and their associated energies (DFT5 Theorem of Chapter 1) can be determined depending on the set of occupation numbers $\{\varepsilon_i(\vec{\gamma})\ \varphi_i^{(\vec{\gamma})}(x)\ ,\ i = 1,...\infty\}$.

It is noteworthy that, based on expression (1.92) of the total energy associated with the electronic system for a given set of occupation numbers, one may formulate the next.

ExtDFT5 Theorem (of Janak, 1978): For a given set of occupational numbers $\vec{\gamma} \equiv (\gamma_1, \gamma_2, ...)$ the total stationary energy is also considered as occupational dependent vector,

$$E(\gamma_1, \gamma_2, ...) := \tilde{E}\left[\varphi_1^{(\vec{\gamma})}, \varphi_2^{(\vec{\gamma})}, ...; \gamma_1, \gamma_2, ...\right] \tag{2.39}$$

and its partial derivatives respecting occupational numbers considered in the set of occupational numbers would correspond to the associated eigen-energy in excited states, and they depend on the entire vector of occupation number:

$$\frac{\partial E(\gamma_1, \gamma_2, ...)}{\partial \gamma_j}\bigg|_{\vec{\gamma}} = \varepsilon_j(\vec{\gamma}) \tag{2.40}$$

Given the importance of choosing the set of occupational numbers we will further illustrate some special elections.

ExtDFT1 Definition: For the non-degenerate case the set of occupation numbers of the fundamental state of a system composed of N electrons is defined by the vector:

$$\vec{\gamma}_0 = \begin{cases} \gamma_i = 1\ \ , i = 1, 2, ..., N \\ \gamma_i = 0\ \ , i > N \end{cases} \tag{2.41}$$

With this definition, the exact fundamental state energy (1.92) is found here

$$E(\vec{\gamma}_0) \equiv E_0 = \sum_{i=1}^{\infty} \gamma_i \varepsilon_i - \frac{1}{2} \iint d\mathrm{r} d\mathrm{r}' \rho(\mathrm{r}) W(\mathrm{r},\mathrm{r}') \rho(\mathrm{r}') + E_{XC}[\rho] - \int \rho(\mathrm{r}) V_{XC}([\rho];\mathrm{r}) d\mathrm{r} \tag{2.42}$$

ExtDFT2 Definition: We define the set of occupation numbers for excited "particle - vacancy" couple for a system consisting of N electrons by the vector:

$$\vec{\gamma}_{qp} = \begin{cases} \gamma_i = 1 \ , i = 1,...,(q-1),(q+1),...,N \ \ \wedge i = p(>N) \\ \gamma_i = 0 \ , i = q \ \wedge \ i > N, i \neq p \end{cases} \tag{2.43}$$

In this case, the approximate energy of the excited states will be given by:

$$\Delta E_{qp} = E(\vec{\gamma}_{qp}) - E(\vec{\gamma}_0) \tag{2.44}$$

but with a numerical disadvantage, because the calculation of this expression must be made through a self-consistent calculation for each set of occupation numbers (2.41) and (2.43). To remove this shortcoming, Slater introduced the concept of "transitional state" formulated by the following definition (Slater, 1974).

ExtDFT3 Definition (of Slater transition state): One defines the set of occupation numbers of excitation by "Slater transition state" of a system composed of N electrons, by the vector:

$$\vec{\gamma}_T = \begin{cases} \gamma_i = 1 \ , i = 1,...,(q-1),(q+1),...,N \\ \gamma_i = \dfrac{1}{2} \ , i = q, i = p \\ \gamma_i = 0 \ , i > N, i \neq p \end{cases} \tag{2.45}$$

In this context the excited state energy is given by the difference:

$$\Delta E_{qp} = E(\vec{\gamma}_T - \Delta\vec{\gamma}) - E(\vec{\gamma}_T + \Delta\vec{\gamma}) \tag{2.46}$$

$$\Delta\vec{\gamma} = \begin{cases} \gamma_q = 1/2 \\ \gamma_p = -1/2 \\ \gamma_i = 0 \ , i \neq q, p \end{cases} \tag{2.47}$$

that through the expansion of both energies around $\vec{\gamma}_T$ and applying Janak's theorem (ExtDFT5 Theorem above) one is let with the result:

$$\Delta E_{qp} = \varepsilon_q(\vec{\gamma}_T) - \varepsilon_p(\vec{\gamma}_T) + O\left((\Delta\vec{\gamma})^3\right) \tag{2.48}$$

in which the terms of order three and above are very small (Slater, 1974).

Therefore, the excited state energy can be well represented by the Slater transition state energy through difference of mono-particles energies.

It may make an extension of the Kohn – Sham methodology for excited states, which will restrict the fundamental state $|\Psi_0\rangle$ and the first excited state $|\Psi_1\rangle$:

ExtDFT6 Theorem (of Kohn – Sham for the first excited state):The wave functions associated to the fundamental and first excited state for a system of N-electrons without Coulomb interaction are Slater determinants $|\Psi_{0,s}\rangle$ and $|\Psi_{1,s}\rangle$ composed of mono-particle orbitals that satisfy the self-consistent set of equations :

$$\left[-\frac{\hbar^2}{2m}\nabla^2 + V_{KS}(\mathrm{r})\right]\varphi_i(\mathrm{r}) = \varepsilon_i\varphi_i(\mathrm{r}),\quad \varepsilon_i \le \varepsilon_i \le \ldots \tag{2.49}$$

$$V_{KS}(\mathrm{r}) = V_0(\mathrm{r}) + \int d\mathrm{r}'W(\mathrm{r},\mathrm{r}')\rho(\mathrm{r}') + V_{XC}([\rho];\mathrm{r}) \tag{2.50}$$

$$V_{XC}^{(\eta)}([\rho];\mathrm{r}) = \frac{\delta E_{XC}^{(\eta)}[\rho]}{\delta\rho_a(\mathrm{r})} \tag{2.51}$$

$$E_{XC}^{(\eta)}[\rho] = F^{(\eta)}[\rho] - \frac{1}{2}\iint d\mathrm{r}d\mathrm{r}'\rho(\mathrm{r})W(\mathrm{r},\mathrm{r}')\rho(\mathrm{r}') - T_{KS}^{(\eta)}[\rho] \tag{2.52}$$

$$\rho(\mathrm{r}) = \sum_{i=1}^{N-1}|\varphi_i(\mathrm{r})|^2 + (1-\eta)|\varphi_N(\mathrm{r})|^2 + \eta|\varphi_{N+1}(\mathrm{r})|^2 \tag{2.53}$$

$$T_{KS}^{(\eta)}[\rho] = \sum_{i=1}^{N-1} t_i + (1-\eta)t_N + \eta t_{N+1} \tag{2.54}$$

Once found the solutions of this self-consistent set of equations, that is $\varphi_i^{(\eta)}(\mathrm{r}),\varepsilon_i(\eta)$, for a density $\rho^{(\eta)}(\mathrm{r})$ for a given value of η, the ensemble energy can be calculated analogous to that of equation (2.24):

$$\begin{aligned} E_0(\eta) &= \sum_{i=1}^{N-1}\varepsilon_i(\eta) + (1-\eta)\varepsilon_N(\eta) + \eta\varepsilon_{N+1}(\eta) - \frac{1}{2}\iint d\mathrm{r}d\mathrm{r}'\rho^{(\eta)}(\mathrm{r})W(\mathrm{r},\mathrm{r}')\rho^{(\eta)}(\mathrm{r}') \\ &+ E_{XC}^{(\eta)}\left[\rho^{(\eta)}\right] - \int d\mathrm{r}\rho^{(\eta)}(\mathrm{r})V_{XC}^{(\eta)}\left([\rho^{(\eta)}];\mathrm{r}\right) \end{aligned} \tag{2.55}$$

Worth noting that for $\eta = 0$ the fundamental energy expression (1.92) is found.

The expression (2.55) allows for the next definition.

ExtDFT4 Definition: The first excited energy is calculated by the expression:

$$\Delta E_1 = \frac{1}{\eta}\left(E_0(\eta) - E_0(0)\right) \tag{2.56}$$

or, more general, by the expression:

$$\Delta E_1 = \frac{dE_0(\eta)}{d\eta} = \varepsilon_{N+1}(\eta) - \varepsilon_N(\eta) + \frac{\partial E_{XC}^{(\eta)}[\rho]}{\partial\eta} \tag{2.57}$$

For very small values of term $\partial/\partial\eta\left(E_{XC}^{(\eta)}[\rho]\right)$, such as the non-degenerate fundamental state and for the first *g*-degenerate excited state with very high *g* (Gross et al., 1988a, 1988b), the relation (2.57) reduces to Slater transition state scheme (2.48), while further improvements to the scheme of Slater transition may be achieved for the better approximations of $E_{XC}^{(\eta)}$ than the term of the Slater exchange itself, see equation (3.50) in Chapter 3.

2.4 DFT REALIZATION OF THE INTERNAL SYMMETRY FOR DEGREE OF FREEDOM

One can setup the axiomatic formalism of the theory of Hohenberg–Kohn–Sham density functional to include the degree of freedom, respectively the internal symmetry invariance properties of multi- electronic through the next theorem systems (Flores and Keller, 1992; Keller, 1986).

ExtDFT7 Theorem (of internal symmetry constraint): The formal structure of the functional density theory can be restrained at three assumptions: (a) total energy of the system in any stationary state is a unique functional of its particle density (b) There is a variational principle for the energy functional for the stationary states of the system, (c) internal symmetry of the system is necessarily part of the theory by the set of parameters $\{c_{i,N}\}$ *and variational equation will look like:*

$$\delta\left\{E\left[\rho;\{c_{i,N}\}\right]-\mu\left(\int\rho\left[\{c_{i,N}\}\right]d\mathrm{r}-N\right)\right\}=0 \tag{2.58}$$

Based on this principle, in the following shall be deducted that the basic KS equation can be amended such that to include the degree of freedom of internal symmetry effects, effects of atomic or molecular electronic states. To this end, one will seek to obtain a square root equation of electronic density, say with the form:

$$h_{eff}\rho^{1/2}(\mathrm{r})=\mu\rho^{1/2}(\mathrm{r}) \tag{2.59}$$

$$h_{eff}=-\nabla^2+V(\mathrm{r})+V_{eff}(\mathrm{r}) \tag{2.60}$$

It starts from an approximate form of the density functional of the kinetic energy of electrons without interaction (KS) system:

$$T_{KS}[\rho]=\int\rho^{1/2}(\mathrm{r})\left(-\nabla^2\right)\rho^{1/2}(\mathrm{r})d\mathrm{r}+T_\theta[\rho] \tag{2.61}$$

in which the $T_\theta[\rho]$ term completes the kinetic density functional expression with specific effects of the electronic state's structure. Taking the variation procedure for equation (2.61) one gets

$$\frac{\delta T_{KS}[\rho]}{\delta\rho(x)}=-\frac{\nabla^2\rho^{1/2}}{\rho^{1/2}}+V_\theta([\rho];\mathrm{r})+V_{SS}([\rho];\mathrm{r}) \tag{2.62}$$

with

$$V_\theta([\rho];\mathrm{r})+V_{SS}([\rho];\mathrm{r})=\frac{\delta T_\theta[\rho]}{\delta\rho(\mathrm{r})} \tag{2.63}$$

acting as effective potential in which the first term should include the potential structure effects function, while the second term is related to the principle of uncertainty, being these reasons potential the whole expression (2.63) is called Pauli potential.

Now, taking also into account the Euler – Lagrange variation one has for the KS energy functional the formal expression:

$$\frac{\delta T_{KS}[\rho]}{\delta\rho(\mathrm{r})}+V_{KS}([\rho];\mathrm{r})=\varepsilon_M \tag{2.64}$$

which along its non-interaction form (2.61) provides the square root equation of the electronic density in the searched form:

$$-\nabla^2\rho^{1/2}+V_{KS}([\rho];\mathrm{r})\rho^{1/2}+\{V_\theta([\rho];\mathrm{r})+V_{SS}([\rho];\mathrm{r})\}\rho^{1/2}=\mu\rho^{1/2} \tag{2.65}$$

ExtDFT5 Definition: One can define the energy correction for the effects of electronic structure of states:

$$\Delta\bar{\varepsilon}(\mathrm{r})=\frac{1}{\rho(\mathrm{r})}\left\{\sum_{i=1}^{N}\varepsilon_i\left|\varphi_{i,KS}(\mathrm{r})\right|^2-\bar{\varepsilon}\rho(\mathrm{r})\right\} \tag{2.66}$$

where:

$$\bar{\varepsilon}=\frac{1}{N}\sum_{i=1}^{N}c_i\varepsilon_i \tag{2.67}$$

corresponds to the summation of the Lagrange ε_i multipliers parameters in the KS procedure associated with auxiliary KS orbitals.

Therefore, $\Delta\bar{\varepsilon}(\mathrm{r})$ expression can be taken as a approximation of V_{SS} obtained from constructing the ExtDFT7 Theorem above.

In these conditions the functional that satisfies the energy variational principle with constraint inclusion of the structure effects of orbital electronic stratification (effects of internal symmetry) can be formulated:

$$\Lambda=E[\rho]-\mu\left(\int\rho(\mathrm{r})d\mathrm{r}-N\right)-\lambda\left(\int\sum_i\varepsilon_i\left|\varphi_i(\mathrm{r})\right|^2d\mathrm{r}-\int\bar{\varepsilon}\rho(\mathrm{r})d\mathrm{r}\right) \tag{2.68}$$

For determining equation that satisfies the constraining effects of internal symmetry one will next consider the system of equations KS:

$$-\nabla^2\varphi_i+[V_{XC}(\mathrm{r})+V_{COUL}(\mathrm{r})+V_{EXT}(\mathrm{r})]\varphi_i=\varepsilon_i\varphi_i\ ,\ \ i=\overline{1,N} \tag{2.69}$$

from which, multiplying to the left with φ_i^* and including all mono-electronic states it yields the expression:

$$\sum_i h_i(\mathrm{r})^{KS}=\sum_i\varepsilon_i\left|\varphi_i\right|^2 \tag{2.70}$$

$$h_i(\mathrm{r})^{KS} = -\varphi_i^* \nabla^2 \varphi_i + \varphi_i^* \left[V_{XC}(\mathrm{r}) + V_{COUL}(\mathrm{r}) + V_{EXT}(\mathrm{r}) \right] \varphi_i \tag{2.71}$$

If both parts of (2.70) are gathered with the terms $N^{-1}\sum_{i=1}^{N}\varepsilon_i \rho^{1/2}$ and $-\nabla^2\rho^{1/2}$, and then dividing the result by $\rho^{1/2}$ the next PROPOSITION can be affirmed.

ExtDFT3 (of Flores–Keller) Proposition: Electronic density and the sum of stationary states of an electronic system that was considered also with the effects of structure on electronic configuration satisfy the following set of self-consistent equations:

$$-\nabla^2 \rho^{1/2} + V^{FK}(\mathrm{r})\rho^{1/2} = \bar{\varepsilon}\rho^{1/2} \tag{2.72}$$

$$V^{FK}(\mathrm{r}) = V_{XC}(\mathrm{r}) + V_{COUL}(\mathrm{r}) + V_{EXT}(\mathrm{r}) + \rho^{-1/2}\nabla^2 \rho^{1/2} + V_\theta([\rho];\mathrm{r}) - \Delta\bar{\varepsilon}(\mathrm{r}) \tag{2.73}$$

Taking into account ExtDFT5 Definition the differential equation (2.72) can be considered also by its integrated form:

$$\int \rho^{1/2}\left(-\nabla^2 \rho^{1/2}\right) d\mathrm{r} + \int \rho^{1/2} V^{FK}(\mathrm{r})\rho^{1/2} d\mathrm{r} = \sum_i c_i \varepsilon_i \tag{2.74}$$

that by numerical methods allows calculating the system's eigen-energies with electronic internal symmetry effects included in the set of parameters $\{c_i\}$ that can improve the results obtained by KS methodology through various approximation of the kinetic term (2.61).

Note that this extended formalism can be applied to stationary excited states as well, when the first variational principle also takes into account (η) parameters:

$$\delta\left\{ E\left[\rho^{(\eta)};\left\{c_{i,N}^{(\eta)}\right\}\right] - \mu\left(\int \rho^{(\eta)}\left[\left\{c_{i,N}^{(\eta)}\right\}\right] d\mathrm{r} - N\right)\right\} = 0 \tag{2.75}$$

In addition, the formalism of the square root of the electronic density certainly has the advantage of reducing the KS system of $N+1$ self-consistent equations (N mono-orbitals equations in addition to the correlation and exchange potential equation – see DFT8 Theorem of Chapter 1) to only two self-consistent equations, yet with the cost of obtaining not the individual stationary energies but their sum along their characteristic electronic density.

2.5 DFT REALIZATION OF THE NON-BORN-OPPENHEIMER STATES

In developing the extension for characterizing molecular states through considering also the nuclear characteristics, as nuclei charges, distances and masses, the starting point is the non-relativistic Hamiltonian containing all inter-electronic, inter-nuclei and electron-nuclei interactions, and also the kinetic energies associated with the system of N electrons and $\{l\}$ nuclei (Capitani et al., 1982):

$$\begin{aligned}\hat{H} &= \hat{T}_e + \hat{T}_n + \hat{V}_{ne} + \hat{W} + \hat{V}_{nn} \\ &= -\frac{1}{2}\sum_{i=1}^{N}\nabla_i^2 - \sum_{a=1}^{l}\frac{1}{2M_a}\nabla_a^2 - \sum_{i=1}^{N}\sum_{a=1}^{l}\frac{Z_a}{|\mathrm{r}_i - \mathrm{R}_a|} + \sum_{i<j}^{N}\frac{1}{|\mathrm{r}_i - \mathrm{r}_j|} + \sum_{a<b}^{l}\frac{Z_a Z_b}{|\mathrm{R}_a - \mathrm{R}_b|}\end{aligned} \tag{2.76}$$

The associated stationary Schrödinger equation, $\hat{H}\Psi = E\Psi$, has therefore the wave function in the overall dependence of the coordinates and spin variables of the electrons and nuclei, $\Psi = \Psi(\{r_i, s_i\}, \{R_a, S_a\})$.

ExtDFT6 Definition: Given the molecular system wave function, there are defined the mono-particle densities respectively for electrons:

$$\rho(r_1) = N \int |\Psi|^2 \frac{d\tau^{N+k}}{d|r_1|} \tag{2.77}$$

and for nuclei:

$$\rho(R_i^a) = k_a \int |\Psi|^2 \frac{d\tau^{N+k}}{dR_1^a} \tag{2.78}$$

in which $d\tau^{N+k}$ denotes the product of the space –spin volume elements for all electrons and all nuclei, while k_a and R_i^a are the number of nuclear species "a" and the position vector of nucleus "i", respectively.

ExtDFT7 Definition: One considers the fundamental state energy density functional for non – Born – Oppenheimer (NBO) extension:

$$E[\rho, \{\rho_a\}] \equiv \min \left\langle \Psi_{\rho,\{\rho_a\}} \middle| \hat{H} \middle| \Psi_{\rho,\{\rho_a\}} \right\rangle \tag{2.79}$$

in which the search of electron densities and the set of nuclear densities for all species, is conducted for Hamiltonian (2.76) adjusted to include specific symmetries (particles' symmetry, the inversion symmetry, Galilean symmetry) for an empirical molecular formula (including isotopes).

Given these definitions the following theorem can be stated for NBO case.

ExtDFT8 Theorem: The variational principle in NBO extension has the form:

$$E[\rho, \{\rho_a\}] \geq E_0 \tag{2.80}$$

$$E[\rho_0, \{\rho_a\}_0] = E_0 \tag{2.81}$$

in which $\rho_0, \{\rho_a\}_0, E_0$ are respectively the electronic density, the densities for the species of nuclei and the total energy density in the molecular fundamental state.

The demonstrations are straightforward. For the first part on will resort to Rayleigh–Ritz principle and ExtDFT7 Definition and immediately writes:

$$E[\rho, \{\rho_a\}] = \min \left\langle \Psi_{\rho,\{\rho_a\}} \middle| \hat{H} \middle| \Psi_{\rho,\{\rho_a\}} \right\rangle \equiv \langle \Phi | \hat{H} | \Phi \rangle \geq \langle \Psi_0 | \hat{H} | \Psi_0 \rangle = E_0 \tag{2.82}$$

For the second part of the theorem one will use the principle of double inequalities to prove (2.81); actually, from equation (2.82) we can write that:

$$E[\rho_0, \{\rho_a\}_0] \geq E_0 \tag{2.83}$$

and on the other side from the property of minimum of the density functional of the total energy we can write that:

$$E\left[\rho_0,\{\rho_a\}_0\right]=\min\left\langle\Psi_{\rho_0,\{\rho_a\}_0}\middle|\hat{H}\middle|\Psi_{\rho_0,\{\rho_a\}_0}\right\rangle\leq E_0 \tag{2.84}$$

Therefore, the last two combined relations render the second theorem relationship equation (2.81) and thus the variational principle is demonstrated also for NBO extension.

ExtDFT9 Theorem: (a) There is a unique functional relationship between the set of mono-particle densities $\rho,\{\rho_a\}$ *and a V external scalar potential; (b) The total molecular energy density functional* $E[\rho,\{\rho_a\}]$ *satisfies the minimum optimization property against the external scalar potentials applied to the molecular system.*

The demonstration for the first affirmation is done by reduction to absurdum method, also used in HK1 theorem; it begins by defining the Hamiltonian extension of this section, equation (2.76), with the action of an external scalar potential embodied:

$$\hat{H}_V=\hat{H}+\sum_{i=1}^{N}V(\mathrm{r}_i)-\sum_{a=1}^{l_a}Z_aV(\mathrm{R}_a) \tag{2.85}$$

There will be now considered two non-equivalent external scalar potentials (that differ by more than a constant) acting on the molecular system, say V and V'. Then, the Rayleigh – Ritz principle as applied for fundamental states associated with wave functions determined from Hamiltonian (2.85), for each external potential action, will have the form:

$$E_{V'}=\left\langle\Psi_{V'}\middle|\hat{H}_{V'}\middle|\Psi_{V'}\right\rangle<\left\langle\Psi_{V}\middle|\hat{H}_{V'}\middle|\Psi_{V}\right\rangle=\left\langle\Psi_{V}\middle|\hat{H}_{V}+V'-V\middle|\Psi_{v}\right\rangle \tag{2.86}$$

Eq. (2.86) equivalents with:

$$E_{V'}<E_V+\int\rho(\mathrm{r})\left[V'(\mathrm{r})-V(\mathrm{r})\right]d\mathrm{r}-\sum_{a=1}^{l_a}Z_a\int\rho_a(\mathrm{R}_a)\left[V'(\mathrm{R}_a)-V(\mathrm{R}_a)\right]d\mathrm{R}_a \tag{2.87}$$

with l_a being the number of groups of identical nuclei. From this point onward the deduction is in a similar way with the reasoning of HK1 theorem bijection proof between externally applied potential to the molecular system and molecular total energy density functional (see Section 1.3.1):

$$E_V\left[\rho,\{\rho_a\}\right]=E\left[\rho,\{\rho_a\}\right]+\int\rho(\mathrm{r})V(\mathrm{r})d\mathrm{r}-\sum_{a=1}^{l_a}Z_a\int\rho_a(\mathrm{R}_a)V(\mathrm{R}_a)d\mathrm{R}_a \tag{2.88}$$

Note that because of dependence on nuclear mass and charge specific to each nuclear species considered, *the functional* $E[\rho,\{\rho_a\}]$ *of equation (2.88) does not have a universal character.*

The second assertion follows directly from considering the variational principle of ExtDFT8 theorem,

$$E_V\left[\rho_{V'},\{\rho_a\}_{V'}\right]>E_V\left[\rho_V,\{\rho_a\}_V\right] \tag{2.89}$$

with the observation that everywhere was made the tacit assumption that the set of densities $\rho,\{\rho_a\}$ is respectively N- and $\{k_a\}$- representative.

Next rise the question of finding the Euler equation for the fundamental state represented by the hypersurface $E = E(N;\{k_a, Z_a, M_a\})$. For this we will start by building an auxiliary variational functional that has embedded all the conservation conditions for the number of particles in the system, electrons and nuclei:

$$\Lambda[\rho,\{\rho_a\}] = E[\rho,\{\rho_a\}] - \mu\left[\int \rho(\mathrm{r})d\mathrm{r} - N\right] - \sum_a \lambda_a\left[\int \rho_a(\mathrm{R}_a)d\mathrm{R}_a - k_a\right] \tag{2.90}$$

Now, the Euler equations for the fundamental state $[\rho_0,\{\rho_a\}_0]$ employ the stationary conditions:

$$\frac{\delta\Lambda}{\delta\rho(\mathrm{r})}\bigg|_{\rho=\rho_0} = \frac{\delta E}{\delta\rho(\mathrm{r})}\bigg|_{\rho=\rho_0} - \mu = 0 \tag{2.91}$$

$$\frac{\delta\Lambda}{\delta\rho_a(\mathrm{R}_a)}\bigg|_{\rho_a=(\rho_a)_0} = \frac{\delta E}{\delta\rho_a(\mathrm{R}_a)}\bigg|_{\rho_a=(\rho_a)_0} - \lambda_a = 0 \quad , a = \overline{1,l} \tag{2.92}$$

to produce the Lagrange multipliers' expressions as resulted by taking into account of the normalized constraints of the functional $\Lambda[\rho,\{\rho_a\}]$ with the significance of the chemical potential for electrons (electronegativity system with changed sign) and nuclear classes, respectively:

$$\mu = \left(\frac{\partial E}{\partial N}\right)_{(k_a, Z_a, M_a)} \tag{2.93a}$$

$$\lambda_a = \left(\frac{\partial E}{\partial k_a}\right)_{N, Z_a, M_a, (k_b, Z_b, M_b)} \tag{2.93b}$$

Observe that the partial derivatives (2.91) and (2.92) above ensure also the passage of the intensive variables of the chemical potential to the corresponding extensive densities. It is therefore permitted to rewrite the functional $\Lambda[\rho,\{\rho_a\}]$ of equation (2.90) as a Legendre transformed functional density:

$$E[\rho,\{\rho_a\}] = \mu N + \sum_{a=1}^{l_a} \lambda_a k_a + Q[\rho,\{\rho_a\}] \tag{2.94}$$

whose total differential is given by:

$$dE = \mu dN + N d\mu + \sum_{a=1}^{l_a} \lambda_a dk_a + \sum_{a=1}^{l_a} k_a d\lambda_a + dQ \tag{2.95}$$

Simultaneously, the total differential of the hyper-surface $E = E(N;\{k_a, Z_a, M_a\})$ will be written taking into account of the electronic and nuclear chemical potential expressions above (2.93a and 2.93b), as follows:

$$dE = \left(\frac{\partial E}{\partial N}\right)_{CT} dN + \sum_{a=1}^{l_a}\left(\frac{\partial E}{\partial k_a}\right)_{CT} dk_a + dR = \mu dN + \sum_{a=1}^{l_a} \lambda_a dk_a + dR \tag{2.96}$$

where

$$dR = \sum_{a=1}^{l_a}\left(\frac{\partial E}{\partial Z_a}\right)_{CT} dZ_a + \sum_{a=1}^{l_a}\left(\frac{\partial E}{\partial M_a}\right)_{CT} dM_a \tag{2.97}$$

Making the difference between the last two expressions, equations (2.95) and (2.96), one obtains an equation analogous to Gibbs–Duhem equation on macroscopic thermodynamic effects:

$$Nd\mu + \sum_{a=1}^{l_a} k_a d\lambda_a + dS = 0 \tag{2.98}$$

with $dS = dQ - dR$. One should remark that through neglecting the dR term and within the approximation of the infinite nuclear masses (i.e., the custom BO approximation) the corresponding Gibbs–Duhem equation for fixed nuclei approximation is also obtained.

2.6 DFT REALIZATION OF THE TEMPORAL DEPENDENCY

The electronic system in atoms and molecules can be seen as a dynamic fluid (Runge and Gross, 1984). Therefore, a very accessible way to implement multi-electronic temporal dependence relays on employing the hydrodynamic approach by related variational principles, from which the principles of energy density functional minimization will arise naturally (Bartolotti, 1981).

ExtDFT8 Definition: Through the time-dependent factorization of molecular wave function in the module and exponential (wave) argument,

$$\Psi(\mathrm{r},t) = \psi(\mathrm{r},t)\exp[iL(\mathrm{r},t)] \tag{2.99}$$

one can define the hydrodynamics functional:

$$J[\psi, V] = \langle\psi|\hat{T} + \hat{W} + \hat{V} + \frac{\partial}{\partial t}L|\psi\rangle \tag{2.100}$$

$$\hat{T} = -\frac{1}{2}\nabla^2 + \frac{1}{2}(\nabla L \cdot \nabla L) \tag{2.101}$$

$$\hat{W} = \frac{1}{2}\sum_{i \neq j}\frac{1}{|\mathrm{r}_i - \mathrm{r}_j|} \tag{2.102}$$

that satisfies the variational equation:

$$\delta J[\psi, L] - \frac{\partial}{\partial t}\langle\psi|\delta L|\psi\rangle = 0 \tag{2.103}$$

ExtDFT9 Definition: The hydrodynamic functional of equation (2.100) will be rewritten in terms of functional energy:

$$J[\psi, L] = E[\psi, L] + \langle\psi|\frac{\partial L}{\partial t}|\psi\rangle \tag{2.104}$$

$$E[\psi, L] = \langle \psi | (\hat{T} + \hat{W} + \hat{V}) | \psi \rangle \tag{2.105}$$

The external potential will be considered as periodic so that one takes the time-average in the hydrodynamic functional:

$$J[\psi, L]_t = \frac{1}{\Delta t} \int_t^{t+\Delta t} J[\psi, L]dt' = E[\psi, L]_t + \langle \psi | \frac{\partial L}{\partial t} | \psi \rangle_t \tag{2.106}$$

leading the variational principle (2.103) with the form:

$$\delta J[\psi, L]_t = 0 \tag{2.107}$$

Considering the independent variations in relation with *y* and *L*, the ExtDFT8 Definition equation is equivalent to the following set of coupled nonlinear differential equations:

$$(\hat{T} + \hat{W} + \hat{V})\psi = -\frac{\partial L}{\partial t}\psi = E\psi \tag{2.108}$$

$$\frac{\partial}{\partial t}\psi\psi + \nabla \cdot (\psi\psi\nabla L) = 0 \tag{2.109}$$

that is, in fact, express the hydrodynamic formulation of quantum mechanics.

On the first of the two equations above is not appropriate to insist, representing – of course – the general Schrödinger equation (2.108). The second equation, equation (2.109), is however reduced to the equation of continuity

$$\frac{\partial}{\partial t}\psi\psi + \nabla \cdot \vec{j} = 0 \tag{2.110}$$

if one takes account of writing the equation of definition ExtDFT8 employing the wave function and probability flux density formulas:

$$\Psi^*\Psi = \psi\psi \tag{2.111}$$

$$\vec{j}(x,t) = \frac{i}{2}(\Psi\nabla\Psi^* - \Psi^*\nabla\Psi) = \psi\psi\nabla L \tag{2.112}$$

Now the expansion of the variational principle for the time depending density functionals can be formulated in an already familiar manner.

ExtDFT10 Theorem: for HK density functional extended by considering the time integral as prescribed by the ExtDFT9 Definition:

$$F[\rho, L]_t = \langle \psi_{\min}^{\rho} | \hat{T} + \hat{W} | \psi_{\min}^{\rho} \rangle_t \tag{2.113}$$

the variational principle is valid for fundamental many-body state under the forms:

$$\text{(a)} \quad F[\rho, L]_t + \int \rho V \, d\sigma \geq E[\psi_0, L_0]_t \tag{2.114}$$

$$\text{(b)} \quad F[\rho_0, L_0]_t + \int \rho_0 V \, d\sigma = E[\psi_0, L_0]_t \tag{2.115}$$

with the notation $d\sigma = (dxdt)/\Delta t$.

The demonstrations are straightforward. For the first part one will consider the relationship derived from writing the HK functional and the Rayleigh – Ritz principle, namely:

$$F[\rho, L]_t + \int \rho V \, d\sigma = \left\langle \psi^{\rho}_{\min} \left| \hat{T} + \hat{W} + \hat{V} \right| \psi^{\rho}_{\min} \right\rangle_t \tag{2.116}$$

$$\left\langle \psi^{\rho}_{\min} \left| \hat{T} + \hat{W} + \hat{V} \right| \psi^{\rho}_{\min} \right\rangle_t \geq E[\psi_0, L_0]_t \tag{2.117}$$

that when summed up resembles the ExtDFT10 Theorem relationship (2.114).

For the second part of ExtDFT10 Theorem one will rewrite the inequality (2.117) only in HK functional components:

$$\left\langle \psi^{\rho_0}_{\min} \left| \hat{T} + \hat{W} \right| \psi^{\rho_0}_{\min} \right\rangle_t \geq \left\langle \psi_0 \left| \hat{T} + \hat{W} \right| \psi_0 \right\rangle \tag{2.118}$$

while, on the other hand, from the ground state wave-function $\psi^{\rho_0}_{\min}$ definition also results the inequality:

$$\left\langle \psi^{\rho_0}_{\min} \left| \hat{T} + \hat{W} \right| \psi^{\rho_0}_{\min} \right\rangle_t \leq \left\langle \psi_0 \left| \hat{T} + \hat{W} \right| \psi_0 \right\rangle \tag{2.119}$$

so the last two inequalities result in the equality:

$$\left\langle \psi^{\rho_0}_{\min} \left| \hat{T} + \hat{W} \right| \psi^{\rho_0}_{\min} \right\rangle_t = \left\langle \psi_0 \left| \hat{T} + \hat{W} \right| \psi_0 \right\rangle \tag{2.120}$$

on which, upon addition of the external potential action to each side, the relation (2.115) of the ExtDFT10 theorem also will result.

Next, given the continuous relationship (2.110) in the form:

$$\frac{\partial \rho}{\partial t} = -\nabla \cdot \vec{j} \tag{2.121}$$

and employing the N – representative density condition:

$$\int \rho \, d\sigma = N \tag{2.122}$$

one can advance the associated density functional:

$$\Lambda[\rho]_t = E[\rho, L]_t - \int \lambda^{\rho} \frac{\partial \rho}{\partial t} d\sigma - \eta^{\rho} \int \rho d\sigma \tag{2.123}$$

whose Euler–Lagrange equations containing the (2.121) and (2.122) conditions imposed by Lagrange multipliers $\lambda^{\rho}, \eta^{\rho}$:

$$\frac{\delta E[\rho, L]_t}{\delta \rho} = \frac{\delta F[\rho, L]}{\delta \rho} + V = -\frac{\partial \lambda^{\rho}}{\partial t} + \eta^{\rho} := \mu \tag{2.124}$$

$$\frac{\partial \rho}{\partial t} + \nabla \cdot (\rho \nabla L) = \frac{\partial \rho}{\partial t} + \nabla \cdot \vec{j} = 0 \tag{2.125}$$

where by the first relationship also the chemical potential from the contribution of total number of particles modified by temporal dynamics of probability current density was defined. These may lead with the following Proposition.

ExtDFT4 Proposition: For a molecular electronic system subject to periodic temporal dynamics coupled with an external potential action, the variational density functional (which satisfies an Euler–Lagrange type equation)of the total electronic energy can be formulated in terms of chemical potential and the Hohenberg–Kohn time-averaged functional as:

$$E[\rho, L]_t = \int \rho\mu \, d\sigma - Q[\rho, L]_t \tag{2.126}$$

with

$$Q[\rho, L]_t = \int \rho \frac{\delta F[\rho, L]_t}{\delta \rho} d\sigma - F[\rho, L]_t \tag{2.127}$$

being these equations with the same form as those obtained for stationary case (Parr et al., 1978).

In the last case, the stationary condition will be re-obtained from the condition $\nabla L = 0$, see for instance equation (2.109), and so the classical formulation Hohenberg – Kohn is resembled, see the term (2.101).

2.7 DFT REALIZATION OF THE THERMAL DEPENDENCY

2.7.1 Mermin Picture

ExtDFT11 Theorem: For the great canonical ensemble at temperature T and given chemical potential m, the sample grand canonical potential,

$$\Omega = E - \mu N - TS \tag{2.128}$$

has the equilibrium minimum values respecting the internal energy E, the number of particles N, and the entropy S variables.

ExtDFT10 Definition: For the calculation of microscopic properties at equilibrium the statistical great canonical operator is defined:

$$\hat{\rho} = \frac{\exp\left[-\beta\left(\hat{H} - \mu\hat{N}\right)\right]}{\mathrm{Tr}\left\{\exp\left[-\beta\left(\hat{H} - \mu\hat{N}\right)\right]\right\}} \tag{2.129}$$

with $\beta = 1/T$ (in atomic units). With this density the great sample canonical potential and the equilibrium density can be calculated by their standard expressions:

$$\Omega = \mathrm{Tr}\left\{\hat{\rho}\left(\hat{H} - \mu\hat{N} + \frac{1}{\beta}\ln\hat{\rho}\right)\right\} \tag{2.130}$$

$$\rho = \mathrm{Tr}\left\{\hat{\rho}\hat{N}\right\} \tag{2.131}$$

In these conditions, the density operator functional of the great canonical potential looks like

$$\Omega_{V-\mu}[\hat{\rho}'] = \mathrm{Tr}\left\{\hat{\rho}'\left(\hat{H} - \mu\hat{N} + \frac{1}{\beta}\ln\hat{\rho}'\right)\right\} \tag{2.132}$$

For a specified inter-particle interaction it will depend only on difference of the applied external potential to the chemical potential for a given temperature. As a consequence, the following theorem can be formulated.

ExtDFT12 Theorem: For a molecular system subject to an external scalar potential action, the application:

$$M : V(\mathrm{r}) - \mu \rightarrow \rho(\mathrm{r}) \tag{2.133}$$

is bijective, so that the electronic density of the fundamental state at finite temperature fixes in a unique way the difference between the applied external potential and internal chemical potential (recalling to the electronegativity of the molecular system, as the electronic availability to enter into an exchange of particles, that is producing the chemical interaction).

The demonstration uses the fact the minimum statistical property at the density functionals' level of the canonical potential, the analogue of the Rayleigh – Ritz principle, as prescribed by the minimum equilibrium affirmed by ExtDFT11 theorem:

$$\Omega_{V-\mu}[\hat{\rho}'] > \Omega_{V-\mu}[\hat{\rho}], \quad \hat{\rho} \neq \hat{\rho}' \tag{2.134}$$

Two different external potentials will be considered, $\hat{H}' = \hat{H} + \hat{V}' - \hat{V}$, and therefore two different "responses of chemical potential" will be recorded so that, using the relations above, one can writes successively:

$$\begin{aligned}
\Omega_{V'-\mu'}[\hat{\rho}'] &= \mathrm{Tr}\left\{\hat{\rho}'\left(\hat{H}' - \mu'\hat{N} + \frac{1}{\beta}\ln\hat{\rho}'\right)\right\} \\
&= \int d\mathrm{r}\left[(V'(\mathrm{r}) - \mu') - (V(\mathrm{r}) - \mu)\right]\rho'(\mathrm{r}) + \mathrm{Tr}\left\{\hat{\rho}'\left(\hat{H} - \mu\hat{N} + \frac{1}{\beta}\ln\hat{\rho}'\right)\right\} \\
&= \int d\mathrm{r}\left[(V'(\mathrm{r}) - \mu') - (V(\mathrm{r}) - \mu)\right]\rho'(\mathrm{r}) + \Omega_{V-\mu}[\hat{\rho}'] \\
&> \int d\mathrm{r}\left[(V'(\mathrm{r}) - \mu') - (V(\mathrm{r}) - \mu)\right]\rho'(\mathrm{r}) + \Omega_{V-\mu}[\hat{\rho}]
\end{aligned} \tag{2.135}$$

Likewise the reverse inequality can also be written:

$$\Omega_{V-\mu}[\hat{\rho}] > \int d\mathrm{r}\left[(V(\mathrm{r}) - \mu) - (V'(\mathrm{r}) - \mu')\right]\rho(\mathrm{r}) + \Omega_{V'-\mu'}[\hat{\rho}'] \tag{2.136}$$

Therefore, if considering the hypothesis the two potential differences $V(\mathrm{r}) - \mu$, $V'(\mathrm{r}) - \mu'$ differ by more than a constant, and they correspond in the two obtained inequalities (2.135) and (2.136) to two densities associated with the same fundamental/ground state then, by making a sum of these relations one gets the obvious contradiction as that of classical theory of Hohenberg–Kohn (HK1 Theorem, Section 1.3.1). Thus, the injection property is proven. On the other way, the surjection is obvious by the way of building the density operator and then the density from the ExtDFT10 Definition, equations (2.129)–(2.131). The uniqueness and inversability of the application (2.133), that is the ExtDFT12 Theorem, is so proved.

One may conclude this way that the statistical operator of the electronic density (2.129) can be considered as an equilibrium functional of the fundamental state density at finite temperature: $\hat{\rho} = \hat{\rho}[\rho_0]$.

Finally, may further formulate the following theorem (Mermin, 1965).

ExtDFT13 Theorem: The density functional of the canonical potential associated to a many-electronic system in the fundamental state at finite temperature can be written:

$$\Omega_{v-\mu}[\rho_0] = \int d\mathrm{r}\left(V(\mathrm{r}) - \mu\right)\rho_0(\mathrm{r}) + F[\rho_0] \tag{2.137}$$

with extended Hohenberg – Kohn functional for finite temperatures written in a statistical fashion,

$$F[\rho_0] = \mathrm{Tr}\left\{\hat{\rho}[\rho_0]\left(\hat{T} + \hat{W} + \frac{1}{\beta}\ln\hat{\rho}[\rho_0]\right)\right\} \tag{2.138}$$

such that it retains the universal nature at any temperature.

2.7.2 Parr Picture

The picture of quantum internal/local temperature begins from a consideration based on the following theorem (Parr, 1980).

ExtDFT14 Theorem: (a) For a molecular system of N electrons, there is not an external thermostat to maintain or provide the electronic fluid with a temperature gradient, as there is no external source for providing a quantum potential – while they are assured by the density distribution itself, with the system temperature having the local dependency. (b) The system's associated entropy is not the entropy associated with an energetic distribution but the microscopic entropy associated with density distribution ρ. A similar statement is valid also for the Gibbs free energy density.

To calculate these quantities we consider the electronic fluid as non-homogeneous and the density functionals will result as the average over all the dynamic variables.

ExtDFT11 Definition: To each probability element of the continuously fluid one associates a distribution function in phase space, $f(\mathrm{r},\mathrm{p})$ which satisfies the properties:

$$\rho(\mathrm{r}) = \int dp f(r,p) \tag{2.139}$$

$$\int d\mathrm{r}\rho(\mathrm{r}) = N \tag{2.140}$$

$$E[\rho] = \int \varepsilon(r,p)dr \tag{2.141}$$

$$\varepsilon(\mathrm{r},\mathrm{p}) = t_{KS}(\mathrm{r},\mathrm{p}) + \varepsilon_{pot}(\mathrm{r},\mathrm{p}) = \int d\mathrm{p} f(\mathrm{r},\mathrm{p})\left[\frac{\mathrm{p}^2}{2} + V(\mathrm{r},\mathrm{p})\right] \tag{2.142}$$

The effective form of distribution function has not mono-determinate character but for the present purpose we will determine its working form from the following definition.

ExtDFT12 Definition: The entropy and the entropy density will be considered in terms of the distribution function $f(\mathrm{r},\mathrm{p})$ *according with the forms:*

$$S = \int d\mathrm{r}\, s(\mathrm{r}) \tag{2.143}$$

$$s(\mathrm{r}) = -\int d\mathrm{p}\, f(\mathrm{r},\mathrm{p})\left[\ln f(\mathrm{r},\mathrm{p}) - 1\right] \tag{2.144}$$

with the units so chosen so that the Boltzmann constant is taken in atomic units (hartrees).

The most probable distribution function is obtained from maximum entropy density given the constraints for the correct density of equation (2.140) and for the kinetic energy density without considering the inter-particle interaction from the ExtDFT11, that is t_{KS} in (2.142); it thus assumes the working form:

$$f(\mathrm{r},\mathrm{p}) = \exp[-\alpha(\mathrm{r})]\exp\left[-\beta(\mathrm{r})\frac{\mathrm{p}^2}{2}\right] \tag{2.145}$$

with $\alpha(\mathrm{r}), \beta(\mathrm{r})$ being r-dependent Lagrange multipliers parameters, so featuring local dependency.

With the probability function (2.145) back into relationships (2.139) and (2.140) from the ExtDFT11 definition, and accounting for the kinetic definition from (2.142) it can be written with the form resembling the ideal gas

$$t_{KS}(\mathrm{r},\rho) = \frac{3}{2}\rho(\mathrm{r})T(\mathrm{r}) \tag{2.146}$$

with the working density:

$$\rho(\mathrm{r}) = \left(\frac{2\pi}{\beta(\mathrm{r})}\right)^{3/2}\exp[-\alpha(\mathrm{r})] \tag{2.147}$$

which proves the first part of theorem ExtDFT14. Equation (2.147) allows also further reformulations of the distribution function and of the electronic density itself, namely:

$$\begin{aligned} f(\mathrm{r},\mathrm{p}) &= \lambda(\mathrm{r})^3\rho(\mathrm{r})\exp\left[-\frac{\mathrm{p}^2}{2T(\mathrm{r})}\right] \\ &= -\frac{1}{2}\sum_{i=1}^{N}\nabla_i^2 - \sum_{a=1}^{I}\frac{1}{2M_a}\nabla_a^2 - \sum_{i=1}^{N}\sum_{a=1}^{I}\frac{Z_a}{|\mathrm{r}_i - \mathrm{R}_a|} + \sum_{i<j}^{N}\frac{1}{|\mathrm{r}_i - \mathrm{r}_j|} + \sum_{a<b}^{I}\frac{Z_a Z_b}{|\mathrm{R}_a - \mathrm{R}_b|} \end{aligned} \tag{2.148}$$

$$\begin{aligned} \rho(\mathrm{r}) &= \lambda(\mathrm{r})^{-3}\exp\left\{-\left[V_{eff}(\mathrm{r},\rho) - \mu(\mathrm{r},\rho)\right]\frac{1}{T(\mathrm{r})}\right\} \\ &= -\frac{1}{2}\sum_{i=1}^{N}\nabla_i^2 - \sum_{a=1}^{I}\frac{1}{2M_a}\nabla_a^2 - \sum_{i=1}^{N}\sum_{a=1}^{I}\frac{Z_a}{|\mathrm{r}_i - \mathrm{R}_a|} + \sum_{i<j}^{N}\frac{1}{|\mathrm{r}_i - \mathrm{r}_j|} + \sum_{a<b}^{I}\frac{Z_a Z_b}{|\mathrm{R}_a - \mathrm{R}_b|} \end{aligned} \tag{2.149}$$

where were considered the notations:

$$\mu(\mathrm{r},\rho)=-\frac{\alpha(\mathrm{r})}{\beta(\mathrm{r})}+V_{eff}(\mathrm{r},\rho) \tag{2.150}$$

$$\lambda(\mathrm{r})=\left[2\pi T(\mathrm{r})\right]^{-1/2} \tag{2.151}$$

Note that the shape of the distribution function (2.148) tells us that the considered electronic fluid is characterized by a local Maxwell–Boltzmann law. However, by replacing the first form of (2.148) into the entropy density (2.144) it acquires the successive equivalent forms:

$$\begin{aligned} s(\mathrm{r}) &= -\rho(\mathrm{r})\left\{\ln\left[\lambda^3\rho(\mathrm{r})\right]-1\right\}+\frac{3}{2}\rho(\mathrm{r}) \\ &= -\rho(\mathrm{r})\ln\rho(\mathrm{r})+\frac{3}{2}\rho(\mathrm{r})\ln T(\mathrm{r})+\frac{1}{2}\rho(\mathrm{r})[5+3\ln(2\pi)] \\ &= \frac{3}{2}\rho(\mathrm{r})\left[\frac{5}{3}+\ln\left(\frac{4\pi}{3}\right)+\ln\left(\frac{3}{2}\frac{\rho T(\mathrm{r})}{\rho^{5/3}(\mathrm{r})}\right)\right] \\ &= \frac{3}{2}\rho(\mathrm{r})\left[ct.+\ln\left(\frac{t_{KS}}{t_{Fermi}}\right)\right] \end{aligned} \tag{2.152}$$

The last relationship includes by two reasons a local addiction: once by the local temperature $T(\mathrm{r})$ and then by Fermi term - belonging by excellence to the local approximation.

Instead, from the first part of the explicit form of electronic density (2.149) can also be inferred immediately the expression for the chemical potential, namely:

$$\mu(\mathrm{r},\rho)=T(\mathrm{r})\ln\left(\lambda^3\rho\right)+V_{eff}(\mathrm{r},\rho) \tag{2.153}$$

in which the first term is the intrinsic chemical potential of a classical system of particles that do not interact. However, when comparing this expression with the corresponding relationship from the Kohn – Sham formalism,

$$\mu(\mathrm{r},\rho)=V_{KS}(\mathrm{r},\rho)+V_{Q}(\mathrm{r},\rho) \tag{2.154}$$

in which the components are written in relation to the equation (2.142) of ExtDFT11 Definition:

$$V_{KS}(\mathrm{r},\rho)=\frac{\delta\varepsilon_{pot}[\rho]}{\delta\rho} \tag{2.155}$$

$$V_{Q}(\mathrm{r},\rho)=\frac{\delta t_{KS}[\rho]}{\delta\rho} \tag{2.156}$$

there can be achieved the effectively potential through a quantum potential correction:

$$V_{eff}(\mathrm{r},\rho)=V_{KS}(\mathrm{r},\rho)+\left[V_Q(\mathrm{r},\rho)-T(\mathrm{r})\ln\left(\lambda^3\rho\right)\right] \tag{2.157}$$

which gives the framework for the following Proposition.

ExtDFT5 Proposition: For An interacting quantum system with the chemical potential m and r density the following two images are equally valid: (a) the auxiliary quantum system without interaction evolves under V_{KS} potential, or (b) the auxiliary classic system without interaction evolves under V_{eff} potential with a local temperature $T(\mathrm{r})$, and it is correlated with KS potential in a relationship such as given by equation (2.157).

Finally, in order to identify the local Gibbs free energy one will equate the internal energy density with ε_{pot} from equation (2.142), while the total energy density will be written in relation to the KS potential, namely with the chemical potential and quantum potential of equation (2.154) by means of a path of integration respecting the variable "*a*" by varying it from 0 to 1 so that $\rho_a \equiv \rho(x,a) = a\rho(x)$:

$$\varepsilon(\mathrm{r},\rho)=\int da\,\rho(\mathrm{r})V_{KS}(\mathrm{r};\rho_a)==\rho(\mathrm{r})\left\{\int_0^1 \mu[\rho_a]da-V_Q\right\} \tag{2.158}$$

Then the Gibbs free energy density will be written:

$$g(\mathrm{r})=\varepsilon_{pot}(\mathrm{r})-T(\mathrm{r})s(\mathrm{r})+P_{KS}(\mathrm{r})=\rho(\mathrm{r})\int_0^1 \mu[\rho_a]da+\rho(\mathrm{r})\left[T\ln\left(\lambda^3\rho\right)-V_Q\right] \tag{2.159}$$

where the kinetic pressure was introduced analogous to the ideal gas kinetic term (2.146),

$$P_{KS}(\mathrm{r},\rho)=\rho(\mathrm{r})T(\mathrm{r}) \tag{2.160}$$

This way a local thermodynamics picture characterizing a local electronic system was built as described by a set of field quantities like $T(\mathrm{r})$, $P_{KS}(\mathrm{r})$, $V_{eff}(\mathrm{r})$, (or $V_Q(\mathrm{r})$) and m. Note that local quantities are not always uniquely defined. For instance, for the choice of kinetic energy density as:

$$t_{KS}=\sum_{i=1}^{N}\frac{1}{8}\frac{\nabla\rho_i\cdot\nabla\rho_i}{\rho_i}-\frac{1}{8}\nabla^2\rho \tag{2.161}$$

then, by taking also into account the equations (2.146) and (2.156), the local quantum potential becomes explicitly:

$$V_Q(\mathrm{r})=\frac{3}{2}T(\mathrm{r})-\frac{1}{8}\frac{\nabla^2\rho(\mathrm{r})}{\rho(\mathrm{r})}-\frac{1}{\rho(\mathrm{r})}\left[\sum_{i=1}^{N}\varepsilon_i\rho_i(\mathrm{r})-\mu\rho(\mathrm{r})\right] \tag{2.162}$$

with the local temperature (for the ground state of hydrogen atom, for example):

$$T(\mathrm{r})=\frac{Z}{3|\mathrm{r}|} \tag{2.163}$$

where ρ_i, ε_i are the density and the KS mono-orbital electronic energy density, respectively.

Note that the equation (2.163) is in accord with the idea of a large electronic temperature near the nucleus, which corresponds to a very large orbital speed to prevent the atomic collapse.

Finally, we should stipulate the fact that the present transcript of the fundamental state DFT in a local thermodynamics allows the redistribution of electronic non-homogenous fluid coming from the two interacting systems into a new one, also non-homogeneous, defining the new binding state, the chemical bond, a molecule. Therefore, the local approach for describing the changing in the thermodynamic quantities of interest (temperature, chemical potential associated with the minus sign of electronegativity, the Gibbs free energy, and so on) are premises of a more realistic modeling picture for a physical-chemical interaction process - quantum information - reactivity - chemical bond.

KEYWORDS

- **Eigen-states**
- **Equi-occupied**
- **Gibbs–Duhem equation**
- **Kohn–Sham formalism**
- **Non-Born-Oppenheimer molecular systems**
- **Quadri-vector density**

3 Popular Density Functionals of Energy

CONTENTS

3.1 INTRODUCTION

The analytical survey of the main workable kinetic, exchange, and correlation density functionals within local and gradient density approximations is undertaken (Putz, 2008a).

3.2 DENSITY FUNCTIONALS OF KINETIC ENERGY

When the electronic density is seen as the diagonal element $\rho(\mathrm{r}_1)=\rho(\mathrm{r}_1,\mathrm{r}_1)$ the kinetic energy may be generally expressed from the Hartree–Fock model, through employing the single determinant $\rho(\mathrm{r}_1,\mathrm{r'}_1)$, as the quantity (Lee and Parr, 1987):

$$T[\rho]=-\frac{1}{2}\int\left[\nabla^2_{\mathrm{r'}_1}\rho(\mathrm{r}_1,\mathrm{r'}_1)\right]_{\mathrm{r}_1=\mathrm{r'}_1}d\mathrm{r}_1 \quad (3.1)$$

it may eventually be further written by means of the thermodynamic (or statistical) density functional (Zhao and Parr, 1992):

$$T_\beta=\frac{3}{2}\int\rho(\mathrm{r})k_B T(\mathrm{r})d\mathrm{r}=\frac{3}{2}\int\rho(\mathrm{r})\frac{1}{\beta(\mathrm{r})}d\mathrm{r} \quad (3.2)$$

that supports various specializations depending on the statistical factor particularization *b*.

For instance, in LDA approximation, the temperature at a point is assumed as a function of the density in that point, $\beta(\mathrm{r})=\beta(\rho(\mathrm{r}))$; this may be easily reached out by employing the scaling transformation to be (Ou–Yang and Levy, 1990)

$$\rho_\lambda(\mathrm{r})=\lambda^3\rho(\lambda\mathrm{r})\Rightarrow T[\rho_\lambda]=\lambda^2 T[\rho],\ \lambda=ct. \quad (3.3)$$

providing that

$$\beta(\mathrm{r}) = \frac{3}{2} C \rho^{-2/3}(\mathrm{r}) \tag{3.4}$$

a result that helps in recovering the traditional (Thomas–Fermi) energetic kinetic density functional form

$$T[\rho] = C \int \rho^{5/3}(\mathrm{r}) d\mathrm{r} \tag{3.5}$$

while the indeterminacy remained is smeared out in different approximation frames in which also the exchange energy is evaluated. Note that the kinetic energy is generally foreseen as having an intimate relation with the exchange energy since both are expressed in Hartree–Fock model as determinant values of $\rho(\mathrm{r}_1, \mathrm{r'}_1)$, see below.

Actually, the different LDA particular cases are derived by equating the total number of particle N with various realization of the integral

$$N = \frac{1}{2} \iint |\rho(\mathrm{r}_1, \mathrm{r'}_1)|^2 \, d\mathrm{r}_1 d\mathrm{r}_1{}' \tag{3.6}$$

by rewriting it within the inter-particle coordinates' frame:

$$\mathrm{r} = 0.5(\mathrm{r}_1 + \mathrm{r}_1{}'), \mathrm{s} = \mathrm{r}_1 - \mathrm{r}_1{}' \tag{3.7}$$

as

$$N = \frac{1}{2} \iint |\rho(\mathrm{r} + \mathrm{s}/2, \mathrm{r} - \mathrm{s}/2)|^2 \, d\mathrm{r} d\mathrm{s} \tag{3.8}$$

followed by spherical averaged expression:

$$N = 2\pi \iint \rho^2(\mathrm{r}) \Gamma(\mathrm{r}, s) d\mathrm{r} s^2 ds \tag{3.9}$$

with

$$\Gamma(\mathrm{r}, s) = 1 - \frac{s}{\beta(\mathrm{r})} + \ldots \tag{3.10}$$

The option in choosing the $\Gamma(\mathrm{r}, s)$ series (3.10) so that to converge in the sense of charge particle integral (3.9) fixes the possible cases to be considered (Lee and Parr, 1987):

1. the Gaussian resummation uses:

$$\Gamma(\mathrm{r}, s) \cong \Gamma_G(\mathrm{r}, s) = \exp\left(-\frac{s^2}{\beta(\mathrm{r})}\right) \tag{3.11}$$

2. the trigonometric (uniform gas) approximation looks like:

$$\Gamma(\mathrm{r}, s) \cong \Gamma_T(\mathrm{r}, s) = 9 \frac{(\sin t - t \cos t)^2}{t^6}, t = s\sqrt{\frac{5}{\beta(\mathrm{r})}} \tag{3.12}$$

In each of (3.11) and (3.12) cases the LDA-b function (3.4) is firstly replaced; then, the particle integral (3.9) is solved to give the constant C and then the respective kinetic energy density functional of (3.5) type is delivered; the results are (Lee and Parr, 1987):

1. in Gaussian resummation:

$$T_{G}^{LDA} = \frac{3\pi}{2^{5/3}} \int \rho^{5/3}(\mathrm{r})d\mathrm{r} \tag{3.13}$$

2. whereas in trigonometric approximation

$$T_{TF}^{LDA} = \frac{3}{10}\left(3\pi^2\right)^{2/3} \int \rho^{5/3}(\mathrm{r})d\mathrm{r} \tag{3.14}$$

one recovers the Thomas–Fermi formula type that closely resembles the original TF (1.70) formulation.

In next one will consider the non-local functionals; this can be achieved through the gradient expansion in the case of slowly varying densities – that is assuming the expansion (Murphy, 1981):

$$\begin{aligned} T &= \int d\mathrm{r}\left[\tau(\rho_{\uparrow}) + \tau(\rho_{\downarrow})\right] \\ &= \int d\mathrm{r}\sum_{m=0}^{\infty}\left[\tau_{2m}(\rho_{\uparrow}) + \tau_{2m}(\rho_{\downarrow})\right] \\ &= \int d\mathrm{r}\sum_{m=0}^{\infty}\tau_{2m}(\rho) \\ &= \int d\mathrm{r}\tau(\rho) \end{aligned} \tag{3.15}$$

The first two terms of the series respectively covers: the Thomas Fermi typical functional for the homogeneous gas

$$\tau_0(\rho) = \frac{3}{10}\left(6\pi^2\right)^{2/3} \rho^{5/3} \tag{3.16}$$

and the Weizsäcker related first gradient correction:

$$\tau_2(\rho) = \frac{1}{9}\tau_W(\rho) = \frac{1}{72}\frac{|\nabla\rho|^2}{\rho} \tag{3.17}$$

They both correctly behave in asymptotic limits:

$$\tau(\rho) = \begin{cases} \tau_0(\rho) = \tau_2(\rho) & \ldots\ \nabla\rho << \ (\textit{far from nucleus}) \\ 9\tau_2(\rho) = \tau_W(\rho) = \dfrac{1}{8}\dfrac{|\nabla\rho|^2}{\rho} & \ldots\ \nabla\rho >> \ (\textit{close to nucleus}) \end{cases} \tag{3.18}$$

However, an interesting resummation of the kinetic density functional gradient expansion series (3.15) may be formulated in terms of the Padé-approximant model (DePristo and Kress, 1987):

$$\tau(\rho) = \tau_0(\rho)P_{4,3}(x) \tag{3.19}$$

with

$$P_{4,3}(x) = \frac{1 + 0.95x + a_2x^2 + a_3x^3 + 9b_3x^4}{1 - 0.05x + b_2x^2 + b_3x^3} \tag{3.20}$$

and where the x-variable is given by

$$x = \frac{\tau_2(\rho)}{\tau_0(\rho)} = \frac{5}{108}\frac{1}{\left(6\pi^2\right)^{2/3}}\frac{|\nabla\rho|^2}{\rho^{8/3}} \tag{3.21}$$

while the parameters a_2, a_3, b_2, and b_3 are determined by fitting them to reproduce Hartree–Fock kinetic energies of He, Ne, Ar, and Kr atoms, respectively (Liberman et al., 1994). Note that Padé function (3.20) may be regarded as a sort of generalized electronic localization function (ELF) susceptible to be further used in bonding characterizations.

3.3 DENSITY FUNCTIONALS OF EXCHANGE ENERGY

Starting from the Hartree–Fock framework of exchange energy definition in terms of density matrix (Levy et al., 1996),

$$K[\rho] = -\frac{1}{4}\iint \frac{|\rho(\mathrm{r}_1, \mathrm{r'}_1)|^2}{|\mathrm{r}_1 - \mathrm{r'}_1|} d\mathrm{r}_1 d\mathrm{r}_1{}' \tag{3.22}$$

within the same consideration as before, we get that the spherical averaged exchange density functional

$$K = \pi \iint \rho^2(\mathrm{r})\Gamma(\mathrm{r}, s) d\mathrm{r} s ds \tag{3.23}$$

takes the particular forms (Lee and Parr, 1987):

1. in Gaussian resummation:

$$K_G^{LDA} = -\frac{1}{2^{1/3}}\int \rho^{4/3}(\mathrm{r}) d\mathrm{r} \tag{3.24}$$

2. and in trigonometric approximation (recovering the Dirac formula):

$$K_D^{LDA} = -\frac{3}{4}\left(\frac{3}{\pi}\right)^{1/3}\int \rho^{4/3}(\mathrm{r}) d\mathrm{r} \tag{3.25}$$

Alternatively, by paralleling the kinetic density functional previous developments the gradient expansion for the exchange energy may be regarded as the density dependent series (Cedillo et al., 1988):

$$\begin{aligned} K &= \sum_{n=0}^{\infty} K_{2n}(\rho) \\ &= \int dr \sum_{n=0}^{\infty} k_{2n}(\rho) \\ &= \int dr k(\rho) \end{aligned} \tag{3.26}$$

while the first term reproduces the Dirac LDA term (Perdew and Yue, 1986; Manoli and Whitehead, 1988):

$$k_0(\rho) = -\frac{3}{2}\left(\frac{3}{4\pi}\right)^{1/3} \rho^{4/3} \tag{3.27}$$

and the second term contains the density gradient correction, with the proposed Becke approximation (Becke, 1986):

$$k_2(\rho) = -b \frac{\dfrac{|\nabla\rho|^2}{\rho^{4/3}}}{\left(1 + d\dfrac{|\nabla\rho|^2}{\rho^{8/3}}\right)^a} \tag{3.28}$$

where the parameters b and d are determined by fitting the k_0+k_2 exchange energy to reproduce Hartree–Fock counterpart energy of He, Ne, Ar, and Kr atoms, and where for the a exponent either 1.0 or 4/5 value furnishes excellent results. However, worth noting that when analyzing the asymptotic exchange energy behavior, we get in small gradient limit (Becke, 1986):

$$k(\rho) \xrightarrow{\nabla\rho<<} k_0(\rho) - \frac{7}{432\pi\left(6\pi^2\right)^{1/3}} \frac{|\nabla\rho|^2}{\rho^{4/3}} \tag{3.29}$$

whereas the adequate large-gradient limit is obtained by considering an arbitrary damping function as multiplying the short-range behavior of the exchange-hole density, with the result:

$$k(\rho) \xrightarrow{\nabla\rho>>} c\rho^{4/5} |\nabla\rho|^{2/5} \tag{3.30}$$

where the constant c depends of the damping function choice.

Next, the Padé-resummation model of the exchange energy prescribes the compact form (Cedillo et al., 1988):

$$k(\rho) = \frac{10}{9} \frac{k_0(\rho)}{P_{4,3}(x)} \tag{3.31}$$

with the same Padé-function (3.20) as previously involved when dealing with the kinetic functional resummation. Note that when $x = 0$, one directly obtains the Ghosh–Parr functional (Ghosh and Parr, 1986):

$$k(\rho) = \frac{10}{9} k_0(\rho) \tag{3.32}$$

Moreover, the asymptotic behavior of Padé exchange functional (3.31) leaves with the convergent limits:

$$k(\rho) = \begin{cases} \frac{10}{9}\left(k_0 + \frac{15}{17} \frac{7}{432\pi\left(6\pi^2\right)^{1/3}} \frac{|\nabla\rho|^2}{\rho^{4/3}} \right) & \ldots\ x \to 0 \\ (SMALL \quad GRADIENTS) & \\ -12\pi \frac{\rho^2}{|\nabla\rho|^2} & \ldots\ x \to \infty \\ (LARGE \quad GRADIENTS) & \end{cases} \tag{3.33}$$

Once again, note that when particularizing small or large gradients and fixing asymptotic long or short range behavior, we are recovering the various cases of bonding modeled by the electronic localization recipe as provided by ELF limits (Bader et al., 1988; Becke, 1990; Cioslowski, 1990; Kohout et al., 2004; Matito et al., 2006; Putz, 2005; Santos et al., 2000; Scemama et al., 2000; Silvi, 2003; Silvi and Savin, 1994; Silvi and Gatti, 2000).

Another interesting approach of exchange energy in the gradient expansion framework was given by Bartolotti through the two-component density functional (Bartolotti, 1982):

$$K[\rho] = C(N)\int \rho(\mathrm{r})^{4/3} d\mathrm{r} + D(N)\int \mathrm{r}^2 \frac{|\nabla\rho|^2}{\rho^{2/3}} d\mathrm{r} \tag{3.34}$$

where the N-dependency is assumed to behave like:

$$C(N) = C_1 + \frac{C_2}{N^{2/3}},\ D(N) = \frac{D_2}{N^{2/3}} \tag{3.35}$$

while the introduced parameters C_1, C_2, and D_2 were fond with the exact values (Alonso and Girifalco, 1978; Perdew et al., 1992; Wang et al., 1990):

$$C_1 = -\frac{3}{4}\pi^{1/3},\ C_2 = -\frac{3}{4}\pi^{1/3}\left[1 - \left(\frac{3}{\pi^2}\right)^{1/3}\right],\ D_2 = \frac{\pi^{1/3}}{729} \tag{3.36}$$

Worth observing that the exchange Bartolotti functional (3.34) has some important phenomenological features: it scales like potential energy, fulfills the non-locality behavior through the powers of the electron and powers of the gradient of the density, while the atomic cusp condition is preserved (Levy and Gorling, 1996).

However, density functional exchange-energy approximation with correct asymptotic (long range) behavior, that is satisfying the limits for the density

$$\lim_{r\to\infty} \rho_\sigma = \exp(-a_\sigma r) \tag{3.37}$$

and for the Coulomb potential of the exchange charge, or Fermi hole density at the reference point r

$$\lim_{r\to\infty} U_X^\sigma = -\frac{1}{r}, \sigma = \alpha(or \uparrow), \beta(or \downarrow) ... spin\ \ states \tag{3.38}$$

in the total exchange energy

$$K[\rho] = \frac{1}{2}\sum_\sigma \int \rho_\sigma U_X^\sigma d\mathrm{r} \tag{3.39}$$

was given by Becke *via* employing the so called semiempirical (SE) modified gradient-corrected functional (Becke, 1986):

$$\begin{aligned} K^{SE} &= K_0 - \beta \sum_\sigma \int \rho_\sigma^{4/3} \frac{x_\sigma^2(\mathrm{r})}{1+\gamma x_\sigma^2(\mathrm{r})} d\mathrm{r}, \\ K_0 &= \int d\mathrm{r} k_0[\rho(\mathrm{r})] \\ x_\sigma(\mathrm{r}) &= \frac{|\nabla \rho_\sigma(\mathrm{r})|}{\rho_\sigma^{4/3}(\mathrm{r})} \end{aligned} \tag{3.40}$$

to the working single-parameter dependent one (Becke, 1988):

$$K^{B88} = K_0 - \beta \sum_\sigma \int \rho_\sigma^{4/3}(\mathrm{r}) \frac{x_\sigma^2(\mathrm{r})}{1+6\beta x_\sigma(\mathrm{r}) \sinh^{-1} x_\sigma(\mathrm{r})} d\mathrm{r} \tag{3.41}$$

where the value $\beta = 0.0042[a.u]$ was found as the best fit among the noble gases (He to Rn atoms) exchange energies; the constant a_σ is related to the ionization potential of the system.

Still, having different exchange approximation energetic functionals as possible worth explaining from where such ambiguity eventually comes. To clarify this, it helps in rewriting the starting exchange energy (3.22) under the formally exact form (Taut, 1996)

$$K[\rho] = \sum_\sigma \int \rho_\sigma(\mathrm{r}) k[\rho_\sigma(\mathrm{r})] g[x_\sigma(\mathrm{r})] d\mathrm{r} \tag{3.42}$$

Where the typical components are identified as:

$$k[\rho] = -A_X \rho^{1/3},\ A_X = \frac{3}{2}\left(\frac{3}{4\pi}\right)^{1/3} \tag{3.43}$$

while the gradient containing correction $g(x)$ is to be determined.

Firstly, one can notice that a sufficiency condition for the two exchange integrals (3.39) and (3.42) to be equal is that their integrands, or the exchange potentials, to be equal; this provides the leading gradient correction:

$$g_0(x) = \frac{1}{2}\frac{U_X(\mathbf{r}(x))}{k[\rho(\mathbf{r}(x))]} \tag{3.44}$$

with $\mathbf{r}(x)$ following from $x(\mathbf{r})$ by (not unique) inversion.

Unfortunately, the above "integrity" condition for exchange integrals to be equal is not also necessary, since any additional gradient correction

$$g(x) = g_0(x) + \Delta g(x) \tag{3.45}$$

fulfills the same constraint if it is chosen so that

$$\int \rho^{4/3}(\mathbf{r})\Delta g(x(\mathbf{r}))d\mathbf{r} = 0 \tag{3.46}$$

or, with the general form:

$$\Delta g(x) = f(x) - \frac{\int \rho^{4/3}(\mathbf{r}) f(x(\mathbf{r}))d\mathbf{r}}{\int \rho^{4/3}(\mathbf{r})d\mathbf{r}} \tag{3.47}$$

being $f(x)$ an arbitrary function.

Nonetheless, if, for instance, the function $f(x)$ is specialized so that

$$f(x) = -g_0(x) \tag{3.48}$$

the gradient correcting function (3.45) becomes:

$$g(x) = -\frac{1}{2A_X}\frac{\int \rho(\mathbf{r})U_X(\mathbf{r})d\mathbf{r}}{\int \rho^{4/3}(\mathbf{r})d\mathbf{r}} \equiv \alpha_X \tag{3.49}$$

recovering the Slater's famous X_α method for exchange energy evaluations (Slater, 1951; Slater and Johnson, 1972)

$$K[\rho] = -\alpha_X A_X \int \rho^{4/3}(\mathbf{r})d\mathbf{r} \tag{3.50}$$

Nevertheless, the different values of the multiplication factor α_X in (3.50) can explain the various forms of exchange energy coefficients and forms above. Moreover, following this conceptual line the above Becke'88 functional (3.41) can be further rearranged in a so called Xα-Becke88 form (Lee and Zhou, 1991)

$$K^{XB88} = \alpha_{XB}\sum_\sigma \int \rho_\sigma^{4/3}(\mathbf{r})\left[2^{1/3} + \frac{x_\sigma^2(\mathbf{r})}{1 + 6\beta_{XB}x_\sigma(\mathbf{r})\sinh^{-1}x_\sigma(\mathbf{r})}\right]d\mathbf{r} \tag{3.51}$$

where the parameters α_{XB} and β_{XB} are to be determined, as usually, throughout atomic fitting; it may lead with a new workable valuable density functional in exchange family.

3.4 DENSITY FUNCTIONALS OF CORRELATION ENERGY

The first and immediate definition of energy correlation may be given by the difference between the exact and Hartree–Fock (HF) total energy of a poly-electronic system (Senatore and March, 1994):

$$E_C[\rho] = E[\rho] - E_{HF}[\rho] \tag{3.52}$$

Instead, in density functional theory the correlation energy can be seen as the gain of the kinetic and electronic repulsion energy between the full interacting ($\lambda = 1$) and non-interacting ($\lambda = 0$) states of the electronic systems (Liu et al., 1999):

$$E_C^{\lambda}[\rho] = \left\langle \psi^{\lambda} \middle| \left(\hat{T} + \lambda\hat{W}\right) \middle| \psi^{\lambda} \right\rangle - \left\langle \psi^{\lambda=0} \middle| \left(\hat{T} + \lambda\hat{W}\right) \middle| \psi^{\lambda=0} \right\rangle \tag{3.53}$$

In this context, taking the variation of the correlation energy (3.53) respecting the coupling parameter λ (Nagy et al., 1999; Ou–Yang and Levy, 1991),

$$\lambda \frac{\partial E_C^{\lambda}[\rho]}{\partial \lambda} = E_C^{\lambda}[\rho] + \int \rho(\mathrm{r})\mathrm{r} \cdot \nabla \frac{\delta E_C^{\lambda}[\rho]}{\delta \rho(\mathrm{r})} d\mathrm{r} \tag{3.54}$$

by employing it through the functional differentiation with respecting the electronic density,

$$\lambda \frac{\partial V_C^{\lambda}[\rho]}{\partial \lambda} - V_C^{\lambda}[\rho] = \mathrm{r} \cdot \nabla V_C^{\lambda} + \int \rho(\mathrm{r}_1)\mathrm{r}_1 \cdot \nabla_1 \frac{\delta^2 E_C^{\lambda}[\rho]}{\delta \rho(\mathrm{r})\delta \rho(\mathrm{r}_1)} d\mathrm{r}_1 \tag{3.55}$$

one obtains the equation to be solved for correlation potential $V_C^{\lambda} = \delta E_C^{\lambda}[\rho]/\delta\rho$; then the correlation energy is yielded by back integration:

$$E_C^{\lambda}[\rho] = \int V_C^{\lambda}(\mathrm{r},[\rho])\rho(\mathrm{r})d\mathrm{r} \tag{3.56}$$

from where the full correlation energy is reached out by finally setting $\lambda = 1$.

When restricting to atomic systems, that is assuming spherical symmetry, and neglecting the last term of the correlation potential equation above, believed to be small (Liu et al., 1999), the equation to be solved simply becomes:

$$\lambda \frac{\partial V_C^{\lambda}[\rho]}{\partial \lambda} - V_C^{\lambda}[\rho] = r\nabla V_C^{\lambda} \tag{3.57}$$

that can really be solved out with the solution:

$$V_C^{\lambda} = A_p \lambda^{p+1} r^p \tag{3.58}$$

with the integration constants A_p and p.

However, since the equation (3.57) is a homogeneous differential one, the linear combination of solutions gives a solution as well. This way, the general form of correlation potential looks like:

$$V_C^{\lambda} = \sum_p A_p \lambda^{p+1} r^p \tag{3.59}$$

This procedure can be then iterated by taking further derivative of (3.55) with respect to the density, solving the obtained equation until the second order correction over above first order solution (3.59),

$$V_C^{\lambda} = \sum_{p1} A_{p1} \lambda^{p1+1} r^{p1} + \sum_{p2} A_{p2} \lambda^{2p2+1} r^{p2} \left\langle r^{p2} \rho \right\rangle \tag{3.60}$$

By mathematical induction, when going to higher orders the K-truncated solution is iteratively founded as:

$$V_C^{\lambda} = \sum_p \sum_{k=1}^{K} A_{pk} \lambda^{pk+1} r^p \left\langle r^p \rho \right\rangle^{k-1} \tag{3.61}$$

producing the λ-related correlation functional:

$$E_C^{\lambda}[\rho] = \sum_p \sum_{k=1}^{K} \frac{1}{k} A_{pk} \lambda^{pk+1} \left\langle r^p \rho \right\rangle^k \tag{3.62}$$

and the associate full correlation energy functional (λ = 1) expression:

$$E_C[\rho] = \sum_p \sum_{k=1}^{K} \frac{1}{k} A_{pk} \left\langle r^p \rho \right\rangle^k \tag{3.63}$$

As an observation, the correlation energy (3.63) supports also the immediate not spherically (molecular) generalization (Liu et al., 1999):

$$E_C[\rho] = \sum_{lmn} \sum_{k=1}^{K} \frac{1}{k} A_{lmnk} \left\langle x^l x^m x^n \rho \right\rangle^k \tag{3.64}$$

Nevertheless, for atomic systems, the simplest specialization of the relation (3.63) involves the simplest density moments $\langle \rho \rangle = N$ and $\langle r\rho \rangle$ that gives:

$$E_C[\rho] = A_{C0} N + A_{C1} \left\langle r\rho \right\rangle \tag{3.65}$$

Unfortunately, universal atomic values for the correlation constants A_{C0} and A_{C1} in (3.65) are not possible; they have to be related with the atomic number Z that on its turn can be seen as functional of density as well. Therefore, with the settings

$$A_{C0} = C_{C0} \ln Z, \; A_{C1} = C_{C1} Z \tag{3.66}$$

the fitting of (3.65) with the HF related correlation energy (3.52) reveals the atomic-working correlation energy with the form (Liu et al., 1999):

$$E_C = -0.16569 N \ln Z + 0.000401 Z \left\langle r\rho \right\rangle \tag{3.67}$$

The last formula is circumvented to the high-density total correlation density approaches rooting at their turn on the Thomas–Fermi atomic theory. Very interesting, the relation (3.67) may be seen as an atomic reflection of the (solid state) high density regime ($r_s < 1$) given by Perdew et al. (Perdew, 1986, 1989; Perdew and Yue, 1986; Perdew et al., 1992, 1996; Seidl et al., 1999; Wang and ; Zhang and Yang, 1988):

$$E_C^{PZ\infty}[\rho] = \int d\mathrm{r}\rho(\mathrm{r})\begin{pmatrix}-0.048 - 0.0116r_s \\ +0.0311\ln r_s + 0.0020r_s \ln r_s\end{pmatrix} \tag{3.68}$$

in terms of the dimensionless ratio

$$r_s = \frac{r_0}{a_0} \tag{3.69}$$

between the Wigner–Seitz radius $r_0 = (3/4\pi\rho)^{1/3}$ and the first Bohr radius $a_0 = \hbar^2 / me^2$. Instead, within the low density regime ($r_s \geq 1$) the first approximation for correlation energy goes back to the Wigner jellium model of electronic fluid in solids thus providing the LDA form (Perdew et al., 1988; Wilson and Levy, 1990):

$$E_C^{W-LDA}[\rho] = \int \varepsilon_C[\rho(\mathrm{r})]\rho(\mathrm{r})d\mathrm{r}\,, \tag{3.70}$$

where

$$\varepsilon_C[\rho(\mathrm{r})] = -\frac{0.44}{7.8 + r_s} \tag{3.71}$$

is the correlation energy per particle of the homogeneous electron gas with density ρ. Further comments and alternative derivations for correlation energy are given in Chapter 6, Section 6.6.2.

However, extended parameterization of the local correlation energy may be unfolded since considering the fit with an LSDA ($\rho_\uparrow$ and $\rho_\downarrow$) analytical expression by *Vosko, Wilk and Nusair* (VWN) (Vosko et al., 1980),

$$E_C^{VWN}[\rho_\uparrow, \rho_\downarrow] = \int \varepsilon_C[\rho_\uparrow(\mathrm{r}), \rho_\downarrow(\mathrm{r})]\rho(\mathrm{r})d\mathrm{r} \tag{3.72}$$

while further density functional *gradient corrected Perdew* (GCP) expansion will look like:

$$E_C^{GCP}[\rho_\uparrow, \rho_\downarrow] = \int d\mathrm{r}\varepsilon_C[\rho_\uparrow(\mathrm{r}), \rho_\downarrow(\mathrm{r})]\rho(\mathrm{r}) + \int d\mathrm{r}B[\rho_\uparrow(\mathrm{r}), \rho_\downarrow(\mathrm{r})]\left|\nabla\rho(\mathrm{r})\right|^2 + \ldots \tag{3.73}$$

where the Perdew recommendation for the gradient integrant has the form (Perdew, 1986):

$$B_C^P[\rho_\uparrow(\mathrm{r}), \rho_\downarrow(\mathrm{r})] = B_C[\rho]\frac{\exp\left(-b[\rho]f\left|\nabla\rho\right|\rho^{-7/6}\right)}{d(m)} \tag{3.74}$$

with

$$B_C[\rho] = \rho^{-4/3}C[\rho] \tag{3.75}$$

being the electron gas expression for the coefficient of the gradient expansion. The normalization in (3.74) is to the spin degeneracy:

$$d(m) = 2^{1/3}\left[\left(\frac{1+m}{2}\right)^{5/3} + \left(\frac{1-m}{2}\right)^{5/3}\right]^{1/2} \tag{3.76}$$

$$m = \frac{\rho_\uparrow - \rho_\downarrow}{\rho}, \rho = \rho_\uparrow + \rho_\downarrow$$

while the exponent containing functional

$$b[\rho] = (9\pi)^{1/6}\frac{C[\rho \to \infty]}{C[\rho]} \tag{3.77}$$

is written as the ratio of the asymptotic long-range density behavior to the current one, and is controlled by the cut-off *f* exponential parameter taking various values depending of the fitting procedures it subscribes: 0.17 for closed shells atoms and 0.11 for Ne particular system (Savin et al., 1986; Savin et al., 1987).

More specifically, we list bellow some nonlocal correlation density functionals in the low density (gradient corrections over LDA) regime:

- the Rasolt and Geldar paramagnetic case ($\rho_\uparrow = \rho_\downarrow = \rho/2$) is covered by correlation energy (Rasolt and Geldart, 1986; Savin et al., 1986):

$$E_C^{RG}[\rho] = c_1 + \frac{c_2 + c_3 r_s + c_4 r_s^2}{1 + c_5 r_s + c_6 r_s^2 + c_7 r_s^3} \tag{3.78}$$

  with $c_1 = 1.667 \cdot 10^{-3}$, $c_1 = 2.568 \cdot 10^{-3}$, $c_3 = 2.3266 \cdot 10^{-2}$, $c_4 = 7.389 \cdot 10^{-6}$, $c_5 = 8.723$, $c_6 = 0.472$, $c_7 = 7.389 \cdot 10^{-2}$ (in atomic units).

- The gradient corrected correlation functional reads as (Savin et al., 1984, 1997):

$$E_C^{GC} = \int d\mathrm{r}\varepsilon_C[\rho_\uparrow,\rho_\downarrow]\rho(\mathrm{r}) + \int d\mathrm{r}B_C^P[\rho_\uparrow,\rho_\downarrow]_{C[\rho]=\sqrt{2}\pi/4(6\pi^2)^{4/3}, f=0.17}|\nabla\rho(\mathrm{r})|^2$$
$$+9\frac{\pi}{4(6\pi^2)^{4/3}}(0.17)^2\int d\mathrm{r}\left(|\nabla\rho_\uparrow|^2\rho_\uparrow^{-4/3} + |\nabla\rho_\downarrow|^2\rho_\downarrow^{-4/3}\right) \tag{3.79}$$

- The *Lee, Yang, and Parr* (LYP) functional within Colle–Salvetti approximation unfolds like (Lee et al., 1988):

$$E_C^{LYP} = -a_C b_C\int d\mathrm{r}\gamma(\mathrm{r})\xi(\mathrm{r})\left(\begin{array}{l}\sum_\sigma\rho_\sigma(\mathrm{r})\sum_i|\nabla\phi_{i\sigma}(\mathrm{r})|^2 - \frac{1}{4}\sum_\sigma\rho_\sigma(\mathrm{r})\Delta\rho_\sigma(\mathrm{r})\\ -\frac{1}{4}|\nabla\rho(\mathrm{r})|^2 + \frac{1}{4}\rho(\mathrm{r})\Delta\rho(\mathrm{r})\end{array}\right)$$
$$-a_C\int d\mathrm{r}\frac{\gamma(\mathrm{r})}{\eta(\mathrm{r})}\rho(\mathrm{r}) \tag{3.80}$$

  where

$$\gamma(\mathrm{r}) = 4\frac{\rho_\uparrow(\mathrm{r})\rho_\downarrow(\mathrm{r})}{\rho(\mathrm{r})^2}, \eta(\mathrm{r}) = 1 + d_C\rho(\mathrm{r})^{-1/3}, \xi(\mathrm{r}) = \frac{\rho(\mathrm{r})^{-5/3}}{\eta(\mathrm{r})}\exp\left[-c_C\rho(\mathrm{r})^{-1/3}\right] \tag{3.81}$$

and the constants: $a_C = 0.04918$, $b_C = 0.132$, $c_C = 0.2533$, $d_C = 0.349$.

- The open-shell (OS) case provides the functional (Wilson and Levy, 1990):

$$E_C^{OS} = \int d\mathrm{r}\frac{a_s\rho(\mathrm{r}) + b_s|\nabla\rho(\mathrm{r})|\rho(\mathrm{r})^{-1/3}}{c_s + d_s\left(|\nabla\rho_\uparrow|\rho_\uparrow^{-4/3} + |\nabla\rho_\downarrow|\rho_\downarrow^{-4/3}\right) + r_s}\sqrt{1-\zeta^2} \tag{3.82}$$

with the spin-dependency regulated by the factor $\zeta = (\rho_\uparrow - \rho_\downarrow)/(\rho_\uparrow + \rho_\downarrow)$, approaching zero for closed-shell case, while the specific coefficients are determined through a scaled-minimization procedure yielding the values: $a_s = -0.74860$, $b_s = -0.06001$, $c_s = 3.60073$, $d_s = 0.900000$.

- Finally, Perdew and Zunger (PZ) recommend the working functional (Perdew and Zunger, 1981):

$$E_C^{PZ0}[\rho] = \int d\mathrm{r}\rho(\mathrm{r})\frac{\alpha_p}{1 + \beta_{1p}\sqrt{r_s} + \beta_{2p}r_s} \tag{3.83}$$

with the numerical values for the fitting parameters founded as: $\alpha_p = -0.1423$, $\beta_{1p} = 1.0529$, $\beta_{2p} = 0.3334$.

3.5 DENSITY FUNCTIONALS OF EXCHANGE-CORRELATION ENERGY

Another approach in questing exchange and correlation density functionals consists in finding them both at once in what was defined as exchange-correlation density functional (1.30). In this regard, following the Lee and Parr approach (Lee and Parr, 1990), the simplest starting point is to rewrite the inter-electronic interaction potential

$$W = \iint\frac{\rho_2(\mathrm{r}_1, \mathrm{r}_2)}{r_{12}}d\mathrm{r}_1 d\mathrm{r}_2 \tag{3.84}$$

and the classical (Coulombic) repulsion

$$J = \frac{1}{2}\iint\frac{\rho(\mathrm{r}_1)\rho(\mathrm{r}_2)}{r_{12}}d\mathrm{r}_1 d\mathrm{r}_2 \tag{3.85}$$

appeared in the formal exchange energy $(W - J)$ in (1.22), by performing the previously introduced coordinate transformation (3.7), followed by integration of the averaged pair and coupled densities (denoted with over-bars) over the angular components of **s**:

$$W = 4\pi\int d\mathrm{r}\int sds\bar{\rho}_2(\mathrm{r}, s) \tag{3.86}$$

$$J = 2\pi\int d\mathrm{r}\int sds\overline{\rho(\mathrm{r} + \mathrm{s}/2)\rho(\mathrm{r} - \mathrm{s}/2)} \tag{3.87}$$

Now, the second order density matrix in (3.86) can be expressed as

$$\bar{\rho}_2(\mathrm{r}, s) = \frac{1}{2}\overline{\rho(\mathrm{r} + \mathrm{s}/2)\rho(\mathrm{r} - \mathrm{s}/2)}\left[1 + F_1(\mathrm{r}, s)\right] \tag{3.88}$$

with the help of the introduced function $F_1(\mathrm{r},s)$ carrying the form

$$F_1(\mathrm{r},s) = -\frac{\exp[-\alpha(\mathrm{r})s]}{1+\alpha(\mathrm{r})}\left\{1+[\alpha(\mathrm{r})s]^2 F_2(\mathrm{r},s)\right\} \quad (3.89)$$

so that the cusp condition for $\overline{\rho}_2(\mathrm{r},s)$

$$\left.\frac{\partial \ln \overline{\rho}_2(\mathrm{r},s)}{\partial s}\right|_{s=0} = 1 \quad (3.90)$$

to be satisfied for a well behaved function of a Taylor series expansion type

$$F_2(\mathrm{r},s) = \sum_{k=0}^{\infty} a_k(\mathrm{r})[\alpha(\mathrm{r})s]^k \quad (3.91)$$

when $\alpha(\mathrm{r})$ stands for a suitable function of r as well, see below.
On the other side, the average $\overline{\rho(\mathrm{r}+\mathrm{s}/2)\rho(\mathrm{r}-\mathrm{s}/2)}$ in (3.87) and (3.88) supports a Taylor expansion (Berkowitz, 1986):

$$\overline{\rho(\mathrm{r}+\mathrm{s}/2)\rho(\mathrm{r}-\mathrm{s}/2)} = \rho^2(\mathrm{r})\left[1-\frac{2\tau_w(\mathrm{r})}{3\rho(\mathrm{r})}s^2+...\right] \quad (3.92)$$

with

$$\tau_w(\mathrm{r}) = \frac{1}{8}\frac{|\nabla\rho(\mathrm{r})|^2}{\rho(\mathrm{r})} - \frac{1}{8}\nabla^2\rho(\mathrm{r}) \quad (3.93)$$

being the Parr modified kinetic energy of Weizsäcker type (Parr and Yang, 1989).

Inserting relations (3.86–3.93) in $(W-J)$ difference it is eventually converted from the "genuine" exchange meaning into practical exchange-correlation energy characterized by the density functional form:

$$\begin{aligned} E_{XC} &= 2\pi\int d\mathrm{r}\int s ds \overline{\rho(\mathrm{r}+\mathrm{s}/2)\rho(\mathrm{r}-\mathrm{s}/2)}F_1(\mathrm{r},s) \\ &= -2\pi\int d\mathrm{r}\frac{\rho^2(\mathrm{r})}{1+\alpha(\mathrm{r})}\int ds\, s\exp[-\alpha(\mathrm{r})s]\times\left\{1-\frac{2\tau_w(\mathrm{r})}{3\rho(\mathrm{r})}s^2+...\right\} \\ &\times\left\{1+[\alpha(\mathrm{r})s]^2\sum_{k=0}^{\infty}a_k(\mathrm{r})[\alpha(\mathrm{r})s]^k\right\} \end{aligned} \quad (3.94)$$

Making use of the two possible multiplication of the series in (3.94), that is either by retaining the $\alpha(\mathrm{r})$ containing function only or by including also the density gradient terms in the first curled brackets, thus retaining also the term containing $\tau_w(\mathrm{r})$ function, the so called *I-XC or II-XC type functionals* are respectively obtained.

Now, laying aside other variants and choosing the simple (however meaningfully) density dependency

$$\alpha(\mathrm{r}) = \kappa\rho^{1/3}(\mathrm{r}),\ \kappa = \text{constant} \tag{3.95}$$

the provided exchange-correlation functionals are generally shaped as (Lee and Parr, 1990):

$$\begin{aligned} E_{XC}^{I} &= -\frac{1}{\kappa^2}\int d\mathrm{r}\rho^{4/3}(\mathrm{r})\frac{A_{XC}(\mathrm{r})}{1+\kappa\rho^{1/3}(\mathrm{r})}, \\ E_{XC}^{II} &= -\frac{1}{\kappa^2}\int d\mathrm{r}\frac{\rho^{4/3}(\mathrm{r})}{1+\kappa\rho^{1/3}(\mathrm{r})}\left[B_{XC}(\mathrm{r})+\frac{2}{3}\frac{\tau_w(\mathrm{r})}{\rho^{5/3}(\mathrm{r})}C_{XC}(\mathrm{r})\right] \end{aligned} \tag{3.96}$$

These functionals are formally exact for any κ albeit the resumed functions $A_{XC}(\mathrm{r})$, $B_{XC}(\mathrm{r})$, and $C_{XC}(\mathrm{r})$ are determined for each particular specialization.

Going now to the specific models, let's explore the type I of exchange-correlation functionals (3.96). Firstly, they can further undergo simplification since the reasonable (atomic) assumption according which

$$\kappa\rho^{1/3}(\mathrm{r}) << 1,\ \forall\mathrm{r} \tag{3.97}$$

Within this frame the best provided model is of $X\alpha$-Padé approximation type, containing N-dependency (Lee and Parr, 1990):

$$E_{XC}^{I(X\alpha)} = -a_0^{X\alpha}\frac{1+a_1^{X\alpha}/N}{1+a_2^{X\alpha}/N}\int\rho^{4/3}(\mathrm{r})d\mathrm{r} \tag{3.98}$$

with $a_0^{X\alpha} = 0.7475$, $a_1^{X\alpha} = 17.1903$, and $a_2^{X\alpha} = 14.1936$ (atomic units).

When the condition (3.97) for κ is abolished the Wigner-like model results, again, having the best approximant exchange-correlation model as the Padé form (Lee and Parr, 1990):

$$E_{XC}^{I(Wig)} = -a_0^{Wig}\frac{1+a_1^{Wig}/N}{1+a_2^{Wig}/N}\int\frac{\rho^{4/3}(\mathrm{r})}{1+\kappa^{I(Wig)}\rho^{1/3}(\mathrm{r})}d\mathrm{r} \tag{3.99}$$

with $a_0^{Wig} = 0.76799$, $a_1^{Wig} = 17.5943$, $a_2^{Wig} = 14.8893$, and $\kappa^{I\,(Wig)} = 4.115\cdot10^{-3}$ (atomic units).

Turning to the II-type of exchange-correlation functionals, the small density condition (3.97) delivers the gradient corrected $X\alpha$ model, taking its best fitting form as the N-dependent Padé approximant (Lee and Parr, 1990):

$$E_{XC}^{II(X\alpha)} = -b_0^{X\alpha}\frac{1+b_1^{X\alpha}/N}{1+b_2^{X\alpha}/N}\int\rho^{4/3}(\mathrm{r})d\mathrm{r} - c_0^{X\alpha}\int\rho^{-1/3}(\mathrm{r})\tau_w(\mathrm{r})d\mathrm{r} \tag{3.100}$$

with $b_0^{X\alpha} = 0.7615$, $b_1^{X\alpha} = 1.6034$, $b_2^{X\alpha} = 2.1437$, and $c_2^{X\alpha} = 6.151\cdot10^{-2}$ (atomic units), while when laying outside the (3.97) condition the gradient corrected Wigner-like best model is proved to be without involving the N-dependency (Lee and Parr, 1990):

$$E_{XC}^{II(Wig)} = -b_0^{Wig} \int \frac{\rho^{4/3}(\mathbf{r})}{1+\kappa^{II(Wig)}\rho^{1/3}(\mathbf{r})} d\mathbf{r} - c_0^{Wig} \int \frac{\rho^{-1/3}(\mathbf{r})\tau_w(\mathbf{r})}{1+\kappa^{II(Wig)}\rho^{1/3}(\mathbf{r})} d\mathbf{r} \tag{3.101}$$

with $b_0^{Wig} = 0.80569$, $c_0^{Wig} = 3.0124\cdot10^{-3}$, and $\kappa^{II(Wig)} = 4.0743\cdot10^{-3}$ (atomic units).

Still, a Padé approximant for the gradient-corrected Wigner-type exchange-correlation functional exists and it was firstly formulated by Rasolt and Geldar (1986) with the working form (Lee and Bartolotti, 1991):

$$E_{XC}^{RG} = E_{XC}^{LDA(or\,X\alpha)} + \int B_{xc}^{RG}[\rho(\mathbf{r})] \frac{|\nabla\rho(\mathbf{r})|^2}{\rho^{1/3}(\mathbf{r})} d\mathbf{r} \tag{3.102}$$

with B_{xc}^{RG} given with the Padé form:

$$B_{XC}^{RG}[\rho(\mathbf{r})] = -1\times10^{-3} c_1^{RG} \frac{1+c_2^{RG} r_s + c_3^{RG} r_s^2}{1+c_4^{RG} r_s + c_5^{RG} r_s^2 + c_6^{RG} r_s^3} \tag{3.103}$$

having the fitted coefficients $c_1^{RG} = 2.568$, $c_2^{RG} = 9.0599$, $c_3^{RG} = 2.877\cdot10^{-3}$, $c_4^{RG} = 8.723$, $c_5^{RG} = 0.472$, and $c_3^{RG} = 7.389\cdot10^{-2}$ (atomic units). Some studies also consider the nonlocal correction in (3.102) pre-multiplied by the 10/7 factor which was found as appropriate procedure for atomic systems.

Finally, worth noting the Tozer and Handy general form for exchange-correlation functionals viewed as a sum of products of powers of density and gradients (Tozer and Handy, 1998):

$$E_{XC}^{TH} = \int F_{xc}\left(\rho_\uparrow, \rho_\downarrow, \varsigma_\uparrow, \varsigma_\downarrow, \varsigma_{\uparrow\downarrow}\right) d\mathbf{r} \tag{3.104}$$

with

$$F_{XC} = \sum_{abcd} \omega_{abcd} R^a S^b X^c Y^d = \sum_{abcd} \omega_{abcd} f_{abcd}(\mathbf{r}) \tag{3.105}$$

where

$$R^a = \rho_\uparrow^a + \rho_\downarrow^a$$

$S^b = m^{2b}$ see equation (3.76) for m definition,

$$X^c = \frac{\varsigma_\uparrow^c + \varsigma_\downarrow^c}{2\rho^{4c/3}}$$

$$Y^d = \left(\frac{\varsigma_\uparrow^2 + \varsigma_\downarrow^2 - 2\varsigma_{\uparrow\downarrow}}{\rho^{8/3}}\right)^d \tag{3.106}$$

and

$$\varsigma_\uparrow = |\nabla\rho_\uparrow|, \varsigma_\downarrow = |\nabla\rho_\downarrow|, \varsigma_{\uparrow\downarrow} = \nabla\rho_\uparrow\cdot\nabla\rho_\downarrow, \rho = \rho_\uparrow + \rho_\downarrow \tag{3.107}$$

The coefficients ω_{abcd} of (3.105) are determined through minimization procedure involving the associated exchange-correlation potentials $V_{XC\uparrow(\downarrow)}^{abcd}(\mathrm{r}) = \delta f_{abcd}(\mathrm{r}) / \delta\rho_{\uparrow(\downarrow)}(\mathrm{r})$ in above (3.104) functional. The results would depend upon the training set of atoms and molecules but presents the advantage of incorporating the potential information in a non-vanishing asymptotical manner, with a semi-empirical value. Moreover, its exact asymptotic exchange-correlation potential equals chemical hardness (Parr and Yang, 1989; Putz, 2003, 2007a, 2007b) for open-shell being less than that for closed shell systems, thus having the merit of including chemical hardness as an intrinsic aspect of energetic approach, a somewhat absent aspect from conventional functionals so far. However, since electronegativity and chemical hardness closely relate with chemical bonding, their relation with the total energy and component functionals is in next at both conceptual and applied levels explored.

KEYWORDS

- **Colle–Salvetti approximation**
- **Electronic density**
- **Electronic localization function**
- **Gaussian resummation**
- **Hartree–Fock model**
- **Padé-resummation model**

4 Chemical Realization of DFT

CONTENTS

4.1 INTRODUCTION

Within the limits of the density functional theory there are introduced and deduced the fundamental chemical descriptors and of their relationships and principles, such as the electronegativity, chemical hardness and softness hierarchies, alongside the chemical action, targeting an unified scenario of the chemical reactivity and bonding phenomenology (Putz, 2003; 2007a, 2008c, 2011a, 2011b, 2011c).

4.2 CHEMICAL REACTIVITY INDICES WITHIN DFT

The essence of the density functional consists in rewriting the total energy of a N-electronic system, as a function of electronic density function of the fundamental state associated to the system (Kohn et al., 1996):

$$E_V[\rho] = \int \rho(\mathrm{r})V(\mathrm{r})d\mathrm{r} + F_{HK}[\rho] \tag{4.1}$$

where V is the external potential applied to the electrons system (for example the potential of nuclei). The size $F_{HK}[\rho]$ is the Hohenberg–Kohn universal functional (exactly unknown as analytic density-functional) expressed as the sum of the density functionals of the kinetic and electronic repulsion energies

$$F_{HK}[\rho] = T[\rho] + V_{ee}[\rho] \tag{4.2}$$

and $\rho(\mathrm{r})$ actually is the uni-electronic density of (free spin) expressed in terms of the wave function of the system (Parr, 1983):

$$\rho(\mathrm{r}) = N\int |\Psi(1,2,...,N)|^2 ds_1 d\tau_2 ... d\tau_N \tag{4.3}$$

with $d\tau_i = ds_i d\mathrm{r}_i$ ($\mathrm{r}_i \in \Re$) being the space-spin volume element, and the number N also interpreted as a density functional:

$$N = N[\rho] = \int \rho(\mathrm{r})d\mathrm{r} \tag{4.4}$$

The univocal relation between the external potential applied to the electronic system and the electronic density, is provided by the Hohenberg–Kohn theorems (1964). Moreover, one of the theorems also statements the inequality relation between the energy, as density function, of any electronic state $E[\rho']$, and the right energy of the fundamental electronic state of the system, $E[\rho]$:

$$E_V[\rho'] \geq E_V[\rho] \tag{4.5}$$

inequality equivalent to the existence of a stationary equations:

$$\delta\{E_V[\rho'] - \mu N[\rho']\} = 0 \tag{4.6}$$

where μ is the Lagrange multiplier associated to the total numbers of the electrons in the system. For the right density of the fundamental state, the multiplier μ represents the chemical potential associated to the system and is expressed by the functional derivative:

$$\mu = \left(\frac{\delta E[\rho]}{\delta \rho(\mathrm{r})}\right)_{\rho=\rho(V)} \tag{4.7}$$

which can be reduced, from the stationarity principle (4.6), to a partial derivative respect to the number of the electrons, in the fundamental state:

$$\mu = \left(\frac{\partial E}{\partial N}\right)_V \tag{4.8}$$

and which, moreover, is associated with the electronegativity with changed sign: $\mu = -\chi$ (Parr et al., 1978). Proceeding with the calculation of the functional derivative in relation with the external potential $V(\mathrm{r})$ in (4.1), is obtained:

$$\rho(\mathrm{r}) = \left(\frac{\delta E[\rho]}{\delta V(\mathrm{r})}\right)_{\rho=\rho(V)} \tag{4.9}$$

wherefrom, together with the relation (4.8), can be expressed the total differential of the functional energy (4.1), $E = E(N, V)$, under the general form (Ayers and Parr, 2001):

$$dE = \mu dN + \int \rho(\mathrm{r})dV(\mathrm{r})d\mathrm{r} \tag{4.10}$$

This is the first equation of the chemical transformations in TFD, because correlates the variation of the total energy of electronic states (atomic or molecular) with the change of the charges to which is subject to, and also with the variation of the potential that governs the electronic state. Applying to the equation (4.10) the Maxwell's relations, results the following string of identities that define the Fukui function $f(\mathrm{r})$:

$$\left(\frac{\delta \mu}{\delta V(\mathrm{r})}\right)_N = \left(\frac{\partial \rho(\mathrm{r})}{\partial N}\right)_V \equiv f(\mathrm{r}) \tag{4.11}$$

Similarly, can be expressed the total differential of the chemical potential, as a function of the number of the electrons in the system and of the applied external potential:

$$d\mu = \left(\frac{\partial \mu}{\partial N}\right)_V dN + \int \left(\frac{\delta \mu}{\delta V(\mathrm{r})}\right)_N \delta V(\mathrm{r}) d\mathrm{r} \tag{4.12}$$

In the second term on the right side of the equation (4.12), is recognized the identity (4.11) and the first term on the right side of the equation (4.12) is recognized as the chemical hardness associated to the electronic system, according to the definitions from Chapter 2:

$$2\eta = \left(\frac{\partial^2 E}{\partial N^2}\right)_V = \left(\frac{\partial \mu}{\partial N}\right)_V \tag{4.13}$$

With the relations (4.11) and (4.13), can be rewritten the total differential (4.12) under the form (Ayers and Parr, 2000):

$$d\mu = 2\eta dN + \int f(\mathrm{r}) dV(\mathrm{r}) d\mathrm{r} \tag{4.14}$$

This is the second equation of the transformations in TFD, equivalent to the first, (4.10), but rewritten at a different level. The equation (4.14) correlates the change of the chemical potential (actually of the electronegativity with changed sign) of an electronic state (atomic or molecular) with the change of charges and with the modification of the potential, through the chemical hardness at this variation, and of the Fukui frontier function, having a promoter role of the variation of the frontier orbital charge (Berkowitz, 1987; Nalewajski, 1998; Yang and Parr, 1985).

If the relation (4.7) is applied in (4.1), is obtained a new Euler–Lagrange equation of the electronic system, but including the term of chemical potential:

$$\mu = \left(\frac{\delta E}{\delta \rho(\mathrm{r})}\right)_V = V(\mathrm{r}) + \frac{\delta F_{HK}[\rho]}{\delta \rho(\mathrm{r})} \tag{4.15}$$

equation that also satisfies the differential form (4.14).

These are the equations of the chemical transformations, in terms of functional densities, which will be processed and converted, in order to apply them to the description of the evolution of the open electronic systems and their potency to participate in chemical reactions.

Further, we will exposed the bond between the local and global sensitivity indices, in TFD, in a way that allows an explicit implementation of the electronic densities. It starts from a generalized form of the Euler–Lagrange equation (4.1)

$$\Xi = \int \rho(\mathrm{r}) V(\mathrm{r}) d\mathrm{r} + F_{HK}[\rho] - \mu \int \rho(\mathrm{r}) d\mathrm{r} = \int \rho(\mathrm{r}) u(\mathrm{r}) d\mathrm{r} + F_{HK}[\rho] \tag{4.16}$$

through which minimize, $\delta \Xi = 0$, compared with the electronic density for the fixed external potential, is obtained a relation equivalent to the expression (4.15)

$$-u(\mathrm{r}) = \mu - V(\mathrm{r}) = \frac{\delta F_{HK}[\rho]}{\delta\rho(\mathrm{r})} \tag{4.17}$$

For an open electronic system, under the influence of the potential $u(\mathrm{r}) = V(\mathrm{r}) - \mu$, which satisfies the equation (4.17), are defined, in analogy with (4.13), but functionally (Baekelandt et al., 1995; Berkowitz and Parr, 1988; Berkowitz et al., 1985; De Proft et al., 1997; Gázquez et al., 1990; Senet, 1996, 1997), the *kernel of the chemical hardness* associated to the system:

$$2\eta(\mathrm{r},\mathrm{r}') = -\frac{\delta u(\mathrm{r})}{\delta\rho(\mathrm{r}')} = \frac{\delta^2 F_{HK}}{\delta\rho(\mathrm{r})\delta\rho(\mathrm{r}')} \tag{4.18}$$

and the *local chemical hardness* for the considered system:

$$\eta(\mathrm{r}) = \frac{1}{N}\int \eta(\mathrm{r},\mathrm{r}')\rho(\mathrm{r}')d\mathrm{r}' \tag{4.19}$$

From the way of definition of the potential $u(\mathrm{r})$, can be observed how it includes all the system properties: both those related to the external potential $V(\mathrm{r})$ where evolve, and those characterized by the chemical potential, the electronegativity with the changed sign of the system. If there is the functional derivative of the electronic density $\rho(\mathrm{r})$ in relation with $u(\mathrm{r})$:

$$\frac{\delta\rho(\mathrm{r})}{\delta u(\mathrm{r})} < \infty, \forall \mathrm{r} \in \Re \tag{4.20}$$

such an existence allows the introduction of new indices, that characterize an electronic system (open).

There can be defined the kernel of chemical softness associated to the electronic system:

$$s(\mathrm{r},\mathrm{r}') = -\frac{\delta\rho(\mathrm{r})}{\delta u(\mathrm{r}')}\theta \tag{4.21}$$

and the *local chemical softness*, that characterizes the electronic system:

$$s(\mathrm{r}) = \int s(\mathrm{r},\mathrm{r}')d\mathrm{r}' \tag{4.22}$$

If combined the definition relations (4.18), (4.19), (4.21), (4.22) can also provide the connection relation between the introduced indices. One considers the integro-differential identity:

$$\int \frac{\delta\rho(\mathrm{r})}{\delta u(\mathrm{r}')}\frac{\delta u(\mathrm{r}')}{\delta\rho(\mathrm{r}'')}d\mathrm{r}' = \delta(\mathrm{r}'' - \mathrm{r}) \tag{4.23}$$

with which, if taken into account (4.18), (4.21), can be obtained also the equivalent expression:

$$2\int s(\mathrm{r},\mathrm{r}')\eta(\mathrm{r},\mathrm{r}'')d\mathrm{r}' = \delta(\mathrm{r}''-\mathrm{r}) \tag{4.24}$$

From this relation it is concluded that, the kernels of the chemical hardness and softness for a considered electronic system, are reverse indices. The relation (4.24) can be also written to another level of localization. If multiplied by $\rho(\mathrm{r}'')$, is integrated over r" and is also considered (4.19), we obtain:

$$2\int s(\mathrm{r},\mathrm{r}')\eta(\mathrm{r}')d\mathrm{r}' = \frac{1}{N}\rho(\mathrm{r}) \tag{4.25}$$

wherefrom, if is again integrated over r, considered (4.22), can be also rewritten the identity:

$$2\int s(\mathrm{r})\eta(\mathrm{r})d\mathrm{r} = 1 \tag{4.26}$$

Moreover, can be also obtained other relations, if considered the sequence of transformations:

$$\begin{aligned} d\rho(\mathrm{r}) &= \int \frac{\delta\rho(\mathrm{r})}{\delta u(\mathrm{r}')} du(\mathrm{r}')d\mathrm{r}' = -\int s(\mathrm{r},\mathrm{r}')du(\mathrm{r}')d\mathrm{r}' \\ &= -\int s(\mathrm{r},\mathrm{r}')\left[dV(\mathrm{r}') - d\mu\right]d\mathrm{r}' = \int s(\mathrm{r},\mathrm{r}')d\mathrm{r}'d\mu - \int s(\mathrm{r},\mathrm{r}')dV(\mathrm{r}')d\mathrm{r}' \end{aligned} \tag{4.27}$$

If considered the equation (4.14) such as:

$$d\mu = 2\eta dN + \int f(\mathrm{r}'')dV(\mathrm{r}'')d\mathrm{r}'' \tag{4.28}$$

the expression (4.27) will have the form:

$$d\rho(\mathrm{r}) = 2s(\mathrm{r})\eta dN + \int\left[-s(\mathrm{r},\mathrm{r}'') + s(\mathrm{r})f(\mathrm{r}'')\right]dV(\mathrm{r}'')d\mathrm{r}'' \tag{4.29}$$

A general expression of the variation of electronic density (in a physicochemical process) can be also written from the functional dependency, $\rho = \rho[N, V]$ correlated with the manner of definition of Fukui function, (4.11)

$$d\rho(\mathrm{r}) = f(\mathrm{r})dN + \int\left(\frac{\delta\rho(\mathrm{r})}{\delta V(\mathrm{r}'')}\right)_N dV(\mathrm{r}'')d\mathrm{r}'' \tag{4.30}$$

By comparing the expressions (4.29), (4.30), are obtained the identities:

$$s(\mathrm{r}) = \frac{f(\mathrm{r})}{2\eta} = f(\mathrm{r})S = \left(\frac{\partial\rho(\mathrm{r})}{\partial N}\right)_V \left(\frac{\partial N}{\partial\mu}\right)_V = \left(\frac{\partial\rho(\mathrm{r})}{\partial\mu}\right)_V, \tag{4.31}$$

$$\kappa(\mathrm{r},\mathrm{r}'') \equiv \left(\frac{\delta\rho(\mathrm{r})}{\delta V(\mathrm{r}'')}\right)_N = -s(\mathrm{r},\mathrm{r}'') + \frac{s(\mathrm{r})s(\mathrm{r}'')}{S} \tag{4.32}$$

In the second identity from (4.31) actually has been introduced the definition of the global chemical softness *S*, and then was used for writing the relationship from (4.32). The relation (4.32) is very important, because it correlates the descriptors of sensitivity, at the level of the density functionals, with the linear response function, $\kappa(\mathrm{r},\mathrm{r}')$.

From the way in which had been introduced the descriptors of sensitivity, is noted the importance of the universal Hohenberg–Kohn functional, $F[\rho]$. However, can be developed a method in order to avoid this dependence (Garza and Robles, 1993), taking into account that an exact analytical expression of this function is not known (yet). This lack can be transposed into the unknowledge of the precise functional relation between the electronic density $\rho(\mathrm{r})$ and the global potential $u(\mathrm{r})$, and this new shortcoming may be eluded by considering the *translation invariant of the external* potential, applied to the electrons system, respectively to their density in the atomic or molecular system. The explicit connection between these two types of variations (of the global electronic potential and of the electronic density), for an open system, may take the forms:

$$\nabla\rho(\mathrm{r}) = \int d\mathrm{r}'\frac{\delta\rho(\mathrm{r})}{\delta u(\mathrm{r}')}\nabla' u(\mathrm{r}') \tag{4.33}$$

$$\nabla u(\mathrm{r}) = \int d\mathrm{r}'\frac{\delta u(\mathrm{r})}{\delta\rho(\mathrm{r}')}\nabla'\rho(\mathrm{r}') \tag{4.34}$$

These relationships can be again transcribed, by virtue of the expressions (4.18) and (4.21), as follows:

$$\nabla\rho(\mathrm{r}) = -\int d\mathrm{r}' s(\mathrm{r},\mathrm{r}')\nabla' u(\mathrm{r}') \tag{4.35}$$

$$\nabla u(\mathrm{r}) = -2\int d\mathrm{r}'\eta(\mathrm{r},\mathrm{r}')\nabla'\rho(\mathrm{r}') \tag{4.36}$$

The expression (4.35) can be easily developed, if considered the kernel of the chemical softness as being (de)composed (in) from a local contribution $L(\mathrm{r})$ and a nonlocal contribution such as:

$$s(\mathrm{r},\mathrm{r}') = L(\mathrm{r}')\delta(\mathrm{r}-\mathrm{r}') + t(\mathrm{r})\rho(\mathrm{r}') \tag{4.37}$$

with which the relation (4.35) becomes:

$$\nabla\rho(\mathrm{r}) = -L(\mathrm{r})\nabla u(\mathrm{r}) - t(\mathrm{r})\int d\mathrm{r}'\rho(\mathrm{r}')\nabla' u(\mathrm{r}') \tag{4.38}$$

The last term from (4.38) is identical to null by virtue of the identity $\nabla V(\mathrm{r}) = \nabla u(\mathrm{r})$ and of the Hellmann–Feynman theorem. Thus, the local contribution $L(\mathrm{r})$ will be provided by the form:

$$L(\mathrm{r}) = -\frac{\nabla\rho(\mathrm{r})\cdot\nabla u(\mathrm{r})}{|\nabla u(\mathrm{r})|^2} = -\frac{\nabla\rho(\mathrm{r})\cdot\nabla V(\mathrm{r})}{|\nabla V(\mathrm{r})|^2} \tag{4.39}$$

In order to establish the non-local contribution, it will be considered firstly the chemical softness (4.37) at the various levels of localization, by successive integration, that is:

$$s(\mathrm{r}) = L(\mathrm{r}) + Nt(\mathrm{r}) \tag{4.40}$$

$$S = \int [L(\mathrm{r}) + Nt(\mathrm{r})]d\mathrm{r} = \int L(\mathrm{r})d\mathrm{r} + N\int t(\mathrm{r})d\mathrm{r} \equiv a + N\int t(\mathrm{r})d\mathrm{r} \tag{4.41}$$

Now can be evaluated also the expression (4.32) and results:

$$\kappa(\mathrm{r},\mathrm{r}') = -L(\mathrm{r}')\delta(\mathrm{r}-\mathrm{r}') - t(\mathrm{r})\rho(\mathrm{r}') + \frac{[L(\mathrm{r}) + Nt(\mathrm{r})][L(\mathrm{r}') + Nt(\mathrm{r}')]}{\int [L(\mathrm{r}) + Nt(\mathrm{r})]d\mathrm{r}} \tag{4.42}$$

If the number of the electrons in the system, N, is established, from the normalization condition of the linear response function

$$\int \kappa(\mathrm{r},\mathrm{r}')d\mathrm{r} = 0 \tag{4.43}$$

applied to the relation (4.42), we will obtained the integral equations

$$\int t(\mathrm{r})d\mathrm{r} = \frac{N}{\rho(\mathrm{r})}t(\mathrm{r}) \tag{4.44}$$

with the simple solution: $t(\mathrm{r}) = \rho(\mathrm{r})$. With this result, the expression of the kernel of chemical softness will have the functional form:

$$s(\mathrm{r},\mathrm{r}') = -\frac{\nabla\rho(\mathrm{r})\cdot\nabla V(\mathrm{r})}{|\nabla V(\mathrm{r})|^2}\delta(\mathrm{r}-\mathrm{r}') + \rho(\mathrm{r})\rho(\mathrm{r}') \tag{4.45}$$

From now on, the analytical implementation is immediate, for an explicit expression of atomic or molecular electronic density.

4.3 INTRODUCING CHEMICAL REACTIVITY PRINCIPLES

Another outstanding *DFT* consequence regards the chemical reactivity principles emerging out from the *HK* theorems. The chemical reactions involve the charge transfer. Therefore is compulsory to derive principles governing charge transfer in *DFT*. If one considers the fundamental relation re-written as the density variation subject to constrain:

$$\int \Delta\rho(\mathrm{r})d\mathrm{r} = \Delta N \tag{4.46}$$

then, the modification in the ground state energy, by means of the Taylor series expansion (assuming $\Delta\rho$ small enough to provide truncation to the second order), writes as (Putz, 2003):

$$E[\rho+\Delta\rho]\cong E[\rho]+\int\left(\frac{\delta E[\rho]}{\delta\rho(\mathrm{r})}\right)_V\Delta\rho(\mathrm{r})d\mathrm{r}$$
$$+\frac{1}{2}\iint\left(\frac{\delta^2 E[\rho]}{\delta\rho(\mathrm{r})\delta\rho(\mathrm{r}')}\right)_V\Delta\rho(\mathrm{r})\Delta\rho(\mathrm{r}')d\mathrm{r}d\mathrm{r}' \tag{4.47}$$

Next, is required that the deviation of energy $E[\rho+\Delta\rho]$ to be minimum. What does mean this in terms of reactivity? It means that the active sites of a reactant molecule are usually placed where the addition or loss of electrons is energetically favorable. Back into mathematical language, the most favorable sites to add or lose electrons further mean that the energy have to be minimized by a function that just accounts for the ground state density variations when electrons are exchanged; this one identifies with the previous introduced Fukui function. Thus, the presence of the Fukui function will produce the minimization of the changed energy, by accounting for the preferred reactive sites, leading the analytical successive equivalences:

$$\min(\Delta E)\cong\int\left(\frac{\delta E[\rho]}{\delta\rho(\mathrm{r})}\right)_V\left(\frac{\partial\rho(\mathrm{r})}{\partial N}\right)_V\Delta\rho(\mathrm{r})d\mathrm{r}$$
$$+\frac{1}{2}\iint\left(\frac{\delta^2 E[\rho]}{\delta\rho(\mathrm{r})\delta\rho(\mathrm{r}')}\right)_V\left(\frac{\partial\rho(\mathrm{r})}{\partial N}\right)_V\left(\frac{\partial\rho(\mathrm{r}')}{\partial N}\right)_V\Delta\rho(\mathrm{r})\Delta\rho(\mathrm{r}')d\mathrm{r}d\mathrm{r}' \tag{4.48}$$
$$=\left(\frac{\partial E}{\partial N}\right)_V\Delta N+\frac{1}{2}\left(\frac{\partial^2 E}{\partial N^2}\right)_V(\Delta N)^2$$
$$=-\chi\,(\Delta N)+\eta\,(\Delta N)^2$$

where the functional derivative rules, together with the standard definitions of electronegativity and of the chemical hardness were considered.

However, the result enlightens two major achievements. The first one states that the electronegativity and chemical hardness are the minimizing global values corresponding to the local Fukui minimizing function for the change in total energy functional, when the system is favorable to exchange electrons around its ground state. The second result shows that for enough small variation in the ground state density the shape of the change in energy displays a parabolic dependence on the exchanged number of electrons. So, the parabolic assumption for the total energy in number of number of electrons, even not demonstrated to be most generally valid, finds however a natural justification, and fits with a wide range of the electronic exchange processes.

It remains to explore the reverse problem, that is: how are optimized the values χ and η during a chemical reaction or through the energy minimization processes? To give a reasonably answer, firstly consider the parabolic shape of the total energy respecting with the total number of electrons, as in Figure 4.1, connecting the three states of interests: $|N-h\rangle$, $|N\rangle$, and $|N+h\rangle$ ones.

Let's next assume that any energetic shape, containing the associated energies for the above considerate electronic states, has a parabolic form, as conceptually already certified.

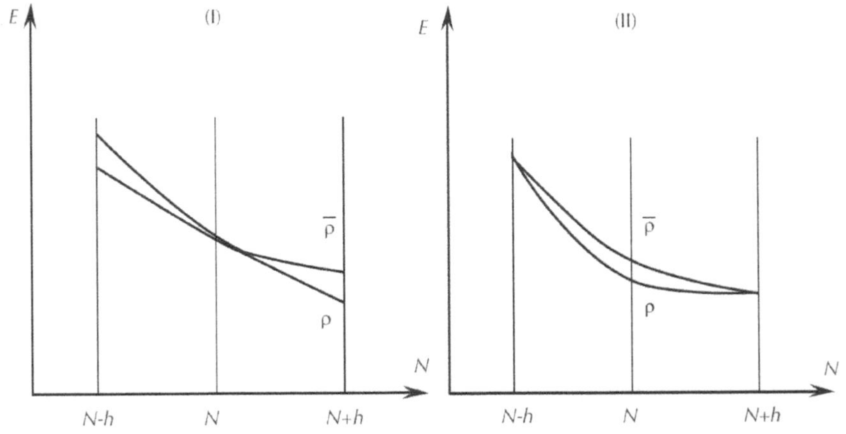

FIGURE 4.1 The two cases of the total energy minimization by distinct acting on the parabolic energetic shape connecting the electronic states $|N-h\rangle$, $|N\rangle$, and $|N+h\rangle$.

Then, one is interested in minimizing the energy throughout all parabolic classes that link those states. Is immediately that such a minimizing procedure can be undertook in two distinct ways: to simultaneously minimize the energetic values of the states $|N-h\rangle$ and $|N+h\rangle$, or to only act on the energy in the "point" $|N\rangle$ of the energetic shape, as in cases (*I*) and (*II*) in Figure 4.1, respectively (Ayers and Parr, 2000).

In the representations (*I*) and (*II*) of Figure 4.1 the negative slope and the convexity of the energetic shape for the state $|N\rangle$ will give information about the behavior of χ and η during the energy minimization, respectively. Firstly, in both analyzed cases the electronegativity approaches its minimum on the right ground density ρ, around the electronic state $|N\rangle$:

$$(I)\,,\,(II):\overline{\chi}\,[\overline{\rho}]\geq\chi\,[\rho] \tag{4.49}$$

whereas the chemical hardness records different optimum values depending with the type of energetic minimization procedure:

$$(I)\;:\overline{\eta}\,[\overline{\rho}]\geq\eta\,[\rho] \tag{4.50}$$

$$(II)\;:\overline{\eta}\,[\overline{\rho}]\leq\eta\,[\rho] \tag{4.51}$$

Now the following dilemma rises: what kind of optimization behavior is the correct one for the chemical hardness, a minimum value, as in case (*I*) or a maximum one like in case (*II*), from Figure 4.1, for the right ground ρ state, respectively? There are two arguments to choose that the right ρ ground state have to have the maximum hardness value, at the end of the energetic minimization (binding) process.

The first argument calls the relation between electronegativity and hardness that prescribes the negative sign between the electronegativity and chemical hardness tendencies when the right ground state is approached. Now, being associated the right ground state, in both above (*I*) and (*II*) cases, with the minimum value in electronegativity, the natural choice for the chemical hardness for ending the minimization process will be therefore its maximum value.

The second argument regards the parabolic energetic shape, that from beginning it was assumed to be maintained as the invariant characteristic during the minimization process; and is clear that in the case (*I*) of Figure 4.1 the minimization of energy tends to deform the parabolic energetic shape into a linear one. Thus, the right way of energy minimization trough the parabolic classes correspond with the case (*II*) which, in turn, prescribes the maximum hardness optimization for the achievement of right ground state.

Ending, worth noting that Pearson had pointed out that *there seems to be a rule of nature that molecules* (or the many-electronic systems in general, n. a.) *arrange themselves* (in their ground state, n. a.) *to be as hard as possible* (Pearson, 1985). This way, was established the Maximum Hardness Principle (MHP) for the equilibrium of the many-electronic systems in their ground states (Pearson, 1985; Chattaraj et al., 1991, 1995; Mineva et al., 1998; Putz, 2008b; Torrent–Sucarrat and Solà, 2006).

Therefore, the global scenario of reactivity, based on electronegativity and hardness principles, implies that there are four stages of bonding (Putz, 2008c, 2011a, 2011b):

(i) approaching stage is dominated by the difference in electronegativity between reactants and is consumed when the electronegativity equalization principle is fulfilled among all constituents of the products; this stage is associated with the charge flow from the more electronegativity regions to the lower electronegativity regions in a molecular formation thus covering the *covalent* binding step;
(ii) even after the chemical equilibrium is attained globally the electrons involved in bonds acts as foreign objects between pairs of regions, at whatever level of molecular partitioning procedure, inducing the appearance of finite difference in adjacent electronegativity of neighbor regions in molecule; it is due to the quantum fluctuations associate with the quantum nature of the bonding electrons and it corresponds to the degree of ionicity occurred in bonds;
(iii) the induced ionicity character of bonds is partially compensated by the chemical forces through the hardness equalization between the pair regions in molecule; the HSAB principle is therefore involved, as a second order effect in charge transfer, being driven by the ionic interaction through bonds;
(iv) still, the quantum fluctuations provides a further amount of finite difference, this time in attained global hardness, that is transposed in relaxation effects among the nuclear and electronic distributions so that the remaining unsaturated chemical forces to be dispersed by stabilization of the molecular structure.

This way, there was proved that electronegativity and hardness provides the minimum set of reactivity indices able to cover the complete process of binding, as a whole.

There is quite surprising that after our best knowledge no systematic studies are reported for linking the hardness with its conceptual source, the electronegativity when applying the HSAB principle. Such link is therefore here advanced based on the very definitions of what soft and hard acid and base are. Still, the right connection can be achieved recalling that electronegativity has the potential nature at the chemical level, so being proportional with the inverse of the radius of atoms or bonding length, $\chi \propto 1/r$. Worth noting that such dependence was the main picture in which one of the most recent atomic electronegativity scale was given (Ghosh, 2005). It can be equally derived from a simple model of charging energy of a conducting sphere of radius r (Ghosh and Biswas, 2002; Pearson, 1988). Yet, electronegativity can be seen as being proportional also with the inverse of the polarizability, $\chi \propto 1/\alpha$, since polarizability α is on its turn proportional with the volume encompassed by the electronic system under discussion (Putz et al., 2003).

With these remarks in hand let's list the main definitions of soft and hard acids and bases, connecting their hardness degree with electronegativity (Putz, 2007a, 2008c):

- a soft base, for example R^- or H^-, is very polarizable and thus with low electronegativity;
- a hard base, for example F^- or OH^-, is not much polarizable and thus with high electronegativity;
- a soft acid, for example RO^+ or HO^+, has usually low positive charge and large size, so posing lower electronegativity;
- a hard acid, for example H^+ or XH (hydrogen bonding molecules), has normally high positive charge and small size, so posing high electronegativity.

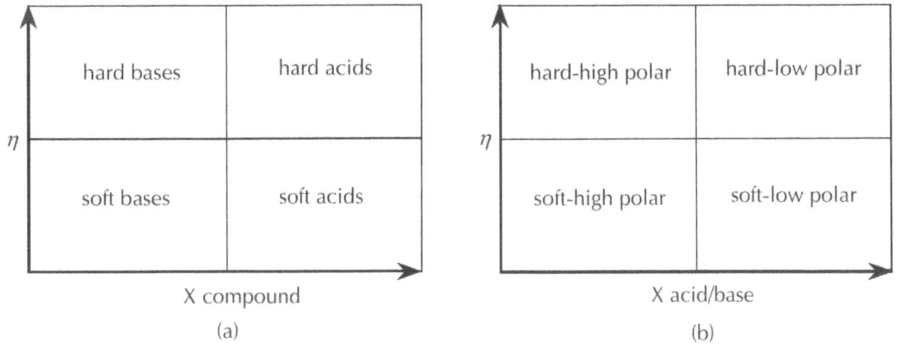

FIGURE 4.2 (a) The diagram of the compound repartition as hard-and-soft acids and bases within the electronegativity-hardness space (χ, η). (b) The diagram of the hard-and-soft nature of the high-and-low polar acids and bases within the electronegativity-hardness space (χ, η) (Putz, 2008c).

More, one can straightforwardly infer from Figure 4.2(a) that the relative position of electronegativities between two reactants can give the acid or base nature of the species since the more basicity the more $-\chi$ is pushed towards positive range.

This observation seems crucial to us and can explain why the consecrated classification of some common compounds to be acids or bases (Pearson, 1997; Sen and Jørgensen, 1987) has not an absolute value and founds some computational disagreements (Drago and Kabler, 1972; Pearson, 1972; Putz et al., 2004), while the relative electronegativities involved in concerned reactions have to as well be taken as the appropriate measure.

Collecting all these ideas in a representative quantum concept we can draw the Figure 4.1 for appropriate indication of the acid/base and hard/soft trends of the chemical species within the electronegativity-hardness chemical space (χ,η).

The Figure 4.2 (a) depicts the phenomenological correlation between electronegativity and hardness for a given chemical species leaving with their natural classification as acids and bases on the electronegativity scale and hard and soft on the hardness scale as their positions are more departed from the (0, 0) origin point within the chemical space (χ,η).

A similar classification can be done in Figure 4.2(b) for a series of acids and bases, separately, with the result in categorizing them as hard-and-soft character with low-and-high polarity as their positions are more departed from the (0, 0) origin point within the chemical space (χ,η). This way, there is provided a new valuable working scheme with the help of which the relative acidic or basic nature as well as the hard and soft strength of the molecules in their state of interaction is analyzed.

4.4 UNIFIED SCENARIO OF CHEMICAL REACTIVITY PRINCIPLES

Since, the special energetic (accuracy about 1 kJ/mol) and space of electronic manifestation of chemical reactivity (the valence state, frontier orbitals) appropriate physico-chemical indicators were advanced in order to better quantify, model, and finally control specific systems (atoms, molecules, atoms in molecules, etc.) and of their properties. Basically, once a system is computed for its energy in various states of donating and accepting electronic the total energy vs. number of electrons $E = E(N)$ is provided, see Figure 4.3, to represent a sort of dynamical mark of the electronic sample under given external potential influence $V(\mathrm{r})$ (Putz, 2011a).

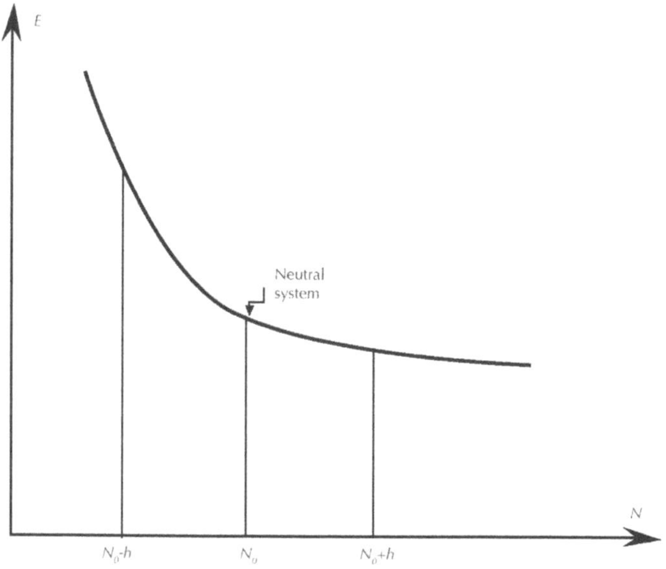

FIGURE 4.3 Schematic representation of the parabolic dependence of the total energy of an N-electronic system, connecting its ionic (± h) states (Parr and Pearson, 1983; Putz, 2003; Putz, 2011a).

Now, under the assumption that such a curve is always convex for Coulombic interacting systems (Ayers and Parr, 2011), one can extract the physico-chemical information from it by considering the first and second derivative on the curve point associated with the neutral system, that is the tangent and the curvature's values on the neutral point eventually provides the prediction for further systems evolution outside of it. This way, the two basic indices of reactivity are at once advanced, namely the electronegativity (EL or χ) or the negative of the chemical potential (Iczkowski and Margrave, 1961; Parr et al., 1978)

$$\chi = -\mu = -\left(\frac{\partial E}{\partial N}\right)_{V(\mathbf{r})} \tag{4.52}$$

and the chemical hardness (HD or η) (Parr and Pearson, 1983)

$$\eta = -\frac{1}{2}\left(\frac{\partial \chi}{\partial N}\right)_{V(\mathbf{r})} = \frac{1}{2}\left(\frac{\partial \mu}{\partial N}\right)_{V(\mathbf{r})} = \frac{1}{2}\left(\frac{\partial^2 E}{\partial N^2}\right)_{V(\mathbf{r})} \tag{4.53}$$

Note that the denominator factor "2", although nowadays it is optional, may have two reasons for being accounted: one is of aesthetical value that is apparent when the two-point-central (2C) finite difference scheme—in accordance with the Koopmans' frozen core theorem (Koopmans, 1934)—is applied to unfold the definitions by derivatives that symmetrical gives (Sen and Jørgensen, 1987; Sen and Mingos, 1993)

$$\chi_{FD} \cong \frac{(E_{N_0-1} - E_{N_0}) + (E_{N_0} - E_{N_0+1})}{2} = \frac{IP + EA}{2} \cong -\frac{\varepsilon_{LUMO} + \varepsilon_{HOMO}}{2} \tag{4.54}$$

$$\eta_{FD} = \frac{1}{2}\left(\frac{\partial^2 E_N}{\partial N^2}\right)_{V(r)} \cong \frac{E_{N_0+1} - 2E_{N_0} + E_{N_0-1}}{2} = \frac{IP - EA}{2} \cong \frac{\varepsilon_{LUMO} - \varepsilon_{HOMO}}{2} \tag{4.55}$$

written in terms of semi-sum and semi-difference of the ionization potential IP and electron affinity EA for the so called "experimental" electronegativity and chemical hardness and also within the approximations of higher occupied and lower unoccupied molecular orbitals, HOMO and LUMO, respectively. Therefore, one can already see also at this frozen core level that (Putz, 2011a)

- the conceptual difference exists between the energetic *level* characterizing the "experimental" electronegativity and the energetic *gap* characterizing the "experimental" chemical hardness, both providing a sort of orthogonal basis $\{\chi,\eta|\chi\perp\eta\}$ for chemical reactivity analysis.

Moreover, under such formulations the direct connection with chemical behavior at frontier is evident, thus emphasizing on the basic role of electronegativity and chemical hardness in reactivity, one establishing the energetic median position and other the energetic interval between the two fundamental tendencies of a chemical system: the electronic accepting (on LUMO) and the electronic donating (from HOMO), respectively.

However, beside the quantitative fundament of EL and HD they also underlie the most celebrated principles of chemical reactivity with which help the qualitative understanding of chemical basic phenomena is possible within the so called *chemical orthogonal space* – COS ($\chi \perp \eta$, see above for their finite difference meaning as energetic level vs. interval). This idea is particularly important since consecrating the need for the EL and HD though their independency (as "orthogonal" quantities) that eventually closes the chemical space of bonding and reactive in two dimensions (2D). However, for having a better view of their reactivity implications we summarize here the main physical-chemical principles they underscore towards reaching the chemical systems equilibrium through interacting with their environment (other atoms, molecules, etc.), this time incorporating also the chemical action intermediary principle as one responsible for DFT density-external potential variational principles, but also relating with related with the electronegativity and chemical hardness ones as will be in next section revealed in a specific chemical variation streamline (Putz, 2011a).

(i) *The electronegativity equalization (EE) principle* ($\Delta\chi = 0$): it was originally stated by the Sanderson (1988) under the assumption that "the molecules in their fundamental state, the electronegativities of different electronic regions in molecule – are equal"; the principle appears as natural since the identification by Parr et al. of electronegativity with the minus of chemical potential (Parr et al., 1978), see equation (1.19), for which the equalization principle is already consecrated from solid state physics (Mortier et al., 1985);

(ii) *Chemical action principle*: by recalling the integral definition of electronegativity as the expectation value of the applied Coulombic on the electronic system due to March (1993)

$$\chi(N,Z) = \left\langle \frac{1}{|\mathrm{r}|} \right\rangle = \int \left\{ \rho(N,Z,\mathrm{r}) \frac{1}{|\mathrm{r}|} \right\} dr \tag{4.56}$$

one may step for further generalization of the electronegativity principle toward assessing the molecular (or chemical) stability/equilibrium with the environment through asking the expectation value of the applied external potential on the ground or valence state density $\rho(\mathrm{r})$

$$C_A \equiv \int \rho(\mathrm{r}) V(\mathrm{r}) dr \tag{4.57}$$

fulfilling the variational principle (Putz, 2007a, 2008c, 2009a, 2009b; Putz and Chiriac, 2008; Putz et al., 2001, 2005)

$$\delta C_A = 0 \tag{4.58}$$

This condition is the analogous of the second Newtonian law for chemical systems, and was recently fully employed to provide the basis of the chemical bonding both within wave-function (Schrödinger) (Putz, 2009b) and the spinorial (Dirac) (Putz, 2010a) frameworks. It nevertheless combines the two main informative quantities of DFT as well, the electronic density and applied potential $\{\rho, V\}$ that constitutes the local Hohenberg–Kohn basis in which the chemical systems are immersed when treated computationally with the aid of DFT.

(iii) *The minimum electronegativity (MinEL) principle*: it may be viewed as a kind of EE principle consequence when the binding process is promoted ($\Delta\chi \neq 0$), firstly formulated by Parr and Yang as (Parr and Yang, 1984, 1989; Yang et al., 1984): "two different sites with generally similar disposition for reacting with a given reagent, the reagent prefers the one which on the reagent's approach is associated with the maximum response of the system's electronegativity. In short, $\Delta\chi \geq 0$ is good for reactivity (n. a.)". Yet, for assessing the chemical stability the other way round form of the last idea will be considered, from where the minimum electronegativity principle $\Delta\chi \leq 0$ immediately results; however, this principle, in order do not conflict with the first one should be seen as a quantum fluctuation remnant effects in system upon the EE was consumed (Tachibana, 1987; Tachibana and Parr, 1992; Tachibana et al., 1999), that is it needs to be minimized as well for the system undergo stable equilibrium;

(iv) *The hard and soft acids and bases (HSAB) principle*: initially formulated by Pearson as "the species with a high chemical hardness prefer the coordination with species that are high in their chemical hardness, and respectively the species with low softness (the inverse of the chemical hardness) will prefer reactions with species that are low in their softness" (Chattaraj and Maiti, 2003; Chattaraj and Schleyer, 1994; Pearson, 1973, 1990, 1997) corresponds in fact with the chemical hardness equalization principle $\Delta\eta = 0$ as the companion principle to that of electronegativity; this principle, since establishing the prerequisite conditions the chemical species will interact is also very useful in addressing also the aromaticity hierarchies; in this respect, the parallelism of the aromaticity concepts and chemical hardness, namely aromatic systems are hard molecules while anti-aromatic systems are on the other hand soft molecules (Parr, 1983), successfully quantified in chemical terms the extra energetic stability of aromatic (benzene as preeminent example) and the lack of stability of anti-aromatic systems, respectively, being this one-to-one rule fundamental to the development of the aromaticity concepts as a special part of chemical reactivity; on these grounds further connections between aromaticity and chemical hardness and electronegativity indices will be also in this review presented.

(v) *The maximum hardness (MaxHD)*: also due to Pearson observation "there seems to be a rule of nature that molecules (or the many-electronic systems in general, n. a.) arrange themselves (in their ground or valence states, n. a.) to be as hard as possible" (Pearson, 1985); however since the inverse of hardness the softness is viewed as direct proportional with the polarizability degree of the system this principles is sometimes referred to as the *minimum polarizability principle* as well; yet, it states as the counterpart of the above minimum electronegativity principle, thus with the analytic realization of inequality $\Delta\eta \geq 0$ (Ayers and Parr, 2000; Chattaraj et al., 1991, 1995; Putz, 2008b); such behavior was both analytically as well computationally proved for various chemical systems (Mineva et al., 1998); nevertheless, the application of this principle to systems with rotational barriers showed that it actually may account also for conformational properties and thus being associated with the steric effects (Torrent–Sucarrat and Solà, 2006); this is not surprising, since in the light of above LUMO-HOMO form directly relates with the frontier orbital concepts; moreover by considering the relaxation effects (beyond the frozen core approximation of Koopmans' theorem) in computing chemical hardness (4.55) the account for the electronic configuration leaves with another route in characterizing the steric effects by its maximum output.

The great use of these principles and indices of reactivity is nowadays still limited by a number of open questions that need be answered for future entrusting in developing physico-chemical methods based on their influences in characterizing complex systems.

4.5 ANALYTICAL NECESSITY OF CHEMICAL REACTIVITY PRINCIPLES

The variational principle of the chemical action density functional of equation (4.58) will be introduced, within DFT, this way proving its necessity in modern conceptual chemistry (Putz, 2011c).

To this end one notes that in physical sciences the total energy variation

$$dE = 0 \tag{4.59}$$

fixes the equilibrium or the evolution towards equilibrium of natural systems. In conceptual DFT framework, however, if one expands the total energy $E = E[N, V(\mathrm{r})]$ such that to contain the electronegativity and chemical hardness appearance relating the number of electrons in first and second order variations in reactivity, respectively, as well as to contain the applied potential, one has the working differential relationship

$$dE = -\chi dN + \eta (dN)^2 + \int \rho(\mathrm{r}) dV(\mathrm{r}) d\mathrm{r} \tag{4.60}$$

Yet, through combining equations (4.59) and (4.60), one has either that:

- there is no action on the system ($dN = 0 = dV$) so that no chemical phenomena is recorded since the physical variational principle (4.59) is fulfilled for whatever electronegativity and chemical hardness values in (4.60);
- or there is no electronic system at all ($\chi = \eta = 0 = \rho(\mathrm{r})$).

Therefore, it seems that the variational physical principle of equation (4.59) do not suffice in order the chemical principles of electronegativity and chemical hardness being encompassed, as above stipulated, see equations (4.49) and (4.50). As a consequence one is forced to perform the double variational procedure on the total energy, that is through applying the additional differentiation on physical energy expansion (4.60), within the so called "chemical variational mode" (and denoted as $\delta[]$) where the total differentiation will be taken only over the scalar-global (extensive χ, η, dN) and local (intensive $\rho(\mathrm{r}), V(\mathrm{r})$) but not over the vectorial (physical – as the coordinate itself $\mathbf{r}$) quantities. This way, one has immediately

$$\delta\left[dE\right]=-\delta[\chi dN]+\delta[\eta(dN)^2]+\int\delta\left[\rho(\mathrm{r})dV(\mathrm{r})\right]d\mathrm{r} \tag{4.61}$$

Now, the chemical variational principle applied to equation (4.61) will look like

$$\delta\left[dE\right]\geq 0 \tag{4.62}$$

In conditions of chemical reactivity or binding certain amount of charge transfer and the system's potential fluctuations (departing from equilibrium) are involved

$$dN=\left|dN\right|=ct.\neq 0,\ dV(\mathrm{r})\neq 0 \tag{4.63}$$

When considering the chemical variation prescribed by equation (4.61) in chemical transformation driven by the condition (4.62), a kind of reactivity towards equilibrium constraint or a sort of reverse Gibbs free energy condition, or a special kind of entropy variation within the second law of thermodynamics are in fact applied, while releasing with the individual reactivity principles:

- for electronegativity contribution we have the condition:

$$-\delta[\chi dN]\geq 0\Leftrightarrow -\left|dN\right|\delta[\chi]\geq 0\Rightarrow\delta[\chi]\leq 0 \tag{4.64}$$

- recovering the electronegativity variational formulation (4.49) for chemical systems toward equilibrium.
- for chemical hardness contribution one gains directly the inequality of equation (4.50)

$$\delta[\eta(dN)^2]\geq 0\Leftrightarrow (dN)^2\delta[\eta]\geq 0\Rightarrow\delta[\eta]\geq 0 \tag{4.65}$$

- whereas for chemical action contribution there is now sufficient in having the exact equality

$$\delta\int\rho(\mathrm{r})dV(\mathrm{r})d\mathrm{r}=0 \tag{4.66}$$

leaving with the successive equivalent forms

$$0=\delta\int\rho(\mathrm{r})\left\{V(\mathrm{r})-V(\mathrm{r}_0)\right\}d\mathrm{r}=\delta\left\{\int\rho(\mathrm{r})V(\mathrm{r})d\mathrm{r}\right\}-\delta\left\{V(\mathrm{r}_0)\int\rho(\mathrm{r})d\mathrm{r}\right\} \tag{4.67}$$

with $V(\mathrm{r}_0)$ being the constant potential at equilibrium. However, through employing the basic DFT relationship for electronic density

$$N=\int\rho(\mathrm{r})d\mathrm{r} \tag{4.68}$$

and by recognizing the total number of electron constancy in the (atoms-in-molecules) system the equation (4.66) produces the so called chemical action principle of equation (4.58) that represents the chemical specialization for the physical variational principle of equation (4.59).

Worth noting that the actual chemical action principle arises along the other chemical reactivity principles of electronegativity and chemical hardness in a natural way, yet by performing special variation of the charge transfer and potential fluctuations around equilibrium rather than by the direct physical implementation of an "action" in searching for the system evolution's "trajectory". Therefore, the present analysis reveals two main features of chemical reactivity principles in general and of that of the chemical action in special:

- the chemical reactivity principles of electronegativity and chemical hardness belong to *chemical* variation principle rather to the physical optimization framework, this way inscribing themselves in the heritage of natural laws provided by Chemistry to complement those of Physics;
- the chemical action principle of equation (4.58), apart representing as well a chemical principle since arising within the same conceptual DFT framework as those of electronegativity and chemical hardness, represents a variational principle that optimizes the relationship between the bijective main quantities of DFT, the electronic density and its driving bare potential, in a convoluted manner over the reactivity or binding space of interest; it may, eventually, produce the chemical global minima of equilibrium in reactivity or bonding.

However, the hierarchy of the electronegativity, chemical hardness and chemical action principles was recently advanced as describing the paradigmatic stages of bonding (Putz, 2008c, 2011a); they may be resumed here through the sequence:

$$\delta\chi = 0 \rightarrow \delta C_A = 0 \rightarrow \Delta\chi < 0 \rightarrow \delta\eta = 0 \rightarrow \Delta\eta > 0 \tag{4.69}$$

as corresponding to the *encountering* (or the electronegativity equality) *stage*, followed by *chemical action minimum variation* (i.e., the global minimum of bonding interaction), then by *the charge fluctuation stage* (due to minimum or residual electronegativity), ending up with *the polarizability stage* (or HSAB) and with the *final steric* (due to maximum or residual hardness) *stage*. Nevertheless, from equation (4.69) one observes the close laying chemical action with electronegativity influence in chemical reactivity and bonding principles.

KEYWORDS

- **Chemical action principle**
- **Electronegativity equalization**
- **Euler–Lagrange equation**
- **Hohenberg–Kohn functional**
- **Minimum electronegativity**
- **Physico-chemical methods**

PART II

PRIMER DENSITY FUNCTIONAL THEORY OF BOSE-EINSTEIN CONDENSATION

5 Basics of the Bose–Einstein Condensation (BEC)

CONTENTS

5.1 INTRODUCTION

Since the nowadays growing interest in Bose–Einstein condensates (BEC) due to the expanded experimental evidence on various atomic systems within optical lattices in weak and strong coupling regimes, its many-body quantum bases and equations are elementarily presented emphasizing on the order parameter, pseudo-potential and mean field approaches, while emphasizing the common places with the Density Functional Theory, like the integral normalization of density or the chemical potential, as the main vehicles of both theories of matter (Putz, 2011d, 2011e).

5.2 THE BACKGROUND PHYSICS OF BEC

Since the Science magazine declared on 22 December, 1995, the Bose condensate as the "molecule of the year", the BEC viewed as *the macroscopic occupation of the same single-particle state in a many-body systems of bosons*, had received new impetus both at theoretical and experimental levels in searching and comprehending new states of matter (Leggett, 2001; Pethick and Smith, 2002; Pitaevskii and Stringari, 2003; Putz, 2011d; Yukalov, 2004).

Historically, the fundaments of BEC were established back to 1925 by the works of Bose (1924) and Einstein (1924, 1925) that predicted the occurrence of a phase-transition in a gas of non-interacting (fixed number of) atoms, based on the analogy with the statistical description of the (not fixed number of) light quanta.

Actually, while considering an ideal Bose gas formed by N undistinguishable non-interacting non-relativistic quantum particles, assumed for simplicity being confined in a space Ω with the volume $|\Omega| = V$, considered in thermal equilibrium at temperature T with a reservoir with chemical potential μ, that is within grand-canonical description, one has the total occupation number given by the bosonic statistics

$$\overline{N} = \sum_{i \geq 0} \frac{1}{\exp\{\beta(\varepsilon_i - \mu)\} - 1} \tag{5.1}$$

associated with the ith excited state energy ε_i of the single-particle Hamiltonian $H^{(1)}$ of the many-body system

$$H = \sum_k H_k^{(1)} = \sum_k \frac{\mathrm{p}_k^2}{2m} \tag{5.2}$$

with each quantum momentum operator

$$\mathrm{p}_k = -i\hbar \nabla_{\mathrm{r}_k} \tag{5.3}$$

acting on the one-body Hilbert space $L^2(\Omega)$ with appropriate boundary conditions. Here, as usual, $\beta = (k_B T)^{-1}$ is the inverse temperature in terms of Boltzmann's constant k_B, while $\hbar$ and m are the Planck's constant and the mass of each particle, respectively. Now, for fixed $\mu < 0$, within the thermodynamic limit one has the *sample density*

$$\frac{\overline{N}}{|\Omega|} \xrightarrow{|\Omega| \to \infty} \frac{1}{(2\pi\hbar)^3} \int_{\Re^3} \frac{1}{\exp\left\{\beta\left(\frac{\mathrm{p}^2}{2m} - \mu\right)\right\} - 1} d\mathrm{p} := \rho \tag{5.4}$$

that is monotonously increasing with μ until reaching its *critical* value ρ_{crit} when bounded by the condition $\mu \to 0$, namely

$$\rho_{\text{crit}} = \frac{1}{(2\pi\hbar)^3} \int_0^\infty \frac{4\pi p^2 dp}{\exp\left\{\frac{\beta}{2m} p^2\right\} - 1} = \left(\frac{m}{2\pi\hbar^2\beta}\right)^{3/2} g_{3/2}(1) \tag{5.5}$$

with

$$g_{3/2}(1) = \frac{2}{\sqrt{\pi}} \int_0^\infty \frac{\sqrt{x}dx}{\exp(x) - 1} = \sum_{l=1}^\infty \frac{1}{l^{3/2}} = \xi(3/2) \cong 2.612 \tag{5.6}$$

where $\xi(z)$ is the Riemann zeta function.

Therefore, *condensation* phenomenon arises when particles exceed the critical density number to all going into the lowest energy state ε_0 with the density of the condensate

$$\rho_{\text{cond}} = \lim_{|\Omega|\to\infty} \frac{\overline{N}_0}{|\Omega|} = \lim_{|\Omega|\to\infty} \frac{1}{|\Omega|} \frac{1}{\exp\left\{\frac{1}{k_B T}\left[\varepsilon_0(\Omega) - \mu(\Omega)\right]\right\} - 1} \neq 0 \tag{5.7}$$

with the temperature bellow the critical one

$$T \leq T_{\text{crit}} = \frac{2\pi\hbar^2}{mk_B}\left[\frac{\rho_{\text{crit}}}{g_{3/2}(1)}\right]^{2/3}_{\rho_{\text{crit}}=\rho} \tag{5.8}$$

The other way around, through separating the contribution from the lowest energy level and approximating the contribution from the remaining terms by an integral, within the thermodynamic limit, one has the net result

$$\rho = \rho_{\text{cond}} + \rho_{\text{crit}} \tag{5.9}$$

Consequently, the critical temperature is the smallest temperature for which $\rho_{\text{cond}} = 0$ that becomes non-vanishing quantity only for lower temperatures! However, worth noting that the above critical temperature criterion may be reshaped as the more usually form of condensation

$$\rho\lambda_{dB}^3 \geq g_{3/2}(1) \cong 2.612 \tag{5.10}$$

in terms of the thermal (or de Broglie) wavelength

$$\lambda_{dB} := \sqrt{\frac{2\pi\hbar^2}{mk_B T}} \tag{5.11}$$

This means that in contrast with other phase transitions (like melting or crystallization) which depend on the interparticle interaction, occurrence of the BEC is subject to such temperature and dilution that the thermal de Broglie wavelength becomes comparable with the interparticle distance $\rho^{-1/3}$! The occupancy in the ground state $\overline{N}_0/\overline{N}$ may be determined from above density equation firstly rewritten as

$$\overline{N} = \overline{N}_0 + |\Omega|\frac{g_{3/2}(1)}{\lambda_{dB}^3} \tag{5.12}$$

from where immediately it follows to be

$$\frac{\overline{N}_0}{\overline{N}} = 1 - \frac{1}{\rho_{\text{crit}}}\frac{g_{3/2}(1)}{\lambda_{dB}^3} = 1 - \left(\frac{T}{T_{\text{crit}}}\right)^{3/2} \tag{5.13}$$

Among many peculiar features for ideal Bose gas, the condensate phase appears to have infinite compressibility and zero entropy, while displaying a first order gas/condensate phase transition (Huang, 2001); in particular, at $T = 0[K]$ there is 100% BEC with all atoms in the ground state φ of the single-particle Hamiltonian $H^{(1)}$ called therefore *the condensate (or macroscopic) wave function.*

Initially considered just as a mathematical curiosity BEC was assumed by London in 1938 as being at the origin of superfluidic manifestation in liquid helium (London, 1938), while being at that time quite controversially with Landau 1941' approach of the superfluid theory in terms of the spectrum of elementary excitations of the fluid (Landau, 1941). The controversy arises because although as bosonic atoms ^{4}He display *strong inter-atomic interaction* that reduces the number of atoms in the zero-momentum state even at absolute zero, while experimentally the detected occupancy is found less than 1/10! Indeed, in the presence of interactions, as naturally occur between trapped bosonic atoms, the many-body Hamiltonian is no longer the sum of single-particle neither Hamiltonians nor the ground state many-body function may be seen as the product of N single-particle wave-functions!

Therefore, the BEC turned to be searched for the weakly interacting Bose gases with a higher condensate fraction, while the first quantitative treatment of interacting Bose fluids was advanced by Bogoliubov (1947) that, although intuitively appealing and correct in many aspects (Zagrebnov and Bru, 2001) still *it does not yield quantum phase transition* (van Oosten et al., 2001).

Due to the experimentally accessible cooling and magneto-optically trapping techniques (Ketterle, 2002), the required temperature and densities were reached to observe BEC in spin-polarized hydrogen, metastable ^{4}He (Fried et al., 1998), vaporous ^{7}Li (Bradley et al., 1995), ^{23}Na (Davis et al., 1995), ^{41}K (Modugno et al., 2001), ^{87}Rb (Anderson et al., 1995), and the list of laser cooled atomic species is still growing. However, the specific conditions and physical parameters are summarized in Table 5.1, conveniently expressed for each quantity.

TABLE 5.1 Typical values and parameters for experimentally acquiring BEC condensation phase.

Parameter	Value
Temperature	50 nK, 50 mK
Number of atoms	$10^2 \div 10^{11}$
Trap size	$10 \div 50$ μm
Mean density (r)	$10^{11} \div 10^{15}$ cm^{-3}
Scattering length (a)	$3 \div 300$ a.u.
Dilution (ρa^3)	$10^{-4} \div 10^{-6}$

Summarizing, the mainstream physical conditions, both theoretically and experimentally, for modeling and obtaining the BEC are basically grounded in the ideal gas picture, however adapted to include weak interactions; they are:

- *Dilute nature of the gas*: allows description via a single parameter, namely the s-wave scattering length (a) of the two-body scattering process, due to the fact all other partial waves may be neglected at lower thermal energy $k_B T$ when BEC takes place as compared with the two-body characteristic scattering energies;
- *Ultra-high dilution*: the scattering length is far smaller than the mean interparticle distance, $\rho a^3 < 1$, allows the Hamiltonian be written as

$$H = \sum_{k=1}^{N}\left(\frac{\mathrm{p}_k^2}{2m} + V\left(\mathrm{r}_k\right)\right) + \sum_{1\le k<h\le N} W\left(\mathrm{r}_k - \mathrm{r}_h\right) \tag{5.14}$$

- with the interparticle potential W having the scattering length a dependency and where V is the trap;
- *The indeterminacy*: bosons in their ground state cannot be localized with respect to each other without enormously changing the kinetic energy; they are thus characterized by the so called *healing length*, $H = 1/\sqrt{8\pi\rho a}$, that completes the ultra-high dilution condition as: $a << \rho^{-1/3} << H$ so that making the BEC possible even though mean density is low;
- *The delocalization picture*: to a very good approximation, at zero temperature the many-body ground state density wave function still may be regarded as a product of N copies of the single-particle wave functions ψ, that is the so called *Hartree–Fock approximation*,

$$\Psi\left(\mathrm{r}_1,\ldots,\mathrm{r}_N\right) \sim \psi(\mathrm{r}_1)\ldots\psi(\mathrm{r}_N) \tag{5.15}$$

- subject it is the solution of the so called Gross–Pitaevsky (GP or non-linear Schrödinger) equation

$$-\frac{\hbar^2}{2m}\Delta\psi(\mathrm{r}) + V(\mathrm{r})\psi(\mathrm{r}) + \frac{4\pi\hbar^2 a}{m}\left|\psi(\mathrm{r})\right|^2\psi(\mathrm{r}) = \mu\psi(\mathrm{r}) \tag{5.16}$$

- while fulfilling the N-body normalization condition

$$\int\left|\psi(\mathrm{r})\right|^2 d\mathrm{r} = N \tag{5.17}$$

- Note that the non-linear term in GP equation accounts for the local (onsite) self-interaction, as each particle would be subjected to an additional potential produced by itself.
- *The quantum depletion of the condensate*: gives the remaining fraction of particles $1-\overline{N}_0/\overline{N}$ remaining outside of the condensate even at zero temperature, due to repulsive interaction, while being enhanced by thermal effects; it gives typically about 1% or less for alkali condensate so one can fairly assume that the with 99% accuracy all atoms have the same single-particle wave function, thus validating above Hartree–Fock assumption (excepting liquid Helium for which the quantum depletion is about 90%).

Overall, the occurrence of the BEC from "imperfect" Bose gas (with some inter-bosonic interaction apart of the inherent statistic interaction) is subject to ultracold framework being driven by the negligible contribution of the $l \neq 0$ quantum waves to the two-body cross section along the ultra-high dilution $\rho a^3 << 1$ criterion that replaces that specific to ideal Bose gas (governed only by the statistical interaction) $\rho\lambda_{dB}^3 \geq 2.612$; as a consequence it will be no more infinitely compressible as the ideal counterpart, while the quantum gas/condensate transition are now regarded to be of higher orders in a/λ_{dB} and $\rho a\lambda_{dB}^2$.

Yet, in order the BEC theory achieves the proper theoretical grounds for whatever coupling interaction is about among the condensing bosons one has to undertake further special framework, see next section (Putz, 2011e) and Section 6.4.

5.3 THE QUANTUM PHENOMENOLOGY OF BEC

5.3.1 N-Body Field Quantization

For an N-identical system of particles with the mass m (fermions or bosons, we treat them unitarily for the moment, or as spineless), say contained in the volume V, not to be confounded with potential that will always have an explicit dependence as $V(...)$ or appropriate indices – see below, whose positions are denoted as $\{\mathrm{r}_1, \mathrm{r}_2, ..., \mathrm{r}_N\}$ or simply as $\{1,2,...,N\}$ one has the working Hamiltonian operator

$$\hat{H} = -\frac{\hbar^2}{2m}\sum_{i=1}^{N}\nabla_i^2 + \sum_{i<j}\underbrace{V(\mathrm{r}_i, \mathrm{r}_j)}_{\equiv V_{ij}} \tag{5.18}$$

the stationary wave function of the N-system $\Psi(1,2,...,N)$ satisfy the Schrödinger equation

$$\hat{H}\Psi(1,2,...,N) = E\Psi(1,2,...,N) \tag{5.19}$$

while being normalized to unity in the volume V

$$1 = \int_V d^N\mathrm{r}\left(\Psi^+(1,2,...,N)\Psi(1,2,...,N)\right) \tag{5.20}$$

Now, while looking to the N particle system state $|\Psi\rangle$ as to the equivalence of the quantized *N-uni-particle fields* $\Psi\left(\mathrm{r}_{i=\overline{1,N}}\right)$ successively acting on the vacuum state $|0\rangle$ one has the following properties of the *many-particle state*

- Normalization

$$1 = \langle\Psi|\Psi\rangle \tag{5.21}$$

- Eigen-equation

$$\hat{H}|\Psi\rangle = E|\Psi\rangle \tag{5.22}$$

- Field Hamiltonian operator

$$\hat{H} = -\frac{\hbar^2}{2m}\int d\mathrm{r}\left(\psi^+(\mathrm{r})\nabla^2\psi(\mathrm{r})\right) + \frac{1}{2}\int d\mathrm{r}_1 d\mathrm{r}_2\left(\psi^+(\mathrm{r}_1)\psi^+(\mathrm{r}_2)V_{12}\psi(\mathrm{r}_2)\psi(\mathrm{r}_1)\right) \tag{5.23}$$

Under the *field commutation rules* for *bosons*

$$\left[\psi(\mathrm{r}), \psi^+(\mathrm{r}')\right] = \delta(\mathrm{r}-\mathrm{r}'), \left[\psi(\mathrm{r}), \psi(\mathrm{r}')\right] = 0, \left[\psi^+(\mathrm{r}), \psi^+(\mathrm{r}')\right] = 0 \tag{5.24}$$

and for *fermions*,

$$\left\{\psi(\mathrm{r}), \psi^+(\mathrm{r}')\right\} = \delta(\mathrm{r}-\mathrm{r}'), \left\{\psi(\mathrm{r}), \psi(\mathrm{r}')\right\} = 0, \left\{\psi^+(\mathrm{r}), \psi^+(\mathrm{r}')\right\} = 0 \tag{5.25}$$

with Hermitian conjugations among the uni-particle fields and with the commutator and anti-commutator basic definitions

$$[A,B] = AB - BA, \{A,B\} = AB + BA \tag{5.26}$$

one immediately sees that the number operator giving the particle number it actually has the form

$$\hat{N} \equiv N = \int dr\left(\psi^{+}(\mathrm{r})\psi(\mathrm{r})\right) \tag{5.27}$$

while verifying the following commutation rules

- Simultaneous diagonalization with the Hamiltonian

$$\left[\hat{H},\hat{N}\right] = 0 \tag{5.28}$$

- Inner-field generating fields

$$\left[\psi(\mathrm{r}),\hat{N}\right] = \psi(\mathrm{r}) \tag{5.29}$$

$$\left[\psi^{+}(\mathrm{r}),\hat{N}\right] = -\psi^{+}(\mathrm{r}) \tag{5.30}$$

The last two properties alongside the properties of the assumed *unique* vacuum state

$$\langle 0|0\rangle = 1, \tag{5.31}$$

$$\hat{H}|0\rangle = 0, \tag{5.32}$$

$$\hat{N}|0\rangle = 0 \tag{5.33}$$

yield in fact the meaning of the direct and conjugated uni-fields as the carriers of the annihilation and creation of one particle respectively

$$\hat{N}\psi(\mathrm{r})|\Psi\rangle = (N-1)\psi(\mathrm{r})|\Psi\rangle \ldots\text{annihilation effect} \tag{5.34}$$

$$\hat{N}\psi^{+}(\mathrm{r})|\Psi\rangle = (N+1)\psi^{+}(\mathrm{r})|\Psi\rangle \cdots \text{ creation effect} \tag{5.35}$$

either for bosonic or fermionic systems.

However, the working with effective quantized field of the N-body system implies introduction of the respective function; to this end it is worth observing first that due to the one-particle annihilation property of function $\psi(\mathrm{r})$ the following so called *field-cancellation* relationship is true

$$0 = \langle \Phi_n|\psi(1)\psi(2)\ldots\psi(N)|\Psi\rangle \tag{5.36}$$

for any N-body state that is different than the vacuum state, $|\Phi_n\rangle \neq |0\rangle$, since systematically recording the decreasing of the number of particles' on the initial state $|\Phi_n\rangle$ as each of the N-uni-field operators comes into action over the previous results, until the final obvious yield $\langle 0|\Psi\rangle = 0$. Now, with this lesson, the other way around, when

$|\Phi_n\rangle = |0\rangle$, due to the unique of the vacuum state one can advance the N-body field wave function under the form (sometimes called also as the Hartree–Fock formulation):

$$\Psi(1,2,...,N) = \frac{1}{\sqrt{N!}}\langle 0|\psi(1)\psi(2)...\psi(N)|\Psi\rangle \tag{5.37}$$

with the property that it is symmetric respecting the exchanging of two coordinate for bosons, and antisymmetric under the same operation for fermions. Firstly, one very important feature of this function is that it looks like a functional, that is as averaging the N-product of N-uni-particle fields between the vacuum and the actual state of the system. Nonetheless, albeit it is formal (and somehow particular or even mysterious definition since its apparent self-defined form) it readily fulfills both the unitary normalization condition and the Schrödinger-equation for the N-body system as it is in next proved.

- The normalization proof is based on inductive iteration of integration over coordinates of successive increasing (or decreasing) particles in the sample, that is

$$N = \int d\mathrm{r}_1 \left(\psi^+(\mathrm{r}_1)\psi(\mathrm{r}_1)\right) \tag{5.38}$$

over the coordinate of N particles;

$$N(N-1) = \int d\mathrm{r}_2 \left\{\psi^+(\mathrm{r}_2)(N)\psi(\mathrm{r}_2)\right\} \tag{5.39}$$

over the coordinates of remaining N-1 particles to integrate;

$$N! = \int d\mathrm{r}_N \left\{\psi^+(\mathrm{r}_N)\left(N(N-1)(N-2)...1\right)\psi(\mathrm{r}_N)\right\} \tag{5.40}$$

over the coordinates of remaining one particle to integrate. So, now we can proceed with the next integral evaluation

$$I = \int d^N\mathrm{r}\left(\Psi^+(1,2,...,N)\Psi(1,2,...,N)\right) \tag{5.41}$$

$$= \frac{1}{N!}\int d^N\mathrm{r}\langle\Psi|\psi^+(N)...\psi^+(1)|0\rangle\langle 0|\psi(1)...\psi(N)|\Psi\rangle \tag{5.42}$$

that it can be generalized with the aid of above field cancellation result as

$$I = \frac{1}{N!}\int d^N\mathrm{r}\sum_n\langle\Psi|\psi^+(N)...\psi^+(1)|\Phi_n\rangle\langle\Phi_n|\psi(1)...\psi(N)|\Psi\rangle \tag{5.43}$$

Next, due to the projection property of the identity operator over the associate N-body Hilbert space

$$\hat{1} = \sum_n |\Phi_n\rangle\langle\Phi_n| \tag{5.44}$$

the last integral further becomes

$$I = \frac{1}{N!}\int d^N\mathrm{r}\langle\Psi|\left(\psi^+(N)...\psi^+(1)\right)\left(\psi(1)...\psi(N)\right)|\Psi\rangle \tag{5.45}$$

that through the iterative distribution of the integrals over less and less coordinate to integrate out, as shown before. The result is now simple to acquire

$$I = \frac{1}{N!}\langle\Psi|\left\{\int d^N\mathrm{r}\left(\psi^+(N)...\psi^+(1)\right)\left(\psi(1)...\psi(N)\right)\right\}|\Psi\rangle = \frac{\langle\Psi|\Psi\rangle}{N!}N! = 1 \tag{5.46}$$

that is proving the normalization condition for the N-body field "function".

- To proof the N-body Schrödinger validity for the above form of the many-field form, one should made recourse to the commutation properties of the uni-particle fields above, alongside of the commutation distribution of the product

$$[AB, C] = [A, C]B + A[B, C] \tag{5.47}$$

distributed while unfolding the eigen-problem for the N-body state

$$\begin{aligned} E\Psi(1,2,...,N) &= \frac{1}{\sqrt{N!}}\langle 0|\psi(1)\psi(2)...\psi(N)\underbrace{E|\Psi\rangle}_{\hat{H}|\Psi\rangle} \\ &= \frac{1}{\sqrt{N!}}\langle 0|\psi(1)\psi(2)...\psi(N)\hat{H}|\Psi\rangle \\ &= \frac{1}{\sqrt{N!}}\langle 0|\left[\psi(1)\psi(2)...\psi(N), \hat{H}\right]|\Psi\rangle \\ &= \frac{1}{\sqrt{N!}}\sum_{j=1}^{N}\langle 0|\psi(1)...\left[\psi(j), \hat{H}\right]...\psi(N)|\Psi\rangle \end{aligned} \tag{5.48}$$

being the before the last term possible due to the Hamiltonian zero action on the vacuum state, see above. Now, either for bosons or fermions (the calculus is made for bosons only but flows in the same manner for fermions as well) the involved commutator evaluates, based on inter-uni-fields commutators rules, as

$$\begin{aligned} &\left[\psi(j), \hat{H}\right] \\ &= -\frac{\hbar^2}{2m}\int dr\left[\psi(\mathrm{r}_j), \psi^+(\mathrm{r})\nabla^2\psi(\mathrm{r})\right] + \frac{1}{2}\int dr_1 dr_2\left[\psi(\mathrm{r}_j), \psi^+(\mathrm{r}_1)\psi^+(\mathrm{r}_2)\right]V(\mathrm{r}_1, \mathrm{r}_2)\psi(\mathrm{r}_2)\psi(\mathrm{r}_1) \\ &= -\frac{\hbar^2}{2m}\int dr\left[\psi(\mathrm{r}_j), \psi^+(\mathrm{r})\right]\nabla^2\psi(\mathrm{r}) \\ &+ \frac{1}{2}\int dr_1 dr_2\left\{\left[\psi(\mathrm{r}_j), \psi^+(\mathrm{r}_1)\right]\psi^+(\mathrm{r}_2) + \psi^+(\mathrm{r}_1)\left[\psi(\mathrm{r}_j), \psi^+(\mathrm{r}_2)\right]\right\}V(\mathrm{r}_1, \mathrm{r}_2)\psi(\mathrm{r}_2)\psi(\mathrm{r}_1) \\ &= -\frac{\hbar^2}{2m}\nabla_j^2\psi(\mathrm{r}_j) + \left(\int dr\psi^+(\mathrm{r})V(\mathrm{r}, \mathrm{r}_j)\psi(\mathrm{r})\right)\psi(\mathrm{r}_j) \\ &= \left[-\frac{\hbar^2}{2m}\nabla_j^2 + \int dr\psi^+(\mathrm{r})V(\mathrm{r}, \mathrm{r}_j)\psi(\mathrm{r})\right]\psi(\mathrm{r}_j) \end{aligned} \tag{5.49}$$

Now, while inserting this expression back into the eigen-value equation,

$$
\begin{aligned}
E\Psi(1,2,...,N) &= \frac{1}{\sqrt{N!}}\sum_{j=1}^{N}\langle 0|\psi(1)...\left[-\frac{\hbar^2}{2m}\nabla_j^2 + V(j)\right]\psi(j)...\psi(N)|\Psi\rangle \\
&= -\frac{\hbar^2}{2m}\sum_{j=1}^{N}\nabla_j^2 \underbrace{\frac{1}{\sqrt{N!}}\langle 0|\psi(1)...\psi(j)...\psi(N)|\Psi\rangle}_{\Psi(1,2,...,N)} \\
&+ \frac{1}{\sqrt{N!}}\sum_{j=1}^{N}\langle 0|(\psi(1)...V(j)\psi(j)...\psi(N))|\Psi\rangle
\end{aligned}
\tag{5.50}
$$

one should use the potential term

$$V(j) = \int d\mathrm{r}\psi^{+}(\mathrm{r})V(\mathrm{r},\mathrm{r}_j)\psi(\mathrm{r}) \tag{5.51}$$

with its immediate properties

$$[\psi(i), V(j)] = V_{ij}\psi(i) \tag{5.52}$$

$$V(j)|0\rangle = 0 \tag{5.53}$$

to firstly produce the transformations via successive (j-1) left-commutations

$$
\begin{aligned}
&\psi(1)...\underline{\psi(j-1)V(j)}\psi(j)...\psi(N) \\
&= \psi(1)...\underline{\psi(j-2)V(j)}\psi(j-1)...\psi(N) + V_{j-1,j}(\psi(1)...\psi(N)) \\
&= \psi(1)...\underline{\psi(j-3)V(j)}\psi(j-2)...\psi(N) + (V_{j-2,j} + V_{j-1,j})(\psi(1)...\psi(N)) \\
&= \left(V(j) + \sum_{i=1}^{j-1} V_{ij}\right)(\psi(1)...\psi(N))
\end{aligned}
\tag{5.54}
$$

leading to the final result

$$
\begin{aligned}
E\Psi(1,2,...,N) &= -\frac{\hbar^2}{2m}\sum_{j=1}^{N}\nabla_j^2\Psi(1,2,...,N) \\
&+ \frac{1}{\sqrt{N!}}\sum_{j=1}^{N}\underbrace{\langle 0|V(j)}_{0}\psi(1)...\psi(N)|\Psi\rangle + \sum_{j=1}^{N}\sum_{i=1}^{j-1}V_{ij}\underbrace{\frac{1}{\sqrt{N!}}\langle 0|(\psi(1)...\psi(N))|\Psi\rangle}_{\Psi(1,2,...,N)} \\
&= \left(-\frac{\hbar^2}{2m}\sum_{j=1}^{N}\nabla_j^2 + \sum_{i<j}V_{ij}\right)\Psi(1,2,...,N)
\end{aligned}
\tag{5.55}
$$

that finally consecrated the Schrödinger equation also for the N-body field quantification.

Yet, one can go further with the quantification picture, since introducing the unifield operators' expansions

$$\psi(\mathrm{r}) = \sum_{\alpha}\hat{a}_{\alpha}u_{\alpha}(\mathrm{r}) \tag{5.56}$$

$$\psi^{+}(\mathrm{r}) = \sum_{\alpha}\hat{a}_{\alpha}^{+}u_{\alpha}^{*}(\mathrm{r}) \tag{5.57}$$

in terms of pure function representing the single particle wave-functions obeying the complete orthonormal set conditions

$$\delta_{\alpha\beta} = \int dr\left(u_{\alpha}^{*}(\mathrm{r})u_{\alpha}(\mathrm{r})\right) \tag{5.58}$$

eventually chosen as for the free particle

$$u_{\mathrm{p}}(\mathrm{r}) = \frac{1}{\sqrt{V}}\exp\left(\frac{i}{\hbar}\mathrm{p}\cdot\mathrm{r}\right) \tag{5.59}$$

being at its turn quantified by its momentum

$$\mathrm{p} = \frac{2\pi\hbar}{L}\mathrm{n} \tag{5.60}$$

that follows from the imposed periodic boundary conditions

$$u_{\mathrm{p}}(\mathrm{r}+\mathrm{n}L) = u_{\mathrm{p}}(\mathrm{r}) \tag{5.61}$$

on each particle confined in the volume V, assumed for convenience being of cubic lattice, that is

$$L = V^{1/3} \tag{5.62}$$

while moving along the vector n.

The new operators appeared are called as: $\hat{a}_{\alpha}$, the annihilation operator for the single state $|\alpha\rangle$; $\hat{a}_{\alpha}^{+}$, the creation operator for the single state $|\alpha\rangle$; they have the basic properties:

- rewrite the commutation rules in a pure operatorial mode for *bosons*

$$\left[\hat{a}_{\alpha},\hat{a}_{\beta}^{+}\right] = \delta_{\alpha\beta}, \left[\hat{a}_{\alpha},\hat{a}_{\beta}\right] = 0, \left[\hat{a}_{\alpha}^{+},\hat{a}_{\beta}^{+}\right] = 0 \tag{5.63}$$

  and for *fermions*,

$$\left\{\hat{a}_{\alpha},\hat{a}_{\beta}^{+}\right\} = \delta_{\alpha\beta}, \left\{\hat{a}_{\alpha},\hat{a}_{\beta}\right\} = 0, \left\{\hat{a}_{\alpha}^{+},\hat{a}_{\beta}^{+}\right\} = 0 \tag{5.64}$$

- produce for their $\hat{a}_{\alpha}^{+}\hat{a}_{\alpha}$ product the following eigen-values

$$\hat{n}_{\alpha} = \hat{a}_{\alpha}^{+}\hat{a}_{\alpha} = \begin{cases} 0,1,2,\ldots \textit{for}\ \ \textit{BOSONS} \\ 0,1\ \ \ldots.\textit{for}\ \ \textit{FERMIONS} \end{cases} \tag{5.65}$$

- allow therefore in reconsidering the number of particle operator with the form

$$\hat{N} = \sum_{\alpha}\hat{a}_{\alpha}^{+}\hat{a}_{\alpha} \tag{5.66}$$

All in all, the associate quantification of the N-body system belongs to the actual Hamiltonian

$$\begin{aligned}\hat{H} &= -\frac{\hbar^2}{2m}\sum_{\alpha,\beta}\hat{a}_\alpha^+\hat{a}_\beta\int d\mathrm{r}\left(u_\alpha^*(\mathrm{r})\nabla^2 u_\beta(\mathrm{r})\right)\\ &+\frac{1}{2}\sum_{\alpha,\beta,\gamma,\lambda}\left(\hat{a}_\alpha\hat{a}_\beta\right)^+\left(\hat{a}_\gamma\hat{a}_\lambda\right)\int d\mathrm{r}_1 d\mathrm{r}_2\left(u_\alpha^*(\mathrm{r}_1)u_\beta^*(\mathrm{r}_2)V_{12}u_\gamma(\mathrm{r}_2)u_\lambda(\mathrm{r}_1)\right)\end{aligned} \tag{5.67}$$

This many-body quantum field (or the so called the *second quantization*) formalism constitutes the main background for correct understanding of condensed phenomena, either for fermions and bosons, as it will be in next presented.

5.3.2 The Bosonic Order Parameter

First of all, it is worth clarifying the discrete-to-continuum transformation when the confined space volume grows asymptotically ($V\to\infty : L\to\infty : \mathrm{p}\to 0$), within the so called *thermodynamic limit*, that gives the highest resolution of the coarse-grain quantum localization picture as based on the common Heisenberg indeterminacy. Actually we have three instances:

- the sum over moments

$$\sum_{\mathrm{p}}[\bullet]\overset{V\to\infty}{\to}\frac{V}{h^3}\int d\mathrm{p}[\bullet] \tag{5.68}$$

- the sum over wave vectors

$$\sum_{\mathrm{k}}[\bullet]\overset{V\to\infty}{\to}\frac{V}{(2\pi)^3}\int d\mathrm{k}[\bullet] \tag{5.69}$$

- the delta-Dirac integrals

$$\int_V (d^D x)\exp(-ik\cdot x)\overset{V\to\infty}{\to}(2\pi)^D\,\delta(k) \tag{5.70}$$

or

$$\frac{1}{(2\pi)^D}\int_V (d^D k)\exp(ik\cdot x)\overset{V\to\infty}{\to}\delta(x) \tag{5.71}$$

Therefore one may consider the Fourier transformations

$$f(x)=\frac{1}{V}\sum_{\mathrm{k}}\tilde{f}(k)\exp(ik\cdot x)\overset{V\to\infty}{\to}\frac{1}{(2\pi)^D}\int_\infty (d^D k)\tilde{f}(k)\exp(ik\cdot x) \tag{5.72}$$

with the inverse Fourier component

$$\tilde{f}(k)=\int_V (d^D x)f(x)\exp(-ik\cdot x) \tag{5.73}$$

Now, one can write down the one particle density $\rho_1(\mathrm{r},\mathrm{r}')$ over the ensemble average $\langle\ \rangle$, see bellow the discussion upon its correct definition in the context of quantum superfluid condensation, which, in the light of above second quantization framework specialized for the free particles' translational invariant system, that is

$$\left\langle \hat{a}_{q}^{+}\hat{a}_{k}\right\rangle = \delta_{q,k}\left\langle \hat{n}_{k}\right\rangle \tag{5.74}$$

successively writes

$$\begin{aligned}\rho_1(\mathrm{r},\mathrm{r}') &= \left\langle \psi^{+}(\mathrm{r})\psi(\mathrm{r}')\right\rangle \\ &= \frac{1}{V}\sum_{k,q}\left\langle \hat{a}_{q}^{+}\hat{a}_{k}\right\rangle \exp\left\{i(\mathrm{k}\cdot\mathrm{r}'\text{-}\mathrm{q}\cdot\mathrm{r})\right\} \\ &= \frac{1}{V}\sum_{k}\left\langle \hat{n}_{k}\right\rangle \exp\left\{i\mathrm{k}\cdot(\mathrm{r}'\text{-}\mathrm{r})\right\} \\ &= \frac{\left\langle \hat{n}_{k=0}\right\rangle}{V} + \frac{1}{(2\pi)^3}\int d\mathrm{k}\left\langle \hat{n}_{k\neq 0}\right\rangle \exp\left\{i\mathrm{k}\cdot(\mathrm{r}'\text{-}\mathrm{r})\right\}\end{aligned} \tag{5.75}$$

But, because the second term, in thermodynamic limit $|\mathrm{r}'\text{-}\mathrm{r}|\to\infty$, behaves like the Dirac function

$$\delta(\mathrm{k}) = \begin{cases} 0 \ldots \mathrm{k}\neq 0 \\ \infty \ldots \mathrm{k}=0 \end{cases} \tag{5.76}$$

it practically vanishes since it belongs only to the $\mathrm{k}\neq 0$ modes, the other being considered out of the sum. This way, one obtains the constant density of zero momentum particles over the whole system

$$\rho_1(\mathrm{r},\mathrm{r}') = \left\langle \psi^{+}(\mathrm{r})\psi(\mathrm{r}')\right\rangle \overset{|\mathrm{r}'\text{-}\mathrm{r}|\to\infty}{\to} \frac{\left\langle \hat{n}_{k=0}\right\rangle}{V} \tag{5.77}$$

that gives the fundamental hint BEC should appear in the context, although not well quantitatively described. A better criterion is acquired when $\langle \hat{n}_{k=0}\rangle / N$ quantity is considered, that can be further refined for the nontranslationally invariant geometry with the help of the so called *superfluid order parameter* (complex number)

$$\left\langle \psi(\mathrm{r})\right\rangle = R(\mathrm{r})\exp\left\{iS(\mathrm{r})\right\} \tag{5.78}$$

towards the modified Penrose–Onsager BEC criteria (Penrose and Osanger, 1956; Anderson, 1966)

$$\rho_1(\mathrm{r},\mathrm{r}') = \left\langle \psi^{+}(\mathrm{r})\psi(\mathrm{r}')\right\rangle \overset{|\mathrm{r}'\text{-}\mathrm{r}|\to\infty}{\to} \left\langle \psi^{+}(\mathrm{r})\right\rangle \left\langle \psi(\mathrm{r}')\right\rangle \tag{5.79}$$

It is clear now that since $R(\mathrm{r}) > 0$ the momentum-space order is formed, that is the Bose–Einstein condensate, that validates the criterion. However, questions remain about the average form of the order parameter $\left\langle \psi(\mathrm{r})\right\rangle$ since the annihilation nature of the uni-field $\psi(\mathrm{r})$ (see above) would imply the cancellation of its expectation value for any of the eigen-states of the N-sample. The point here is that the basic *global gauge invariance* (symmetry) of the Hamiltonian respecting the field transformation

$$\psi(\mathrm{r}) \to \psi(\mathrm{r})\exp(i\theta),\ \forall\theta \in \Re \tag{5.80}$$

that preserves the total number of particles, is eventually *spontaneously broken*, under the action of the auxiliary (external) field $\xi(\mathrm{r})$, coupled with $\psi(\mathrm{r})$, that practically does not exist (Gunton and Buckingham, 1968), while having only the mathematical-conceptual (fictitious) role in explaining the transition of a gas to its condensed phase; therefore, the correct definition of the order parameter is as the *ensemble average in the thermodynamic + external zero filed limits*

$$\langle\psi(\mathrm{r})\rangle := \lim_{\xi\to 0}\lim_{V\to\infty}\frac{\mathrm{Tr}\left\{\psi(\mathrm{r})\exp\left(-\beta E[\psi,\xi]\right)\right\}}{\mathrm{Tr}\left\{\exp\left(-\beta E[\psi,\xi]\right)\right\}},\ \beta = \frac{1}{k_B T} \tag{5.81}$$

with the energy functional casting as

$$E[\psi,\xi] := H - \mu N - \int d\mathrm{r}\left\{\psi(\mathrm{r})\xi(\mathrm{r}) + \psi^{+}(\mathrm{r})\xi^{+}(\mathrm{r})\right\} \tag{5.82}$$

However, as a matter of fact, the modulus of the superfluid order parameter at absolute zero for liquid ^{4}He is about $|\langle\psi(\mathrm{r})\rangle| = \sqrt{\langle\hat{n}_{\mathrm{k}=0}\rangle / N} \cong \sqrt{0.08}$ (Huang, 1987), thus having non-zero influence on the bosonic condensation that acquires the density

$$\rho(\mathrm{r}) \cong |\langle\psi(\mathrm{r})\rangle|^2 \tag{5.83}$$

As such the order parameter features its most important property – shaping the bosonic condensate density. However, the next discussion will address the practical forms of the effective Hamiltonian for the imperfect gases (i.e., ideal gases perturbed) at low temperature.

5.3.3 The Pseudo-Potential Picture for the Imperfect Bosonic Gas

The next problem is how to model from the Hamiltonian perspective the bosonic interaction in the condensate, that is how looks like the potential responsible for bosonic correlation – so to speak. The basic model, initially advanced by Fermi (1936), then refined by Huang and Yang (1957), assumes the basic idea that at low temperature the inter-particle potential depends on a single parameter – the scattering length "a". Therefore, at extremely low temperature the effective parameters that determine the motion of *the imperfect Bose gas* (i.e., the bosonic gas is sufficiently diluted towards the ideal one, i.e., occupying the whole available space – conceptually asymptotically ad infinitum, yet with some small inter-particle interaction that basically relaying on scattering) are:

- the thermal wave-length (Putz, 2009c)

$$\lambda_{dB} = \sqrt{2\pi\beta\hbar^2 / m} \tag{5.84}$$

- that plays in condensation the de Broglie wavelength role of the collapsing particles, see the Figure 5.1 for the semi-qualitative picture of BEC.
- the average interparticle separation

$$v^{1/3} = (V / N)^{1/3} \tag{5.85}$$

- the scattering length usually implemented as the diameter of condensing atoms,

$$a \cong 2R \tag{5.86}$$

That is, the scattering length plays the role in the interacting bosonic potential itself. This picture is completed when the problem of *N*-body hard sphere interaction is replaced, in analogy with custom Poisson electrostatics' method of condensed charges on a sphere, by the multipole potentials (producing scattered S waves, P waves, D waves, etc.), further represented by the collection of points constrained by the geometric loci of the *binary* $i \leftrightarrow j$ collision

$$\left|\mathrm{r}_i - \mathrm{r}_j\right| \leq a \tag{5.87}$$

This is the picture of the pseudo-potentials, here restricted to the two-body pseudo-potentials $\delta\left(\mathrm{r}_i - \mathrm{r}_j\right)$, whose extension to the all particle sin the system generates the so called "three-like hypersurface" in the $3N$- configuraton space, see Figure 5.2.

Although not completed, since neglecting the 3-, 4-, and finally the *N*- body interacting effects, the present 2-body pseudopotential picture seems adequate representing also the whole *N*-body problem when the interparticle potential is a finite-ranged potential (as is usually the case) instead of the hard-sphere potential. Therefore, while considering the (bosonic) two-particle mass-center wave function $\psi(\mathrm{r})$, $\mathrm{r} = \mathrm{r}_2 - \mathrm{r}_1$ the Schrödinger equation in the center of mass system for the bosonic condensed mode, that is when $\mathrm{k} = 0$ see the previous section, it looks like the extended Poisson equation in electrostatics (at the first sight in the atomic units)

$$\nabla^2 \psi(\mathrm{r}) = 4\pi a \delta(\mathrm{r}) \tag{5.88}$$

yielding the Hamiltonian contribution (dimensional) term of *inter-bosonic potential*

$$\left(\frac{\hbar^2}{m}\right) \times 4\pi a \delta(\mathrm{r}) \tag{5.89}$$

thus providing the working Hamiltonian for the bosonic condensate of *N*-particles as

$$H_{Bosonic} = \sum_{i=1}^{N} \left(-\frac{\hbar^2}{2m} \nabla_i^2 + V(\mathrm{r}_i) \right) + \frac{4\pi a \hbar^2}{m} \sum_{i<j} \delta\left(\mathrm{r}_i - \mathrm{r}_j\right) \tag{5.90}$$

Worth noting that, being the inter-bosonic strength (Dalfovo et al., 1999; Pethick and Smith, 2002)

$$g = \frac{4\pi a \hbar^2}{m_B} \tag{5.91}$$

an interaction term will be customarily associated with the product wave-functions of the form $(\psi^{+}\psi)^{2}$, in any functional development of condensation theory, see next section.

Note that, within the Hartree–Fock approximation, the total wave-function of the N-bosonic system it can be taken as the product of the single-particle fields, meaning that all bosons are in their single states, that is

$$\Psi(\mathrm{r}_1,\mathrm{r}_2,...\mathrm{r}_N)=\psi(\mathrm{r}_1)\psi(\mathrm{r}_2)...\psi(\mathrm{r}_N) \tag{5.92}$$

rooting in the above second quantization many-body treatment, see Section 5.3.1.

However, now we have all basic bosonic condensate features in hands so that to advance the specific equation of motion, in a similar way the Schrödinger equation describes the fermionic evolution.

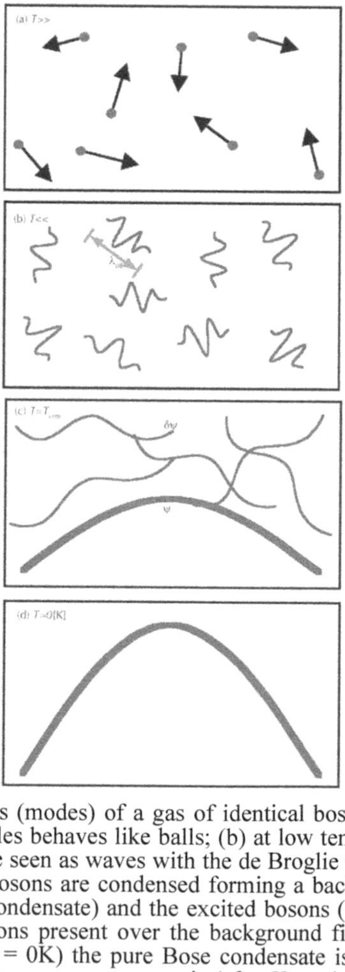

FIGURE 5.1 Various phases (modes) of a gas of identical bosons at various temperatures: (a) at high temperature particles behaves like balls; (b) at low temperature the quantum effect are dominant and particles are seen as waves with the de Broglie wavelength λ_{dB}; (c) at critical (transition) temperature the bosons are condensed forming a background wave-length (in the ground state of the bosonic condensate) and the excited bosons (with a non-zero momentum) due to the quantum fluctuations present over the background field (see Section 5.2); (d) at absolute zero temperature (T = 0K) the pure Bose condensate is formed with a giant wave-length and the wave function becomes macroscopic (after Ketterle, 2002).

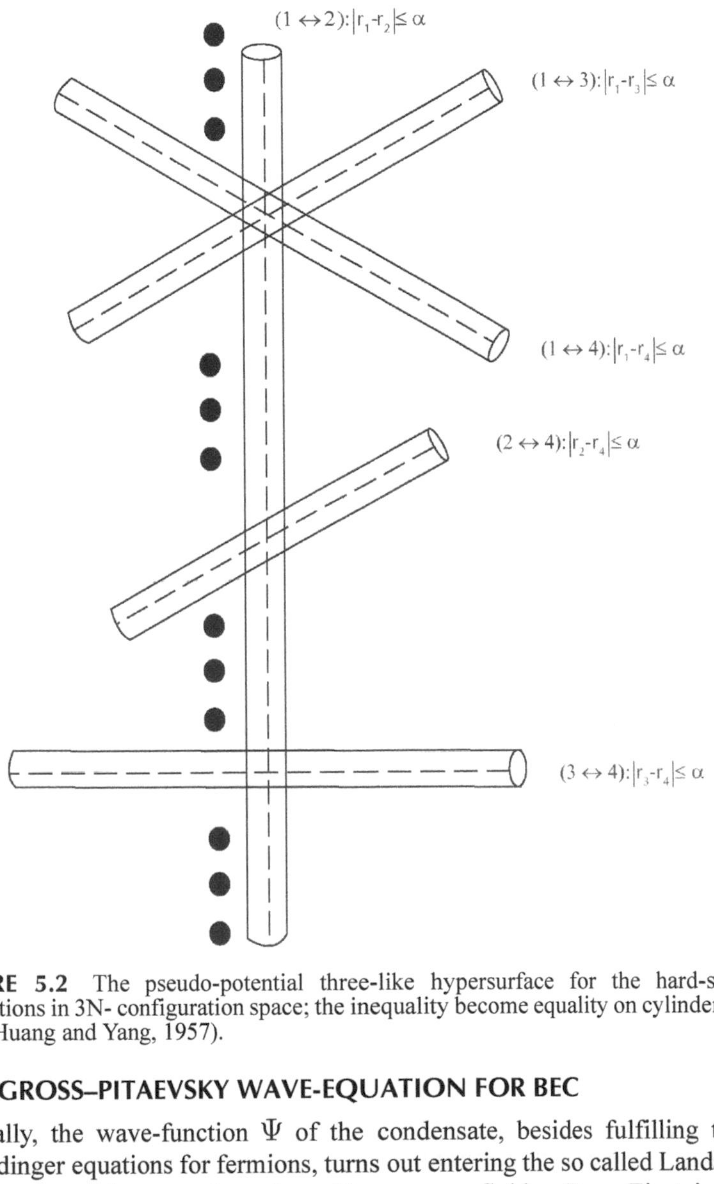

FIGURE 5.2 The pseudo-potential three-like hypersurface for the hard-sphere binary interactions in 3N- configuration space; the inequality become equality on cylinders' boundaries (after Huang and Yang, 1957).

5.4 GROSS–PITAEVSKY WAVE-EQUATION FOR BEC

Basically, the wave-function Ψ of the condensate, besides fulfilling the ordinary Schrödinger equations for fermions, turns out entering the so called Landau free energy when describing a condensation, either as superfluid or Bose–Einstein condensate; yet, through collecting all previous unfolded information about energetic (interacting) terms in a condensate, its functional energy looks like (Huang, 2001)

$$E[\psi,\psi^{+}]=\int d\mathrm{r}\left[\frac{\hbar^{2}}{2m_{B}}|\nabla\psi(\mathrm{r})|^{2}+\left(V(\mathrm{r})-\mu\right)\psi^{+}(\mathrm{r})\Psi(\mathrm{r})+\frac{g}{2}\left(\psi^{+}(\mathrm{r})\psi(\mathrm{r})\right)^{2}\right] \tag{5.93}$$

in terms of

- m_B : the mass of particles that condense (the *bondons* in the case of chemical bonding)
- $V(\mathrm{r})$: the external potential
- μ : the chemical potential, so it associates/fixes (in grand canonical ensemble) the macroscopic observable as the total number of particles in condensate – see the Section 5.3.1 above

$$N = \int d\mathrm{r} \left|\psi(\mathrm{r})\right|^2 \tag{5.94}$$

- thus making the connection also with the density functional basic requirement in terms of density

$$N = \int d\mathrm{r} \rho(\mathrm{r}) \tag{5.95}$$

- g: the strength of the condensate, modulating the interaction between two particles of the condensate, with the meaning of "scattering length" between particles in condensate, on the custom size of an atomic hard-sphere equivalent diameter, see the Section 5.3.3 above.

One can see nevertheless that, while performing the energy minimization respecting the background field ψ^+

$$0 = \frac{\partial E}{\partial \psi^+} \tag{5.96}$$

while considering the Gauss integration theorem over the kinetic term

$$\begin{aligned} \int d\mathrm{r} \left|\nabla\psi\right|^2 &= \int d\mathrm{r} \nabla\psi^+ \cdot \nabla\psi = \int d\mathrm{r} \nabla\left[\left(\nabla\psi\right)\psi^+\right] - \int d\mathrm{r}\left(\nabla^2\psi\right)\psi^+ \\ &= \underbrace{\int d\Sigma_\infty \left[\left(\nabla\psi\right)\psi^+\right]}_{0} - \int d\mathrm{r}\left(\nabla^2\psi\right)\psi^+ \\ &= -\int d\mathrm{r}\left(\nabla^2\psi\right)\psi^+ \end{aligned} \tag{5.97}$$

one has in fact to perform the integral differentiation

$$0 = \frac{\partial}{\partial \psi^+} \int d\mathrm{r} \left[-\frac{\hbar^2}{2m_B}\left(\nabla^2\psi\right)\psi^+ + \left(V(\mathrm{r}) - \mu\right)\psi^+\psi + \frac{g}{2}\left(\psi^+\psi\right)^2 \right] \tag{5.98}$$

which immediately leaves with the result

$$0 = -\frac{\hbar^2}{2m_B}\left(\nabla^2\psi\right) + \left(V(\mathrm{r}) - \mu\right)\psi + g\left(\psi^+\psi\right)\psi \tag{5.99}$$

that can be rearranged under the so called GP equation (Gross, 1961; Pitaevsky, 1961)

$$\left[-\frac{\hbar^2}{2m_B}\nabla^2 + V(\mathrm{r}) + g\left|\psi\right|^2 \right]\psi = \mu\psi \tag{5.100}$$

It can also be generalized to the time-dependent equation

$$\left[-\frac{\hbar^2}{2m_B}\nabla^2+V(\mathrm{r})+g|\psi|^2\right]\psi=i\hbar\frac{\partial\psi}{\partial t} \tag{5.101}$$

as being viewed like the *non-linear Schrödinger equation* (NLSE) that reduces to the stationary case under the traditional wave-function ansatz

$$\psi(\mathrm{r},t)=\psi(\mathrm{r},t)\exp\left(-\frac{i}{\hbar}\mu t\right) \tag{5.102}$$

Nonetheless, the present picture is assumed as valid within the next fulfilling criteria:

- The existence of an *imperfect Bose-* gas, that is the dilute gas that is the ideal gas + weak interactions, such that the interactions are described via the single parameter of the length of the two-body scattering process with s-waves: thus the existence of "a" !
- The conditions of the *ultra-high dilution*, acquired for the cases when the interparticle distance is far larger than the scattering length: $v^{1/3}>>a$. This is equivalent to saying that no many-body interactions other than two-body interactions are present, and this leads for thermodynamic limit $N,V\to\infty$ (see also the Thomas–Fermi approximation in the Chapter 9, Section 9.3) with the condensate ground state energy proportional with the bosonic interaction term that urns to be

$$E_0\sim\frac{4\pi a\hbar^2}{m_B}\rho N\sim NE[\psi,\psi^+], \tag{5.103}$$

with

$$\rho=\frac{N}{V} \tag{5.104}$$

- Fulfilling the *healing (or indeterminacy) condition*, when scattering wavelength is largely over-passed by the thermal (de Broglie) length, $a<<\lambda_{dB}$, for assuring the losing of bosonic identity in the condensate, since this way, in the low density regime ($a<<\rho^{-1/3}$) it is impossible to localize the particles relative to each other; as a consequence, bosons in the ground state (being characterized by the mean field or the order parameter) are smeared out over large distance (or wavelength, see Figure 5.1) compared with the mean particle distance: $\rho^{-1/3}<<\lambda_{dB}$.

However, due to the particular chemical potential role played in forming the condensate through determining the number of particle in the condensate since the above normalization appears as a an evolution constant, the grand partition function of the condensate

$$Z=\int(D\psi)(D\psi^+)\exp\{-\beta E[\psi,\psi^+]\} \tag{5.105}$$

with the weighting Boltzmann factor $\beta=(k_BT)^{-1}$, should also be seen at the origin of the bosonic condensate or as its marker in establishing some conceptual condensate properties as we will see in what follows.

5.4 MODELING ULTRACOLD ATOMIC BOSONS TRAPPED IN OPTICAL LATTICES

Since, the recent experimental possibility in creating artificial solids by optical lattices, by cooling the atoms trapped as a gas in ultrahigh vacuum to nano-Kelvins temperature then captured on the nodes of the standing waves created by combinations of laser beams, see Figure 5.3, there emerged the ideal conditions for better understanding the quantum mechanical phase transitions when strong interactions between particles are tuned (Bakr et al., 2010). This way, the atoms play the role of electrons (for fermionic atoms) or of the superconducting pairs (for bosonic atoms) in a solid, whereas the optical lattice stands for the underlying ionic crystal (and potential). On the other side, the physics behind such systems may relay on the celebrated *Hubbard model* which qualitatively advances two basic ideas (DeMarco, 2010):

- Cold atoms can hop between sites by quantum tunneling;
- Cold atoms can interact with each other only if they are on the same site.

Yet, one key features of this model is the appearance of the so called Mott insulator with electrons (or cold atoms) prevented to exercise their conductivity by strong inter-particle interaction – thus experiencing a quantum phase transition.

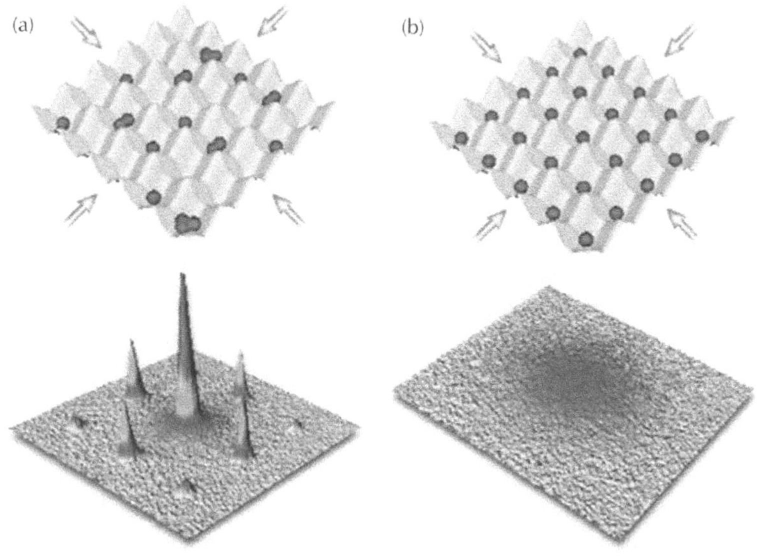

FIGURE 5.3 Laser based optical lattice; (a) the instance with low lattice potential allows for the superfluid state with fluctuating number of atoms per site and with the largest interference at zero momentum peak; (b) the instance with high lattice potential produces the insulating (Mott) state that having exactly one atom per site produces blurring in momentum distribution (Bloch, 2004, 2008).

However, depending on fermionic or bosonic type of particles under consideration there are two versions of Hubbard model and of the associate phenomena, namely Fermi–Hubbard (FH) and Bose–Hubbard (BH) realizations. The FH model is used if atoms are fermions (Schneider et al., 2008; Jördens et al., 2008) and may constitute the analytical framework for modeling key technological materials such as high-temperature cuprates. Actually, it works with the Hamiltonian (Bernier et al., 2009; Paiva et al., 2010):

$$\hat{H}_{FH} = -J \sum_{\substack{<ij>, \\ \sigma\in\{\uparrow,\downarrow\}}} \left(\hat{a}^{+}_{j\sigma}\hat{a}_{i\sigma} + h.c.\right) + U\sum_{i}\hat{n}_{i\uparrow}\hat{n}_{i\downarrow} + \sum_{i}\varepsilon_i\hat{n}_i \tag{5.106}$$

having essentially the two parts accounting for above two features of the Hubbard description: the *hopping energy* driven by the tunneling matrix elements (*J*) between nearest neighbors <*ij*>, and the *onsite interaction energy* given (*U*) driven by the particle number operators $\hat{n}_{i\sigma} = \hat{a}^{+}_{i\sigma}\hat{a}_{i\sigma}$ on magnetic sublevels $\sigma \in \{\uparrow, \downarrow\}$, while the ε_i is the offset energy associated with an atom on *i*-site of lattice due to the specific confining potential (cubic, harmonic, etc.) or the (minus) chemical potential μ.

On the other side, the Bose–Hubbard model is used if atoms are bosons as ^{87}Rb atoms are, for instance, currently applied on modeling granular superconductors and Josephson junctions; the corresponding Hamiltonian it takes the form (Fischer et al., 1989; Greiner et al., 2002; Jaksch et al., 1998):

$$\hat{H}_{BH} = -J\sum_{<ij>}\hat{a}^{+}_{j\sigma}\hat{a}_{i\sigma} + \hat{H}_0 \tag{5.107}$$

with the on-site part

$$H_0 = \sum_{i} H_i, \quad H_i = \frac{1}{2}U\hat{n}_i\left(\hat{n}_i - 1\right) - \mu\hat{n}_i \tag{5.108}$$

The two side effects are modeled by the working expression for the strength of tunneling

$$J = -\int d\mathrm{r} w(\mathrm{r}-\mathrm{r}_i)\left(-\frac{\hbar^2\nabla^2}{2m} + V_{lat}(\mathrm{r})\right)w(\mathrm{r}-\mathrm{r}_j) \tag{5.109}$$

associated with the hopping matrix element through the single Wannier functions $w(\mathrm{r}-\mathrm{r}_i)$, $w(\mathrm{r}-\mathrm{r}_j)$ localized on adjacent sites (*i,j*), respectively, with $V_{lattice}(\mathrm{r})$ referring to the optical lattice potential; and by the onsite repulsion between cold atoms quantified by the onsite interaction matrix element

$$U = g\int d\mathrm{r}\left|w(\mathrm{r})\right|^4 \tag{5.110}$$

with

$$g = \frac{4\pi\hbar^2 a}{m} \tag{5.111}$$

the coupling strength in terms of the scattering length of an atom (a)– a reminiscence from the "imperfect" Bosonic gas treatment through GP equation, see equation (5.16) in the Section 5.2.

However, the experimental possibility in varying the optical lattice potential depth allows for tuning the system to behave as superfluid (or BEC) when tunneling dominates ($U = 0$) or as a Mott insulator phase when the phase coherence is no more prevalent in the system ($J = 0$) but strong correlation of atomic numbers per lattice site (n) is recording. Analytically, for a homogeneous lattice system ($\mu = ct$) with M sites and N trapped cold atoms, one may write the many-body wave ground state function for the two quantum regimes distinctly as

$$|\Psi\rangle \propto \begin{cases} \left(\sum_{i=1}^{M} \hat{a}_i^+\right)^N |0\rangle ... SUPERFLUID : U = 0 \\ \prod_{i=1}^{M} \left(\hat{a}_i^+\right)^n |0\rangle ... MOTT\ INSULATOR : J = 0 \end{cases} \tag{5.112}$$

while having now the U/J ratio as the main descriptor for the quantum phase transition. This is the third level in characterizing the quantum phase transitions, along the above presented gas-to-ideal and gas-to-imperfect (or diluted) BEC states, here as superfluid-to-Mott insulator matter manifestation.

However, one may notice that while in the superfluid case the Hartree–Fock approximation hold widely, as in the diluted BEC case – see equation (5.15), in the Mott quantum state it is no longer valid, with the consequence that it will be no longer the subject of the GP equation for instance. Fortunately, the late case (including the prediction of the Mott insulators) turns out to be still tractable through GP picture yet amended as following (Kleinert et al., 2004, 2005):

(i) Considering the GP Hamiltonian for a grand-canonical ensemble of Bose particles with repulsive two-particle d-function interaction

$$\hat{H}_{GP} = \hbar\partial_\tau + \varepsilon_k(-i\hbar\nabla) - \mu + \frac{g}{2}|\psi(\mathrm{r},\tau)|^2 \tag{5.113}$$

to the Euclidian action

$$A[\psi,\psi^+] = \int_0^{\hbar\beta} d\tau \int d\mathrm{r}\psi^+(\mathrm{r},\tau)\hat{H}_{GP}\psi(\mathrm{r},\tau) \tag{5.114}$$

entering the functional integral of for the partition function

$$\mathcal{Z} = \oint D\psi \oint D\psi^+ \exp\left\{-\frac{1}{\hbar}A[\psi,\psi^+]\right\} \tag{5.115}$$

with ε_k denoting the one-particle energies (latter to be considered either as free spectra or, in optical lattice, restricted to the first Brillouin zone, for instance), while the bosons being represented by the complex fields $\psi^+(\mathrm{r},\tau), \psi(\mathrm{r},\tau)$ periodic on the imaginary time interval $\tau \in [0, \hbar\beta]$.

(ii) The background field method is implemented by expanding the Bose fields

$$\psi(\mathrm{r},\tau)=\Psi+\delta\psi(\mathrm{r},\tau), \tag{5.116a}$$

$$\psi^{+}(\mathrm{r},\tau)=\Psi^{+}+\delta\psi^{+}(\mathrm{r},\tau) \tag{5.116b}$$

around a constant background field or the main field Ψ giving the BEC density

$$\rho_0=\Psi^{+}\Psi \tag{5.117}$$

with the fluctuations $\delta\psi(\mathrm{r},\tau),\delta\psi^{+}(\mathrm{r},\tau)$ neglected in the first order (the methods' prescription) and restrained to only harmonic ones.

(iii) Calculating the density of the depleted particles $\rho-\rho_0$ in the small interaction (diluted) regime by the aid of extremizing the background field dependent effective potential (or action)

$$\Gamma[\Psi,\Psi^{+}]=-\frac{\ln \mathcal{Z}}{\beta}\xrightarrow{\quad OPTIMIZATION \quad}\Gamma[\mu,T] \tag{5.118}$$

and computing the temperature dependent sample density

$$\rho(\mu,T)=-\frac{1}{|\Omega|}\frac{\partial\Gamma[\mu,T]}{\partial\mu}\bigg|_{T} \tag{5.119}$$

(iv) Applying the powerful variational perturbation theory – VPT (Kleinert, 2003; Janke et al., 2001; Kleinert et al., 2002) on the previous one-loop (or Popov) approximation leading to a self-consistent approximation of it, $\rho-\rho_0=f(\rho)$, that then transposed into the parametric equation in the plane on the *effective scattering length – temperature* ($a_{eff}-T$) eventually produces the high-loop expansion

$$\frac{T}{T_C}=1+c_1 a_{eff}\rho^{1/3}+O(a_{eff}^{2}n^{2/3}) \tag{5.120}$$

with the phase diagram draw either for homogeneous dilute Bose gas (Figure 5.4(a)) or for the optical lattice (Figure 5.4(b)) in the plane

$$\frac{T}{T_C}=F(\gamma) \tag{5.121}$$

in terms of the ratio

$$\gamma\equiv\frac{U}{J}\propto a_{eff}\rho^{1/3} \tag{5.122}$$

that generalize for the optical lattice environment the earlier critical BEC conditions, see Section 5.2.

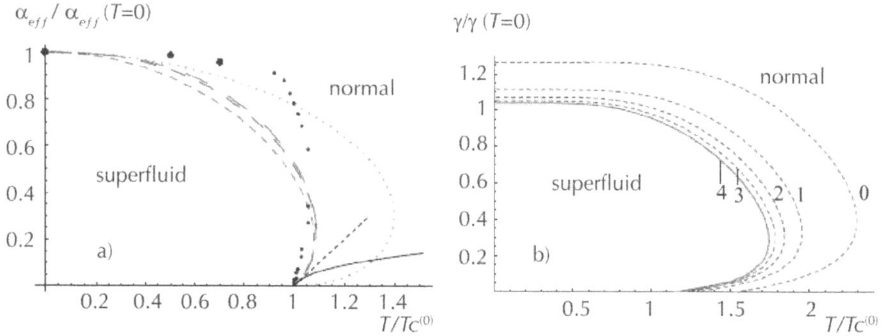

FIGURE 5.4 Quantum phase diagrams for (a) homogeneous dilute Boson gas (normal) – to BEC (superfluid) transition and (b) optical boson lattice – to Mot insulator (normal) transition illustrating the realization of equations. (5.120) and (5.121), in various variational improved one-loop approximations and expansions in hopping order, respectively (Kleinert et al., 2004).

However, although the success of this approach there is still the open question whether is possible working out a theory of strong interaction of cold bosonic atoms in optical lattice without recourse to variational perturbation theory applied on GP (valid for dilute bosonic gases) yet resembling the Bose–Hubbard performances in modeling the superfluid-Mott insulator quantum phase transitions. Such promise comes from the most celebrated in Chemistry Density Functional Theory and the main outline of the project is in the Chapter 6 exposed.

5.5 CHALLENGING CHEMICAL BONDING WITH BEC

Since, the nowadays growing interest in BEC due to the expanded experimental evidence on various atomic systems within optical lattices in weak and strong coupling regimes, the connection with Density Functional Theory is firstly advanced within the mean field framework at three levels of comprehension: the many-body normalization condition, Thomas–Fermi limit, and the chemical hardness closure with the inter-bosonic strength and universal Hohenberg–Kohn functional.

As first application on Chemistry, the traditional Heitler–London quantum mechanical description of the chemical bonding for homopolar atomic systems is reloaded within the non-linear Schrödinger (here the Gross–Pitaevsky) Hamiltonian; the results show that a fermionic-bosonic energetic gap is registered, with the bosonic contribution being driven by the square of the order parameter for the BEC density in free motion, while the bonding and antibonding wave functions remaining the same.

Further application may be unfolded for the diatomic heteropolar molecules, as well as for hydrocarbons and aromatic complexes; in this regard, one should note that the consecrated dichotomy between bonding and antibonding, or highest occupied molecular orbital-HOMO and the lowest unoccupied molecular orbital-LUMO orbitals, of chemical bonding respectively, may be employed in terms of the *electronic density* $\rho(\mathrm{r})$, eventually written as the condensate (bonding) + critical (atoms basins in molecule, AIM)

$$\rho(\mathrm{r}) = \rho_{bond}(\mathrm{r}) + \rho_{AIM}(\mathrm{r}) \tag{5.123}$$

so that resembling the basic mean-field approach of BEC, according to which the condensate itself is represented by the mean field (of bonding), while the critical regime (from where the condensate had spring out) is modeled by quantum field fluctuations of atoms in molecule (AIM).

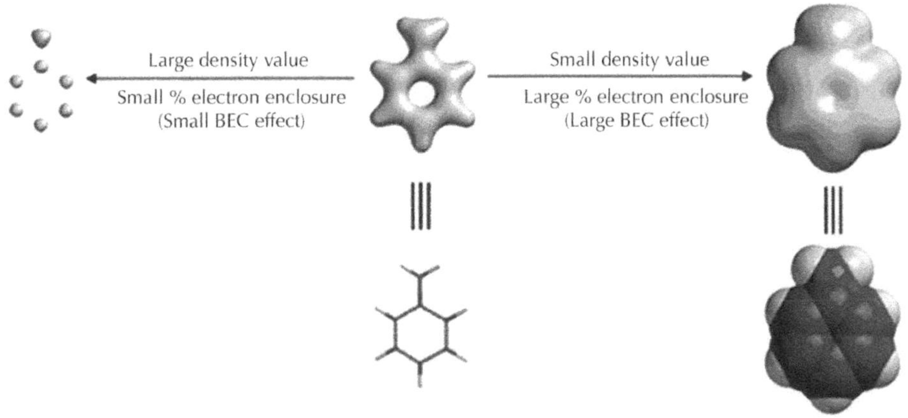

FIGURE 5.5 The structure information (aniline $C_6H_5NH_2$ case) enclosed by electronic density (Putz, 2001e).

Finally, it is worth remarking also the striking behavior of electronic cloud representations in molecules, regarding the electronic density-electronic enclosure inverse relation - see Figure 5.5 that further encourages the study of bosonic manifestation of delocalized electrons to represent the molecular sample as a whole bosonic system. Yet this is an open subject that may leave with further advancement in chemical reactivity theory in terms of redefining the frontiers of AIM, beyond the current (Bader) quantum approach (Bader, 1990, 1994, 1998). Further DFT-BEC combined approach of the chemical bond will be provided in the Chapter 9.

KEYWORDS

- **Atoms in molecule**
- **Bose–Einstein condensates**
- **Bosonic atoms**
- **Density Functional Theory**
- **Eigen-value equation**
- **Gross–Pitaevsky equation**
- **Many-body systems**
- **N-bosonic system**

6 Ψ-Density Functional Theory of Bose–Einstein Condensation

CONTENTS

6.1 INTRODUCTION

The basics of Density Functional Theory (DFT) and of Bose–Einstein Condensation (BEC) are merged such that the DFT enriches with the order parameter equation from the BEC modeling, while BEC acquires the new equation for the condensate density as provided by the DFT Hohenberg-KS analysis. Together, BEC–DFT may constitute as the next level of comprehension in modeling condensed matter in physics and chemical bond in chemistry, in particular, while providing the clue how the unified theory of physical-chemical phenomena should look like (Albus et al., 2003; Argaman and Band, 2011; Brand, 2004; Kim and Vetter, 1997; Nunes, 1999; Putz, 2011d; Zubarev, 2003).

6.2 Ψ-DFT-BEC VARIATIONAL EQUATIONS

Apart of the above noted N-integration connection between DFT and BEC one likes to see whether DFT may be eventually related with the more canonical treatment of BEC in term of the *order parameter* (Putz, 2011d)

$$\Psi(\mathrm{r}) \equiv \langle \psi(\mathrm{r}) \rangle_T \tag{6.1}$$

providing the condensation density realization

$$\rho_{\mathrm{cond}} \equiv |\Psi(\mathrm{r})|^2 \tag{6.2}$$

as part of the *overall density* of the superfluid

$$\rho(\mathrm{r}) \equiv \langle \psi^+(\mathrm{r})\psi(\mathrm{r}) \rangle_T \tag{6.3}$$

One should see the relation between the two types of densities given by equations (6.2) and (6.3) in the light of the so called Penrose–Osanger criteria for BEC (Penrose and Osanger, 1956):

$$\rho(\mathrm{r}) = \langle \psi^+(\mathrm{r})\psi(\mathrm{r'}) \rangle_T \xrightarrow[\textit{LIMIT}]{\textit{THERMODYNAMIC}} \langle \psi^+(\mathrm{r}) \rangle_T \langle \psi(\mathrm{r'}) \rangle_T = \rho_{\mathrm{cond}} \tag{6.4}$$

respecting the *thermodynamic sample* average

$$\langle A \rangle_T = \frac{1}{Z}\mathrm{Tr}[Ae^{-\beta E[\psi]}] = \mathrm{Tr}[\rho_Z A] \tag{6.5}$$

$$Z = \mathrm{Tr}[e^{-\beta E[\psi]}],\ \rho_Z = \frac{e^{-\beta E[\psi]}}{Z} = \frac{e^{-\beta E[\psi]}}{\mathrm{Tr}[e^{-\beta E[\psi]}]} \tag{6.6}$$

with the quantum average included

$$E[\psi] = \langle \hat{H} \rangle_\psi = \int d\mathrm{r}\psi^+(\mathrm{r})\hat{H}\psi(\mathrm{r}) \tag{6.7}$$

Yet, in order to avoid the cancellation of the order parameter $\langle \psi(\mathrm{r}) \rangle$ by taking the expectation of the annihilation uni-field $\psi(\mathrm{r})$ the global gauge U(1) invariance of the working Hamiltonian is spontaneously broken by considering the additional source terms $J(\mathrm{r}), J^*(\mathrm{r})$, as a reminiscence of the Ginzburg–Landau (GL) treatment of superconductivity, which are then considered to the vanishing limit for consistency (Ginzburg, 2004). Therefore it reads as

$$\begin{aligned} E_{GL} &= \langle \hat{H}_{GL} \rangle_\psi = \int d\mathrm{r}\psi^+(\mathrm{r})\left[-\frac{\hbar^2\nabla^2}{2m} + V(\mathrm{r}) - \mu\right]\psi(\mathrm{r}) \\ &+ \frac{1}{2}\int d\mathrm{r}\int d\mathrm{r'}\psi^+(\mathrm{r})\psi^+(\mathrm{r'})W(\mathrm{r},\mathrm{r'})\psi(\mathrm{r'})\psi(\mathrm{r}) \\ &+ \int d\mathrm{r}\left[\psi^+(\mathrm{r})J(\mathrm{r}) + J^*(\mathrm{r})\psi(\mathrm{r})\right] \end{aligned} \tag{6.8}$$

In these conditions, one may ask how the original Hohenberg–Kohn (HK) potential-density bijection theorem looks like. As a general recipe for DFT treatment of BEC one should note that any analysis has to include information on both the order parameter

$\Psi(\mathrm{r})$ and overall density $\rho(\mathrm{r})$; actually, the potential–density bijection is transposed in the uniqueness of the functionals $(V(\mathrm{r})-\mu)[\Psi(\mathrm{r}),\rho(\mathrm{r})]$ and $J[\Psi(\mathrm{r}),\rho(\mathrm{r})]$. Note that the potential fields $J(\mathrm{r}),J^*(\mathrm{r})$ were included to achieve the breaking symmetry for the particle number conservation, see the discussion on Section 5.3.2, so that they couple with the condensate fields $\Psi^+(\mathrm{r}),\Psi(\mathrm{r})$ as one may immediately lay down by the grand-potential functional

$$\begin{aligned}\Omega[\Psi,\rho] &:= \left\langle E_{GL}-TS\right\rangle_T = \left\langle E_{GL}+\frac{\ln\rho_{\neq}}{\beta}\right\rangle_T \\ &= F_{HK}[\Psi,\rho]+\int d\mathrm{r}\left[V(\mathrm{r})-\mu\right]\rho(\mathrm{r})+\int d\mathrm{r}\left[\Psi^+(\mathrm{r})J(\mathrm{r})+J^*(\mathrm{r})\Psi(\mathrm{r})\right]\end{aligned} \tag{6.9}$$

through introducing the HK universal functional

$$F_{HK}[\Psi,\rho]=\left\langle T+W+\frac{\ln\rho_{\neq}}{\beta}\right\rangle_T \tag{6.10}$$

Then, the above bijection transposes into the variational principle, either as the inequality

$$\Omega^{V',J'}[\Psi',\rho']>\Omega^{V,J}[\Psi,\rho] \tag{6.11}$$

in the case of

$$\left[\Psi'(\mathrm{r}),\rho'(\mathrm{r})\right]\neq\left[\Psi(\mathrm{r}),\rho(\mathrm{r})\right] \tag{6.12}$$

or as the optimal decopupled equations

$$\begin{cases}\left.\dfrac{\delta\Omega^{V',J'}[\Psi',\rho']}{\delta\Psi'}\right|_{\substack{\Psi'=\Psi\\ \rho'=\rho}}=0\\[2ex] \left.\dfrac{\delta\Omega^{V',J'}[\Psi',\rho']}{\delta\rho'}\right|_{\substack{\Psi'=\Psi\\ \rho'=\rho}}=0\end{cases} \tag{6.13}$$

So, up to now, worth remarking that, unlike the "standard" DFT way, the BEC adapted DFT uses also the statistical average in order to properly account for the order parameter and superfluidic density that are custom for quantum condensation phenomena. Then, also note that the two averages are combined in the statistical average $\langle\,\rangle_T$ since the appearance of the quantum average $\langle\,\rangle_\psi$ for evaluation of the energy functional; this may be called as the specific "double average" for BEC modeling.

The next step is to consider the non-interacting particle system (called afterwards as "KS-system") due to the Kohn–Sham (KS) specialization (Kohn and Sham, 1965):

$$\begin{aligned}E_{KS} &= \left\langle\hat{H}_{KS}\right\rangle_\psi = \int d\mathrm{r}\psi^+(\mathrm{r})\left[-\frac{\hbar^2\nabla^2}{2m}+V_{KS}(\mathrm{r})-\mu\right]\psi(\mathrm{r}) \\ &+\int d\mathrm{r}\left[\psi^+(\mathrm{r})J_{KS}(\mathrm{r})+J^*_{KS}(\mathrm{r})\psi(\mathrm{r})\right]\end{aligned} \tag{6.14}$$

where V_{KS} is formed such as not destroying the overall external (applied) potential – being therefore considered as the an *effective* one

$$\hat{V}_{KS} = \hat{V} + \hat{W} + (\hat{T} - \hat{T}_{KS}) \tag{6.15}$$

It may however seen under the more detailed unfolded form

$$\begin{aligned} V_{KS}(\mathrm{r}) &= V(\mathrm{r}) + \int d\mathrm{r}'\rho(\mathrm{r}')W(\mathrm{r},\mathrm{r}') + V_{XC}[\rho(\mathrm{r})] \\ &\equiv V_{ext}(\mathrm{r}) + V_{Hartree}(\mathrm{r}) + \frac{\delta E_{XC}[\rho(\mathrm{r})]}{\delta\rho} \end{aligned} \tag{6.16}$$

where we recognized the exchange and correlation potential-energy relationship (1.30).

However, the last equation may constitute the first key in considering the computational DFT for BEC since appropriately interpreting of its components, namely:

- The first term may associate in the case of optical lattices with the *lattice potential*;
- The second term is usually referred to as the *Hartree term* describing the inter-particle interaction; it thus may be considered as of the δ-delta form + other dispersion terms –see the Stuttgart experiment (Griesmaier et al., 2005), this way accounting for the *onsite effect* for cold atoms trapped in optical lattices;
- The third term it is generally called as the *exchange-correlation (XC) contribution*, and for the bosonic treatment will play combining role for the *hope and onsite* effects, for its exchange and correlation parts, respectively; it is implemented by the custom local–density approximation (LDA) and of its generalized gradient approximation (GGA) where the functionals $F_{XC}[\rho(\mathrm{r})]$ depends on the density at the coordinate + the gradient + second order density (Laplacian based) extensions as already employed by quantum chemistry (Putz, 2008a), with the possibility being applied for bosonic treatment when the inter-electronic Coulombic potential is replaced by the London inter-atomic potential, for instance.

Very interesting, the DFT-BEC correspondence seems naturally since the condensation phenomenon implies "no inter-particle interaction" so recovering the KS-condition at once. Therefore, the BEC order parameter and the superfluidic density have to resemble those associated with KS system of DFT:

$$\Psi(\mathrm{r}) \overset{!}{=} \Psi_{KS}(\mathrm{r}) := \left\langle \psi(\mathrm{r}) \right\rangle_T^{KS} \tag{6.17}$$

$$\rho(\mathrm{r}) \overset{!}{=} \rho_{KS}(\mathrm{r}) := \left\langle \psi^+(\mathrm{r})\psi(\mathrm{r}) \right\rangle_T^{KS} \tag{6.18}$$

under the actual KS-condition of performing the statistical average

$$\left\langle A \right\rangle_T^{KS} = \frac{\mathrm{Tr}[Ae^{-\beta E_{KS}[\psi]}]}{Z_{KS}} = \mathrm{Tr}[\rho_Z^{KS} A] \tag{6.19}$$

$$\mathcal{Z}_{KS} = \mathrm{Tr}[e^{-\beta E_{KS}[\psi]}],\ \rho_{\mathcal{Z}}^{KS} = \frac{e^{-\beta E_{KS}[\psi]}}{\mathcal{Z}_{KS}} = \frac{e^{-\beta E_{KS}[\psi]}}{\mathrm{Tr}[e^{-\beta E_{KS}[\psi]}]} \tag{6.20}$$

Yet, accordingly with DFT the KS-density has the simple form of superposition of the particles' wave functions

$$\rho(\mathrm{r}) = \rho_{KS}(\mathrm{r}) = \sum_{i}^{N} |\psi_i(\mathrm{r})|^2 \tag{6.21}$$

However, they are called as the KS uni-particle orbitals ψ_i that reproduce the density of the original many-body system through solving the KS equations of the auxiliary non-interacting system

$$\left[-\frac{\hbar^2}{2m}\nabla^2 + V_{KS}(\mathrm{r}) - \mu\right]\psi_i(\mathrm{r}) = \varepsilon_i\psi_i(\mathrm{r}) \tag{6.22}$$

All in all, since both Hartree and XC term depend on $\rho(\mathrm{r})$ which depends on the ψ_i by equation (6.13), which in turn depend on V_{KS} by the KS equation of equation (1.32) - it appears to be solved in a self-consistent (iterative) way; usually one starts with an initial guess (trial) for $\rho(\mathrm{r})$, then calculates the corresponding V_S (in a given hopping-onsite picture for trapped bosons in optical lattices) and solves the KS equations for the ψ_i. From these one calculates a new density and starts again. This procedure is then repeated until convergence is reached in providing the ground state density – the BEC density (equally with that of KS system) and the energy (of KS type here) as well.

However, now the HK theorem has the analytical specialization:

$$\begin{cases} \left.\dfrac{\delta\Omega_{KS}^{V'_{KS},J'_{KS}}[\Psi',\rho']}{\delta\Psi'}\right|_{\substack{\Psi'=\Psi\\ \rho'=\rho}} = 0 \\ \left.\dfrac{\delta\Omega_{KS}^{V'_{KS},J'_{KS}}[\Psi',\rho']}{\delta\rho'}\right|_{\substack{\Psi'=\Psi\\ \rho'=\rho}} = 0 \end{cases} \tag{6.23}$$

by employing the KS-grand potential

$$\begin{aligned} \Omega_{KS}[\Psi,\rho] &:= \left\langle E_{KS} + \frac{\ln \rho_{\mathcal{Z}}^{KS}}{\beta} \right\rangle_T^{KS} \\ &= F_{KS}[\Psi,\rho] + \int d\mathrm{r}\left[V_{KS}(\mathrm{r}) - \mu\right]\rho(\mathrm{r}) + \int d\mathrm{r}\left[\Psi^+(\mathrm{r})J_{KS}(\mathrm{r}) + J_{KS}^*(\mathrm{r})\Psi(\mathrm{r})\right] \end{aligned} \tag{6.24}$$

through having adapted also the universal HK functional

$$F_{KS}[\Psi,\rho] = \left\langle T_{KS} + \frac{\ln \rho_{\mathcal{Z}}^{KS}}{\beta} \right\rangle_T^{KS} \tag{6.25}$$

In these conditions, through noting that the thermal average provides the so called Hartree-energy

$$\begin{aligned}\langle W\rangle_T &= \left\langle \frac{1}{2}\int dr \int dr' \psi^+(r)\psi^+(r')W(r,r')\psi(r')\psi(r)\right\rangle_T \\ &= \frac{1}{2}\int dr \int dr' \rho(r)W(r,r')\rho(r') := E_{Hartree}[\rho]\end{aligned} \tag{6.26}$$

one may see the original grand-canonical potential may be rewritten in terms of its KS form, as

$$\begin{aligned}&\Omega[\Psi,\rho] = F_{KS}[\Psi,\rho] + \int dr[V(r)-\mu]\rho(r) \\ &+\int dr\left[\Psi^+(r)J(r) + J^*(r)\Psi(r)\right] + \frac{1}{2}\int dr \int dr' \rho(r)W(r,r')\rho(r') + E_{XC}[\Psi,\rho]\end{aligned} \tag{6.27}$$

where the XC energy intervenes with the general definition

$$\begin{aligned}E_{XC}[\Psi,\rho] &:= F_{HK}[\Psi,\rho] - F_{KS}[\Psi,\rho] - E_{Hartree}[\rho] \\ &= \left\langle T + W + \frac{\ln\rho_{\mathcal{Z}}}{\beta}\right\rangle_T - \left\langle T_{KS} + \frac{\ln\rho_{\mathcal{Z}}^{KS}}{\beta}\right\rangle_T^{KS} - \langle W\rangle_T \\ &= \left\langle T + \frac{\ln\rho_{\mathcal{Z}}}{\beta}\right\rangle_T - \left\langle T_{KS} + \frac{\ln\rho_{\mathcal{Z}}^{KS}}{\beta}\right\rangle_T^{KS}\end{aligned} \tag{6.28}$$

Finally, we are in position combining the two formulations of the variational principle under HK and KS versions, employing in fact the system of equations

$$\begin{cases}\dfrac{\delta\Omega^{V,J}[\Psi,\rho]}{\delta\Psi} = \dfrac{\delta\Omega_{KS}^{V_{KS},J_{KS}}[\Psi,\rho]}{\delta\Psi} \\ \dfrac{\delta\Omega^{V,J}[\Psi,\rho]}{\delta\rho} = \dfrac{\delta\Omega_{KS}^{V_{KS},J_{KS}}[\Psi,\rho]}{\delta\rho}\end{cases} \tag{6.29}$$

Note that in these equations the last form in terms of KS and XC functionals is convenient to be used for performing the functional derivatives of the original grand-canonical potential, aiming appropriate cancellations, as follows.

Actually, the realization of the first equation of the above system gives

$$\frac{\delta F_{KS}[\Psi(r),\rho(r)]}{\delta\Psi(r)} + J^*(r) + \frac{\delta E_{XC}[\Psi(r),\rho(r)]}{\delta\Psi(r)} = \frac{\delta F_{KS}[\Psi(r),\rho(r)]}{\delta\Psi(r)} + J^*_{KS}(r) \tag{6.30}$$

from where, while defining the XC-source current

$$J^*_{XC}(r) = \frac{\delta E_{XC}[\Psi(r),\rho(r)]}{\delta\Psi(r)} \tag{6.31}$$

it yields the KS-form of the conjugated condensation field's current source

$$J^{*}_{KS}(\mathrm{r}) = J^{*}(\mathrm{r}) + J^{*}_{XC}(\mathrm{r}) \tag{6.32}$$

In the similar manner, one obtains for the direct current source of the KS (condensed) system the expression

$$J_{KS}(\mathrm{r}) = J(\mathrm{r}) + J_{XC}(\mathrm{r}) \tag{6.33}$$

yet, with the conjugated definition of the XC-current

$$J_{XC}(\mathrm{r}) = \frac{\delta E_{XC}[\Psi(\mathrm{r}), \rho(\mathrm{r})]}{\delta \Psi^{*}(\mathrm{r})} \tag{6.34}$$

Turning to the realization of the second equation in the HK–KS system above one has

$$\frac{\delta F_{KS}[\Psi(\mathrm{r}), \rho(\mathrm{r})]}{\delta \rho(\mathrm{r})} + V(\mathrm{r}) - \mu + \int d\mathrm{r}' W(\mathrm{r}, \mathrm{r}')\rho(\mathrm{r}') + \underbrace{\frac{\delta E_{XC}[\Psi(\mathrm{r}), \rho(\mathrm{r})]}{\delta \rho(\mathrm{r})}}_{V_{XC}[\Psi(\mathrm{r}),\rho(\mathrm{r})]} \tag{6.35}$$

$$= \frac{\delta F_{KS}[\Psi(\mathrm{r}), \rho(\mathrm{r})]}{\delta \rho(\mathrm{r})} + V_{KS}(\mathrm{r}) - \mu$$

from where, through introducing the so called *the Hartree potential*

$$V_{Hartree}(\mathrm{r}) = \int d\mathrm{r}' W(\mathrm{r}, \mathrm{r}')\rho(\mathrm{r}') \tag{6.36}$$

and the recognizing the above defined XC-potential, the result gives the expression of the KS-potential as a sum of the terms of the original (externally applied) one, Hartree and XC- potentials as well

$$V_{KS}(\mathrm{r}) = V(\mathrm{r}) + V_{Hartree}(\mathrm{r}) + V_{XC}[\Psi(\mathrm{r}), \rho(\mathrm{r})] \tag{6.37}$$

to be used in the self-consistent procedure above.

Therefore, there seems that through conjunction of BEC with DFT two main additions appears to both theories:

- DFT influences BEC in providing the so called KS-currents that educes exactly to those of exchange and correlation, while the external currents are to be set to zero for assuring the physical equations of motion

$$J(\mathrm{r}) \overset{!}{=} 0,\ J^{*}(\mathrm{r}) \overset{!}{=} 0 \tag{6.38}$$

- BEC influences DFT in respecting the extended version of effective KS potential that now depend on a corrected XC-potential defined in terms of superfluid density

So, in whatever direction, the exchange and correlation effects are present, and its knowledge further requires appropriate formulation of the order parameter and of the superfluidic density itself, that nevertheless includes the condensation density part. Being so important dealing with appropriate density formulation – this subject will be therefore approached in the next section with which occasion also the condensate (the KS system) energy will be reformulated according with the Bogoliubov theory of superfluidity.

6.3 Ψ-DFT-BEC DENSITY AND ENERGY

While remembering that the HK optimization procedure on BEC implies double-averaging (upon ψ and T) followed by *variational* technique, one may like to see how the condensate density and energy behave under appropriate perturbation-like treatment of the bosonic wave function itself (Putz, 2011d). As such, the mean fields

$$\Psi(\mathrm{r}) = \langle \psi(\mathrm{r}) \rangle_T \tag{6.39}$$

$$\Psi^+(\mathrm{r}) = \langle \psi^+(\mathrm{r}) \rangle_T \tag{6.40}$$

responsible for the order parameter through T-averaging and then for the condensate density by squaring are here augmented by a superposition of the *single-particle basis wave-functions*, the so called spectral or Bogoliubov expansion or *excitation spectra* directly related with the creation and annihilation operators

$$\tilde{\psi}(\mathrm{r}) = \sum_i \varphi_i(\mathrm{r})\hat{a}_i \tag{6.41}$$

$$\tilde{\psi}^+(\mathrm{r}) = \sum_i \varphi_i^*(\mathrm{r})\hat{a}_i^+ \tag{6.42}$$

to produce the so called Bogoliubov transformations of the overall superfluidic wave-function

$$\psi(\mathrm{r}) = \Psi(\mathrm{r}) + \tilde{\psi}(\mathrm{r}) \tag{6.43}$$

$$\psi^+(\mathrm{r}) = \Psi^+(\mathrm{r}) + \tilde{\psi}^+(\mathrm{r}) \tag{6.44}$$

Here, the introduced canonical functions $\varphi_i(\mathrm{r}), \varphi_i^*(\mathrm{r})$ have the properties

$$\sum_i \varphi_i^*(\mathrm{r})\varphi_i(\mathrm{r'}) = \delta(\mathrm{r}-\mathrm{r'}),\ \int d\mathrm{r}\varphi_i^*(\mathrm{r})\varphi_j(\mathrm{r}) = \delta_{ij} \tag{6.45}$$

that are seen as a reminiscence of the dilute bosonic gas (compatible with the Gross-Pitaevsky treatment), while for the creation and annihilation of bosonic operators the ordinary commutation rules apply

$$\left[\hat{a}_i, \hat{a}_j^+\right] = \delta_{ij} \tag{6.46}$$

$$\left[\hat{a}_i, \hat{a}_j\right] = 0 = \left[\hat{a}_i^+, \hat{a}_j^+\right] \tag{6.47}$$

One has therefore similar relationships at the level of field operators, namely

$$\left[\psi(\mathrm{r}), \psi^+(\mathrm{r'})\right] = \delta(\mathrm{r}-\mathrm{r'}) \tag{6.48}$$

$$\left[\psi(\mathrm{r}), \psi(\mathrm{r'})\right] = 0 = \left[\psi^+(\mathrm{r}), \psi^+(\mathrm{r'})\right] \tag{6.49}$$

However, in order the above mean field "legs" be not affected in the condensate region, there will be imposed the constrain the basis function will vanish under T-average in the KS system

$$\left\langle \tilde{\psi}(\mathrm{r}) \right\rangle_T^{KS} = 0 \tag{6.50}$$

In other terms, this condition assures both the condensation regime and the critical regime outside of it. To be more explicit, let's take the overall density above and transform it according with the present Bogoliubov relations; one gets successively:

$$\begin{aligned}
\rho(\mathrm{r}) &= \rho_{KS}(\mathrm{r}) = \left\langle \psi^+(\mathrm{r})\psi(\mathrm{r}) \right\rangle_T^{KS} \\
&= \left\langle \left[\Psi^+(\mathrm{r}) + \tilde{\psi}^+(\mathrm{r})\right]\left[\Psi(\mathrm{r}) + \tilde{\psi}(\mathrm{r})\right] \right\rangle_T^{KS} \\
&= \left\langle \Psi^+(\mathrm{r})\Psi(\mathrm{r}) \right\rangle_T^{KS} + \left\langle \Psi^+(\mathrm{r})\tilde{\psi}(\mathrm{r}) \right\rangle_T^{KS} + \left\langle \tilde{\psi}^+(\mathrm{r})\Psi(\mathrm{r}) \right\rangle_T^{KS} + \left\langle \tilde{\psi}^+(\mathrm{r})\tilde{\psi}(\mathrm{r}) \right\rangle_T^{KS} \\
&= \left|\Psi(\mathrm{r})\right|^2 + \Psi^+(\mathrm{r})\underbrace{\left\langle \tilde{\psi}(\mathrm{r}) \right\rangle_T^{KS}}_{0} + \underbrace{\left\langle \tilde{\psi}^+(\mathrm{r}) \right\rangle_T^{KS}}_{0}\Psi(\mathrm{r}) + \left\langle \tilde{\psi}^+(\mathrm{r})\tilde{\psi}(\mathrm{r}) \right\rangle_T^{KS} \\
&= \left|\Psi(\mathrm{r})\right|^2 + \left\langle \tilde{\psi}^+(\mathrm{r})\tilde{\psi}(\mathrm{r}) \right\rangle_T^{KS} \\
&= \left|\Psi(r)\right|^2 + \sum_{i\neq 0\neq j} \varphi_i^*(r)\varphi_j(r)\left\langle \hat{a}_i^+\hat{a}_j \right\rangle_T^{KS}
\end{aligned} \tag{6.51}$$

Now, in the last relation appears the statistical average of the number of (bosonic) particle operator that can be immediately identified with the consecrated Bose–Einstein distribution (Vetter, 1997)

$$\left\langle \hat{a}_i^+\hat{a}_j \right\rangle_T^{KS} = \delta_{ij} f_{BE}(E_i) = \delta_{ij}\frac{1}{\exp(\beta E_i) - 1} \tag{6.52}$$

Thus the superfluidic density writes indeed as condensed plus critical regime density contributions

$$\begin{aligned}
\rho(\mathrm{r}) &= \left|\Psi(\mathrm{r})\right|^2 + \sum_{i,j} \varphi_i^*(\mathrm{r})\varphi_j(\mathrm{r})\delta_{ij} f_{BE}(E_i) \\
&= \left|\Psi(\mathrm{r})\right|^2 + \sum_{i\neq 0} \left|\varphi_i(\mathrm{r})\right|^2 \frac{1}{\exp(\beta E_i) - 1} \\
&= [CONDENSED + CRITICAL\,/\,THERMAL\,/\,FLUCTUATING\ REGIME]densities
\end{aligned} \tag{6.53}$$

Next application of the Bogoliubov transformations is on the KS system itself, that is on the condensate regime, aiming the diagonalization (aka optimization) of the KS-energy in terms of the eigen-energies above (E_i), so with the form

$$\hat{\mathrm{I}}E_{KS} = E_0 + \sum_{i\neq 0} E_i\hat{a}_i^+\hat{a}_i \tag{6.54}$$

accounting, like the density before, for the contribution within the condensed phase (the term E_0) and outside of it in the critical regime (the term depending on creation and annihilation operators $\hat{a}_i^+, \hat{a}_i$).

The proof of such expression is circular and makes use of its desired form to see firstly the working relationships as consequences; then inserted in the KS-energy with KS-current sources, along the Bogoliubov transformation, has to recover this form indeed; this way proofing the effectiveness of Bogoliubov transformation in provided.

However, one starts employing the KS-BEC energy ansatz to compute the commutator

$$\begin{aligned}
\left[\left(\hat{1}E_{KS}\right),\hat{a}_i\right] &= \left[E_0+\sum_j E_j\hat{a}_j^+\hat{a}_j,\hat{a}_i\right]=\left[\sum_j E_j\hat{a}_j^+\hat{a}_j,\hat{a}_i\right]\\
&=\sum_j E_j\left[\hat{a}_j^+\hat{a}_j,\hat{a}_i\right]=\sum_j E_j\{\hat{a}_j^+\underbrace{\left[\hat{a}_j,\hat{a}_i\right]}_{=0}+\underbrace{\left[\hat{a}_j^+,\hat{a}_i\right]}_{=-\delta_{ij}}\hat{a}_j\}=-\sum_j E_j\delta_{ij}\hat{a}_j\\
\left[\left(\hat{1}E_{KS}\right),\hat{a}_i\right] &= -E_i\hat{a}_i
\end{aligned}\tag{6.55}$$

Then take the more interesting commutation

$$\begin{aligned}
\left[\left(\hat{1}E_{KS}\right),\psi(\mathrm{r})\right] &= \left[\left(\hat{1}E_{KS}\right),\Psi(\mathrm{r})+\sum_i\varphi_i(\mathrm{r})\hat{a}_i\right]=\sum_i\varphi_i(\mathrm{r})\left[\left(\hat{1}E_{KS}\right),\hat{a}_i\right]\\
\left[\left(\hat{1}E_{KS}\right),\psi(\mathrm{r})\right] &= -\sum_i E_i\varphi_i(\mathrm{r})\hat{a}_i
\end{aligned}\tag{6.56}$$

We keep in mind this result, especially the right hand part of the equation, and unfold in what follows the left hand side with the help of KS-explicit energy terms, namely

$$\begin{aligned}
&\left[\left(\hat{1}E_{KS}\right),\psi(\mathrm{r})\right]\\
&=\left[\int d\mathrm{r}'\psi^+(\mathrm{r}')\hat{h}_{KS}(\mathrm{r}')\psi(\mathrm{r}'),\psi(\mathrm{r})\right]+\left[\int d\mathrm{r}'\psi^+(\mathrm{r}')J_{KS}(\mathrm{r}'),\psi(\mathrm{r})\right]\\
&+\left[\int d\mathrm{r}'J^*_{KS}(\mathrm{r}')\psi(\mathrm{r}'),\psi(\mathrm{r})\right]\\
&=\int d\mathrm{r}'\psi^+(\mathrm{r}')\hat{h}_{KS}(\mathrm{r}')\underbrace{\left[\psi(\mathrm{r}'),\psi(\mathrm{r})\right]}_{0}+\int d\mathrm{r}'\psi^+(\mathrm{r}')\underbrace{\left[\hat{h}_{KS}(\mathrm{r}'),\psi(\mathrm{r})\right]}_{0}\psi(\mathrm{r}')\\
&+\int d\mathrm{r}'\underbrace{\left[\psi^+(\mathrm{r}'),\psi(\mathrm{r})\right]}_{-\delta(\mathrm{r}-\mathrm{r}')}\hat{h}_{KS}(\mathrm{r}')\psi(\mathrm{r}')+\int d\mathrm{r}'\underbrace{\left[\psi^+(\mathrm{r}'),\psi(\mathrm{r})\right]}_{-\delta(\mathrm{r}-\mathrm{r}')}J_{KS}(\mathrm{r}')\\
&+\int d\mathrm{r}'J^*_{KS}(\mathrm{r}')\underbrace{\left[\psi(\mathrm{r}'),\psi(\mathrm{r})\right]}_{0}\\
&=-\hat{h}_{KS}(\mathrm{r})\psi(\mathrm{r})-J_{KS}(\mathrm{r})\\
&\left[\left(\hat{1}E_{KS}\right),\psi(\mathrm{r})\right]=-\hat{h}_{KS}(\mathrm{r})\Psi(\mathrm{r})-\sum_i\hat{h}_{KS}(\mathrm{r})\varphi_i(\mathrm{r})\hat{a}_i-J_{KS}(\mathrm{r})
\end{aligned}\tag{6.57}$$

Note that in the last expression the Bogoliubov transformation was considered for the field $\psi(\mathrm{r})$, while the "small" KS–Hamiltonian of BEC was considered through the notation

$$\hat{h}_{KS}(\mathrm{r}) = -\frac{\hbar^2\nabla_{\mathrm{r}}^2}{2m} + V_{KS}(\mathrm{r}) - \mu \tag{6.58}$$

Now we are ready to compare (term-by-term) the last two expressions for the commutator $\left[\left(\hat{1}E_{KS}\right), \psi(\mathrm{r})\right]$ and get the identities:

$$\begin{cases} \hat{h}_{KS}(\mathrm{r})\varphi_i = E_i\varphi_i \\ \hat{h}_{KS}(\mathrm{r})\Psi(\mathrm{r}) + J_{KS}(\mathrm{r}) = 0 \end{cases} \tag{6.59}$$

The first identity recovers in fact the traditional KS-DFT treatment when no BEC current sources are considered into the system; yet it gives the valuable information *the basis functions* $\{\varphi_i\}$ *from the Bogoliubov transformations are in fact identified with the KS-orbitals*. However the second relationship is a specific DFT-BEC connection and will be next used in either direct or conjugated forms

$$\begin{cases} \hat{h}_{KS}(\mathrm{r})\Psi(\mathrm{r}) = -J_{KS}(\mathrm{r}) \\ \Psi^+(\mathrm{r})\hat{h}_{KS}(\mathrm{r}) = -J^*_{KS}(\mathrm{r}) \end{cases} \tag{6.60}$$

In these conditions, one may finally consider the Bogoliubov transformations directly to the KS-BEC energy to equivalently yield

$$\begin{aligned}
\hat{1}E_{KS} &= \int d\mathrm{r}\psi^+(\mathrm{r})\hat{h}_{KS}(\mathrm{r})\psi(\mathrm{r}) + \int d\mathrm{r}\left[\psi^+(\mathrm{r})J_{KS}(\mathrm{r}) + J^*_{KS}(\mathrm{r})\psi(\mathrm{r})\right] \\
&= \int d\mathrm{r}\Psi^+(\mathrm{r})\underbrace{\hat{h}_{KS}(\mathrm{r})\Psi(\mathrm{r})}_{-J_{KS}(\mathrm{r})} + \int d\mathrm{r}\,\underbrace{\Psi^+(\mathrm{r})\hat{h}_{KS}(\mathrm{r})}_{-J^*_{KS}(\mathrm{r})}\tilde{\psi}(\mathrm{r}) \\
&+ \int d\mathrm{r}\tilde{\psi}^+(\mathrm{r})\underbrace{\hat{h}_{KS}(\mathrm{r})\Psi(\mathrm{r})}_{-J_{KS}(\mathrm{r})} + \int d\mathrm{r}\tilde{\psi}^+(\mathrm{r})\hat{h}_{KS}(\mathrm{r})\tilde{\psi}(\mathrm{r}) \\
&+ \int d\mathrm{r}\Psi^+(\mathrm{r})J_{KS}(\mathrm{r}) + \int d\mathrm{r}\tilde{\psi}^+(\mathrm{r})J_{KS}(\mathrm{r}) \\
&+ \int d\mathrm{r}J^*_{KS}(\mathrm{r})\Psi(\mathrm{r}) + \int d\mathrm{r}J^*_{KS}(\mathrm{r})\tilde{\psi}(\mathrm{r}) \\
&= \int d\mathrm{r}J^*_{KS}(\mathrm{r})\Psi(\mathrm{r}) + \int d\mathrm{r}\tilde{\psi}^+(\mathrm{r})\hat{h}_{KS}(\mathrm{r})\tilde{\psi}(\mathrm{r}) \\
&= \int d\mathrm{r}J^*_{KS}(\mathrm{r})\Psi(\mathrm{r}) + \int d\mathrm{r}\sum_{i,j}\varphi^*_i(\mathrm{r})\hat{a}^+_i\underbrace{\hat{h}_{KS}(\mathrm{r})\varphi_j(\mathrm{r})}_{E_j\varphi_j(\mathrm{r})}\hat{a}_j \\
&= \int d\mathrm{r}J^*_{KS}(\mathrm{r})\Psi(\mathrm{r}) + \sum_{i,j}E_j\underbrace{\left[\int d\mathrm{r}\varphi^*_i(\mathrm{r})\varphi_j(\mathrm{r})\right]}_{\delta_{ij}}\hat{a}^+_i\hat{a}_j \\
&= E_0 + \sum_{i\neq 0}E_i\hat{a}^+_i\hat{a}_i
\end{aligned} \tag{6.61}$$

where we recognize now the KS-energy of the condensed phase be of the form

$$E_0 = \int d\mathrm{r}J^*_{KS}(\mathrm{r})\Psi(\mathrm{r}) \tag{6.62}$$

that closes and proofs the present BEC Bogoliubov transformation approach as being consonant with KS-DFT picture, although enriching it accordingly (Vetter, 1997).

Now, one widely recognizes the DFT approach readily provides the bosonic density formulation in the way it describes the superfluid condensate + critical regime, in accordance with the basic formulation of BEC, while the associate KS-energy may virtually contain the basic information of *hoping* and *onsite* interactions as demanding by Bose-Hubbard model, for instance.

6.4 Ψ-DENSITY FUNCTIONALS OF BEC

The peculiar feature of the DFT in combining information coming from the natural system, hereafter called HK system, and from the auxiliary non-interacting KS one seems to offer the ideal framework that makes the difference between the thermal (system at a finite temperature) and condensed state (system at low or ultralow temperature where the inter-particle interaction, are at least as it is understood in the thermal case, eventually ceases to be manifested). Therefore, one may think that the difference between the HK and KS systems' information may provide the route of introducing the density functionals for the BEC . For this reason the perturbation theory at finite temperature main concepts are firstly reviewed then leading with the quested Y-density functionals.

6.4.1 Perturbation Theory at Finite Temperature

As a general rule, the interplay between the *quantum mechanics* and the *quantum statistics* is assure by the Wick rotation interplay

$$it \to \tau \to \beta\hbar \tag{6.63}$$

and this will be often considered whenever necessary. Consider a free statistical system with Hamiltonian $\hat{H}_0$ placed in an external influence driven by the Hamiltonian $\hat{H}_1$. For the total Hamiltonian

$$\hat{H} = \hat{H}_0 + \hat{H}_1 \tag{6.64}$$

the grand canonical partition function looks like

$$\mathcal{Z} = \mathrm{Tr}\left(e^{-\beta\hat{H}}\right) \tag{6.65}$$

with the grand canonical potential (free energy as the great thermodynamic potential)

$$\Omega = -\frac{1}{\beta}\ln \mathcal{Z} \tag{6.66}$$

Noting that for the unperturbed system we have the similar relationships

$$\mathcal{Z}_0 = \mathrm{Tr}\left(e^{-\beta\hat{H}_0}\right) \tag{6.67}$$

$$\Omega_0 = -\frac{1}{\beta}\ln \mathcal{Z}_0 \tag{6.68}$$

one can look on the perturbation amount through the free-energy difference

$$\Delta\Omega = \Omega - \Omega_0 = -\frac{1}{\beta}\ln\frac{\mathcal{Z}}{\mathcal{Z}_0} \tag{6.69}$$

Now, in order to proper evaluate (6.69) one has to learn how to deal with the ratio $\mathcal{Z}/\mathcal{Z}_0$ for finite temperature. For that one firstly notes the grand partition functions may be written in terms of free energy (grand potential) out of equations (6.66) and (6.68) so that essentially we have the identities

$$\mathcal{Z} = e^{-\beta\Omega} = \mathrm{Tr}\left(e^{-\beta\hat{H}}\right) \tag{6.70}$$

$$\mathcal{Z}_0 = e^{-\beta\Omega_0} = \mathrm{Tr}\left(e^{-\beta\hat{H}_0}\right) \tag{6.71}$$

The idea is to relate equations (6.70) and (6.71), eventually with the aid of the so called evolution operator $\hat{U}(\tau,0) = \hat{U}(\beta\hbar,0)$ such that to can write

$$e^{-\beta\hat{H}} = \hat{U}(\beta\hbar,0)e^{-\beta\hat{H}_0} \tag{6.72}$$

and have to evaluate

$$e^{-\beta F} = Tr\left(\hat{U}(\beta\hbar,0)e^{-\beta\hat{H}_0}\right) \tag{6.73}$$

However, from (6.72) one gets for finite times and temperature the evolution operator at finite time

$$\hat{U}(\tau,0) = e^{-\frac{(\hat{H}-\hat{H}_0)\tau}{\hbar}} = e^{-\frac{\hat{H}_1\tau}{\hbar}} \tag{6.74}$$

that can be immediate generalized under the form

$$\hat{U}(\tau,\tau_0) = e^{-\frac{(\hat{H}-\hat{H}_0)\tau}{\hbar}} e^{\frac{(\hat{H}-\hat{H}_0)\tau_0}{\hbar}} \tag{6.75}$$

Equivalently with the more familiar one

$$\hat{U}(\tau,\tau_0) = e^{\frac{\hat{H}_0\tau}{\hbar}} e^{-\frac{\hat{H}(\tau-\tau_0)}{\hbar}} e^{-\frac{\hat{H}_0\tau_0}{\hbar}} \tag{6.76}$$

which assures the conditions for it, namely:

- Identity:

$$\hat{U}(\tau_0,\tau_0) = 1 \tag{6.77}$$

- Unitary:

$$\hat{U}^{+}(\tau,\tau_0)\hat{U}(\tau,\tau_0) = \hat{U}(\tau,\tau_0)\hat{U}^{+}(\tau,\tau_0) = 1 \tag{6.78}$$

- Group Property:

$$\hat{U}(\tau_1,\tau_2)\hat{U}(\tau_2,\tau_3)=\hat{U}(\tau_1,\tau_3) \tag{6.79}$$

- Idempotency:

$$\hat{U}(\tau,\tau_0)\hat{U}(\tau_0,\tau)=1\Rightarrow\hat{U}(\tau_0,\tau)=\hat{U}^{+}(\tau,\tau_0) \tag{6.80}$$

Next, one would like to express the evolution operator such that the various perturbative degrees to be well evidenced; to this end one switches from the Schrödinger picture of wave function evolution

$$\begin{cases}-\hbar\dfrac{\partial}{\partial\tau}\left|\Psi_S(\tau)\right\rangle=\hat{H}\left|\Psi_S(\tau)\right\rangle\\ \left|\Psi_S(\tau)\right\rangle=e^{-\frac{\hat{H}(\tau-\tau_0)}{\hbar}}\left|\Psi_S(\tau_0)\right\rangle\end{cases} \tag{6.81}$$

to the *interaction picture* where the wave-function connects with that of the Schrödinger case though the application of the evolution amplitude operator

$$\left|\Psi_I(\tau)\right\rangle=e^{\frac{\hat{H}_0\tau}{\hbar}}\left|\Psi_S(\tau)\right\rangle \tag{6.82}$$

through the successive time-derivations:

$$\begin{aligned}-\hbar\frac{\partial}{\partial\tau}\left|\Psi_I(\tau)\right\rangle&=-\hat{H}_0e^{\frac{\hat{H}_0\tau}{\hbar}}\left|\Psi_S(\tau)\right\rangle+e^{\frac{\hat{H}_0\tau}{\hbar}}\underbrace{(-1)\hbar\frac{\partial}{\partial\tau}\left|\Psi_S(\tau)\right\rangle}_{\hat{H}\left|\Psi_S(\tau)\right\rangle}\\&=e^{\frac{\hat{H}_0\tau}{\hbar}}\left(-\hat{H}_0+\hat{H}_0+\hat{H}_1\right)\underbrace{\left|\Psi_S(\tau)\right\rangle}_{\exp\left(-\frac{\hat{H}_0\tau}{\hbar}\right)\left|\Psi_I(\tau)\right\rangle}\\&=e^{\frac{\hat{H}_0\tau}{\hbar}}\hat{H}_1e^{-\frac{\hat{H}_0\tau}{\hbar}}\left|\Psi_I(\tau)\right\rangle\end{aligned} \tag{6.83}$$

so that providing the interaction equation

$$\begin{cases}-\hbar\dfrac{\partial}{\partial\tau}\left|\Psi_I(\tau)\right\rangle=\hat{H}_1(\tau)\left|\Psi_I(\tau)\right\rangle\\ \left|\Psi_I(\tau)\right\rangle=\hat{U}(\tau,\tau_0)\left|\Psi_I(\tau_0)\right\rangle\end{cases} \tag{6.84}$$

that is formally identically with that of Schrödinger (19) excepting the fact it refers to the interacting Hamiltonian only and that is becoming time-evolution, with the working expression

$$\hat{H}_1(\tau)=e^{\frac{\hat{H}_0\tau}{\hbar}}\hat{H}_1e^{-\frac{\hat{H}_0\tau}{\hbar}} \tag{6.85}$$

while the second condition of (6.84) is equivalently with (6.82) once the evolution amplitude (6.76) is implemented.

There is obvious now that the interaction picture, in general, and consequently also the evolution operator in special are closely related with the interaction Hamiltonian, and carrying its information therefore. Moreover, the interaction picture transforms the Schrödinger operators in temporal ones upon the recipe:

$$\hat{O}_I(\tau) = e^{\frac{\hat{H}_0\tau}{\hbar}} \hat{O} e^{-\frac{\hat{H}_0\tau}{\hbar}} \tag{6.86}$$

Even more, through combining the two equations of (6.84) one obtains the time-evolution operatorial equation

$$-\hbar\frac{\partial}{\partial\tau}\hat{U}(\tau,\tau_0) = \hat{H}_1(\tau)\hat{U}(\tau,\tau_0) \tag{6.87}$$

whose solution

$$\hat{U}(\tau,\tau_0) - \underbrace{\hat{U}(\tau_0,\tau_0)}_{1} = -\frac{1}{\hbar}\int_{\tau_0}^{\tau} d\tau'\hat{H}_1(\tau')\hat{U}(\tau',\tau_0) \tag{6.88}$$

may be set up in an iteratively way, in what become known as *Neumann-Liouville expansion* or *Dyson series*, since noting its self-consistency

$$\begin{aligned}
\hat{U}(\tau,\tau_0) &= 1 - \frac{1}{\hbar}\int_{\tau_0}^{\tau} d\tau'\hat{H}_1(\tau')\hat{U}(\tau',\tau_0) \\
\hat{U}(\tau,\tau_0) &= 1 + \frac{(-1)}{\hbar}\int_{\tau_0}^{\tau} d\tau'\hat{H}_1(\tau') + \left(\frac{-1}{\hbar}\right)^2 \int_{\tau_0}^{\tau} d\tau' \int_{\tau_0}^{\tau'} d\tau''\hat{H}_1(\tau')\hat{H}_1(\tau'') + \ldots
\end{aligned} \tag{6.89}$$

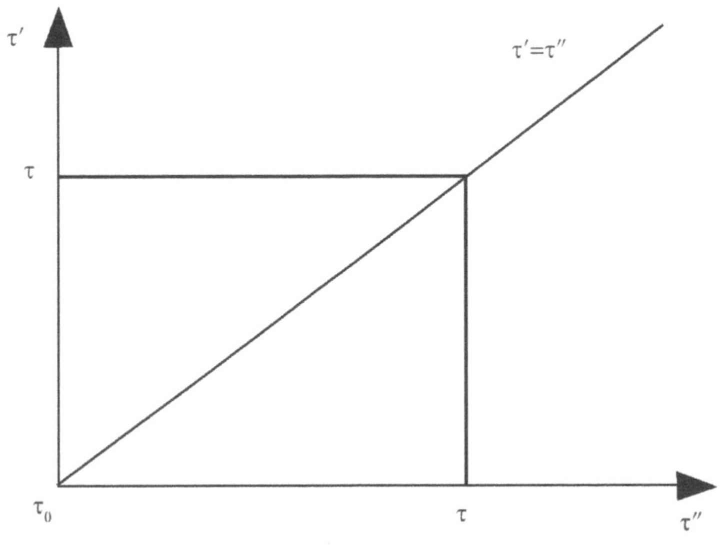

FIGURE 6.1 Integration regions for time ordering process in equation (6.89).

In order to understand how to generalize the solution (6.89) one focuses on equivalent forms of the third term of the right-hand-side of (6.89) and successively obtains respecting the time-ordering integrals and contributions (see Figure 6.1):

$$\begin{aligned}
&\int_{\tau_0}^{\tau} d\tau' \int_{\tau_0}^{\tau'} d\tau'' \hat{H}_1(\tau')\hat{H}_1(\tau'') \\
&= \frac{1}{2}\int_{\tau_0}^{\tau} d\tau' \int_{\tau_0}^{\tau'} d\tau'' \hat{H}_1(\tau')\hat{H}_1(\tau'') + \frac{1}{2}\int_{\tau_0}^{\tau} d\tau'' \int_{\tau''}^{\tau} d\tau' \hat{H}_1(\tau')\hat{H}_1(\tau'') \\
&= \frac{1}{2}\int_{\tau_0}^{\tau} d\tau' \int_{\tau_0}^{\tau'} d\tau'' \hat{H}_1(\tau')\hat{H}_1(\tau'') + \frac{1}{2}\int_{\tau_0}^{\tau} d\tau' \int_{\tau'}^{\tau} d\tau'' \hat{H}_1(\tau'')\hat{H}_1(\tau') \\
&= \frac{1}{2}\int_{\tau_0}^{\tau} d\tau' \int_{\tau_0}^{\tau} d\tau'' \left[\hat{H}_1(\tau')\hat{H}_1(\tau'')\theta(\tau'-\tau'') + \hat{H}_1(\tau'')\hat{H}_1(\tau')\theta(\tau'-\tau'')\right] \\
&= \frac{1}{2}\int_{\tau_0}^{\tau} d\tau' \int_{\tau_0}^{\tau} d\tau'' \hat{\mathrm{T}}\left[\hat{H}_1(\tau')\hat{H}_1(\tau'')\right]
\end{aligned} \tag{6.90}$$

where one introduced the Heaviside step function

$$\theta(\tau'-\tau'') = \begin{cases} 1 \ldots \tau' > \tau'' \\ 0 \ldots \tau' < \tau'' \end{cases} \tag{6.91}$$

and the time-ordering product of operators accounting for that operators do not necessarily commute even with itself at different times

$$\hat{\mathrm{T}}\left[\hat{A}(\tau')\hat{B}(\tau'')\right] = \begin{cases} \hat{A}(\tau')\hat{B}(\tau'') \ldots \tau' > \tau'' \\ \mp\hat{B}(\tau'')\hat{A}(\tau') \ldots \tau'' > \tau' \end{cases} \tag{6.92}$$

with the "–" sign for fermions and "+" sign for bosons.

With this rule, the series solution (6.89) can be resumed under the working form

$$\hat{U}(\tau,\tau_0) = \sum_{n=0}^{\infty}\left(-\frac{1}{\hbar}\right)^n \frac{1}{n!}\int_{\tau_0}^{\tau} d\tau_1 \ldots \int_{\tau_0}^{\tau} d\tau_n \hat{\mathrm{T}}\left[\hat{H}_1(\tau_1)\ldots\hat{H}_1(\tau_n)\right] \tag{6.93}$$

equivalently rewritten as

$$\hat{U}(t,t_0) = \sum_{n=0}^{\infty}\left(-\frac{i}{\hbar}\right)^n \frac{1}{n!}\int_{t_0}^{t} dt_1 \ldots \int_{t_0}^{t} dt_n \hat{\mathrm{T}}\left[\hat{H}_1(t_1)\ldots\hat{H}_1(t_n)\right] \tag{6.94}$$

or

$$\hat{U}(\hbar\beta,0) = \sum_{n=0}^{\infty}\left(-\frac{1}{\hbar}\right)^n \frac{1}{n!}\int_{0}^{\hbar\beta} d\tau_1 \ldots \int_{0}^{\hbar\beta} d\tau_n \hat{\mathrm{T}}\left[\hat{H}_1(\tau_1)\ldots\hat{H}_1(\tau_n)\right] \tag{6.95}$$

in the virtue of equation (6.63). The form (6.95) can be now used for evaluation of (6.69), yet under the more convenient exponential resumed series

$$\hat{U}(\hbar\beta,0)=\exp\left\{\hat{\mathrm{T}}\left[-\frac{1}{\hbar}\int_0^{\hbar\beta}d\tau\hat{H}_1(\tau)\right]\right\} \tag{6.96}$$

Firstly, through recalling (6.73) one has

$$\begin{aligned}\frac{\mathcal{Z}}{\mathcal{Z}_0}&=\frac{\mathrm{Tr}\left(\hat{U}(\beta\hbar,0)e^{-\beta\hat{H}_0}\right)}{\mathrm{Tr}\left(e^{-\beta\hat{H}_0}\right)}=\mathrm{Tr}\left(\hat{U}(\beta\hbar,0)\frac{e^{-\beta\hat{H}_0}}{\mathrm{Tr}\left(e^{-\beta\hat{H}_0}\right)}\right)\\&=\sum_{n=0}^{\infty}\left(-\frac{1}{\hbar}\right)^n\frac{1}{n!}\int_0^{\hbar\beta}d\tau_1...\int_0^{\hbar\beta}d\tau_n\mathrm{Tr}\left(\hat{\rho}_0\hat{\mathrm{T}}\left[\hat{H}_1(\tau_1)...\hat{H}_1(\tau_n)\right]\right)\\&=\exp\left\{\mathrm{Tr}\left(\hat{\rho}_0\hat{\mathrm{T}}\left[-\frac{1}{\hbar}\int_0^{\hbar\beta}d\tau\hat{H}_1(\tau)\right]\right)\right\}\end{aligned} \tag{6.97}$$

then finally:

$$\Delta\Omega=-\frac{1}{\beta}\ln\frac{\mathcal{Z}}{\mathcal{Z}_0}=-\frac{1}{\beta}\mathrm{Tr}\left(\hat{\rho}_0\hat{\mathrm{T}}\left[-\frac{1}{\hbar}\int_0^{\hbar\beta}d\tau\hat{H}_1(\tau)\right]\right)=\sum_{n=1}^{\infty}\Omega_n \tag{6.98}$$

with

$$\Omega_n=-\frac{1}{\beta}\left(-\frac{1}{\hbar}\right)^n\int_0^{\hbar\beta}d\tau_1...\int_0^{\hbar\beta}d\tau_n\mathrm{Tr}\left(\hat{\rho}_0\hat{\mathrm{T}}\left[\hat{H}_1(\tau_1)...\hat{H}_1(\tau_n)\right]\right) \tag{6.99}$$

after re-considering the full thermal times' decomposition.

Yet, we remain with the problem of evaluating the Trace-operation in (6.99)

While liking to evaluate the trace in (6.97) one needs specific field theory tools as eventually presented elsewere (Fetter and Walecka, 2003). For the present purpose, however, we focus on the use of the (6.98) in DFT of Bose–Einstein condensates aiming the determination for the XC functional.

6.4.2 Ψ-Exchange-Correlation Functional

For beginning one needs some pre-requisites regarding the DFT contact with the statistical mechanics of Bosonic condensation . The main point is that in Bosonic condensates all energetic functionals are functionals of the couple $[\rho(\mathrm{r}),\Psi(\mathrm{r})]$ in terms of the overall *super-fluidic density* $\rho(\mathrm{r})$ and of the *order parameter* $\Psi(\mathrm{r})$,

$$\begin{cases}\rho(\mathrm{r})\equiv\left\langle\psi^+(\mathrm{r})\psi(\mathrm{r})\right\rangle_T\\\Psi(\mathrm{r})\equiv\left\langle\psi(\mathrm{r})\right\rangle_T\end{cases} \tag{6.100}$$

both defined through the N-particles' ensemble statistical average

$$\begin{cases} \langle \bullet \rangle_T = \frac{1}{\mathcal{Z}} \mathrm{Tr}[\bullet e^{-\beta E[\psi]}] == \mathrm{Tr}[\rho_Z \bullet] \\ \rho_Z = \frac{e^{-\beta E[\psi]}}{\mathcal{Z}} = \frac{e^{-\beta E[\psi]}}{\mathrm{Tr}[e^{-\beta E[\psi]}]} \end{cases} \tag{6.101}$$

of single-particle field $\psi(\mathrm{r})$. Note that the squares of this particle wave-function and of its thermal average (order parameter) are related with the total number of particle in the system and with the condensate density, see also eq. (5.17), namely as:

$$\begin{cases} N = \int |\psi(\mathrm{r})|^2 \, dr \\ \rho_{\mathrm{cond}} \equiv |\Psi(\mathrm{r})|^2 \end{cases} \tag{6.102}$$

In these conditions, one can advance the main DFT-BEC idea according which the real many-body system is represented by the so called HK grand canonical potential (free energy)

$$\Omega_{HK}[\Psi,\rho] = F_{HK}[\Psi,\rho] + \int dr [V_{ext}(\mathrm{r}) - \mu] \rho(\mathrm{r}) + D_{ext}[\Psi,\rho] \tag{6.103}$$

with

$$\begin{cases} F_{HK}[\Psi,\rho] = \left\langle T_{HK} + W + \frac{\ln \rho_{\mathcal{Z}}}{\beta} \right\rangle_T \\ D_{ext}[\Psi,\rho] = \int dr \left[\Psi^+(\mathrm{r}) J(\mathrm{r}) + J^*(\mathrm{r}) \Psi(\mathrm{r}) \right] \end{cases} \tag{6.104}$$

where F_{HK} stays for the HK functional, $V(\mathbf{r})$ is the external applied potential, μ is the chemical potential, W- the inter-particle pair interaction, T_{HK}: the kinetic energy of the real system, and $J(\mathrm{r})$ the external source that assure the observable character of the DFT-BEC picture and will be further analytically unfolded.

In the same manner, if one considers the KS systems, as a non-interacting many body system, so fortunately carrying by its very construction the same density as that of the (HK) super-fluid system, yet the free energy will have look differently like is

$$\Omega_{KS}[\Psi,\rho] = F_{KS}[\Psi,\rho] + \int dr [V_{KS}(\mathrm{r}) - \mu] \rho(\mathrm{r}) + D_{KS}[\Psi,\rho] \tag{6.105}$$

with the following components:

- The KS functional

$$F_{KS}[\Psi,\rho] = \left\langle T_{KS} + \frac{\ln \rho_{\mathcal{Z}}^{KS}}{\beta} \right\rangle_T^{KS} \tag{6.106}$$

- employing the actual KS-statistical average

$$\begin{cases} \langle \bullet \rangle_T^{KS} = \dfrac{1}{\mathcal{Z}_{KS}} \mathrm{Tr}[\bullet e^{-\beta E_{KS}[\psi]}] == \mathrm{Tr}[\rho_{\mathcal{Z}}^{KS} \bullet] \\ \rho_{\mathcal{Z}}^{KS} = \dfrac{e^{-\beta E_{KS}[\psi]}}{\mathcal{Z}_{KS}} = \dfrac{e^{-\beta E_{KS}[\psi]}}{\mathrm{Tr}[e^{-\beta E_{KS}[\psi]}]} \end{cases} \tag{6.107}$$

- The KS functional casts as

$$V_{KS}(\mathrm{r}) = V_{ext}(\mathrm{r}) + V_{Hartree}(\mathrm{r}) + [T(\mathrm{r}) - T_{KS}(\mathrm{r})] \tag{6.108}$$

- with the introduced Hartree-potential

$$V_{Hartree}(\mathrm{r}) = \int d\mathrm{r}'\rho(\mathrm{r}')W(\mathrm{r},\mathrm{r}') \tag{6.109}$$

- and the XC potential

$$D_{KS}[\Psi,\rho] = \int d\mathrm{r}\left[\Psi^{+}(\mathrm{r})J_{KS}(\mathrm{r}) + J_{KS}^{*}(\mathrm{r})\Psi(\mathrm{r})\right] \tag{6.110}$$

- The KS external current potential

$$D_{KS}[\Psi,\rho] = \int d\mathrm{r}\left[\Psi^{+}(\mathrm{r})J_{KS}(\mathrm{r}) + J_{KS}^{*}(\mathrm{r})\Psi(\mathrm{r})\right] \tag{6.111}$$

All in all one may see that KS grand canonical functional (6.105) with all its components (6.106–6.111) formally resembles the real KH functional (6.103) but for different level of interaction. Therefore, the interaction comes in through the perturbation form that is their difference corresponds to the perturbation interaction that brings the non-interacting system into the interacted one; it reads:

$$\Omega_{HK}[\Psi,\rho] = \Omega_{KS}[\Psi,\rho] + \Delta\Omega[\Psi,\rho] \tag{6.112}$$

Through employing in two ways the information encompassed in this functional one can find out the expression from XC functional.

In the first instance one can rewrite (6.112) at the level of corresponding Hamiltonians

$$\begin{aligned} H_{HK} &= \left(T + W + \frac{\ln \rho_{\mathcal{Z}}}{\beta}\right) + (V - \mu N) + D_{ext} \\ &= \left(T + \frac{\ln \rho_{\mathcal{Z}}}{\beta}\right) + \underbrace{V + V_{Hartree} + V_{XC}}_{V_{KS}} - \mu N + D_{KS} + W - V_{Hartree} - V_{XC} - D_{XC} \\ &= H_{KS} + H_1 \end{aligned} \tag{6.113}$$

where we identified the KS Hamiltonian

$$H_{KS} = \left(T + \frac{\ln \rho_{\mathcal{Z}}}{\beta}\right) + V_{KS} - \mu N + D_{KS} \tag{6.114}$$

and the perturbation Hamiltonian

$$H_1 = W - V_{Hartree} - V_{XC} - D_{XC} \tag{6.115}$$

along the introduced XC source energy

$$D_{XC} = D_{KS} - D_{ext} \tag{6.116}$$

On the other side, one may rearrange the HK grand canonical energy in terms of KS functional through the relationship:

$$\begin{aligned}&\Omega_{HK}[\Psi,\rho]\\&= F_{KS}[\Psi,\rho] + \int (V_{ext}(\mathrm{r}) - \mu)\rho(\mathrm{r})d\mathrm{r} + D_{ext}[\Psi,\rho] + E_{Hartree}[\Psi,\rho] + E_{XC}[\Psi,\rho]\end{aligned} \tag{6.117}$$

while one recognizes the Hartree energy has to have the form

$$E_{Hartree}[\Psi,\rho] = \frac{1}{2}\int d\mathrm{r}d\mathrm{r}'\rho(\mathrm{r})W(\mathrm{r},\mathrm{r}')\rho(\mathrm{r}') \tag{6.118}$$

and the XC energy has to have the formal form

$$E_{XC}[\Psi,\rho] = F_{HK}[\Psi,\rho] - E_{Hartree}[\Psi,\rho] \tag{6.119}$$

Now the difference $\Delta\Omega[\Psi,\rho]$ looks like

$$\begin{aligned}\Delta\Omega[\Psi,\rho] &= \Omega_{HK}[\Psi,\rho] - \Omega_{KS}[\Psi,\rho]\\&= \left\{F_{KS}[\Psi,\rho] + \int (V_{ext}(\mathrm{r}) - \mu)\rho(\mathrm{r})d\mathrm{r} + D_{ext}[\Psi,\rho] + E_{Hartree}[\Psi,\rho] + E_{XC}[\Psi,\rho]\right\}\\&-\left\{F_{KS}[\Psi,\rho] + \int (V_{KS}(\mathrm{r}) - \mu)\rho(\mathrm{r})d\mathrm{r} + D_{KS}[\Psi,\rho]\right\}\\&= \int (V_{ext}(\mathrm{r}) - V_{KS}(\mathrm{r}))\rho(\mathrm{r})d\mathrm{r} - D_{XC}[\Psi,\rho] + E_{Hartree}[\Psi,\rho] + E_{XC}[\Psi,\rho]\end{aligned} \tag{6.120}$$

from where the XC potential gets out

$$\begin{aligned}&E_{XC}[\Psi,\rho]\\&= \Delta\Omega[\Psi,\rho] + \int (V_{Hartree}(\mathrm{r}) + V_{XC}(\mathrm{r}))\rho(\mathrm{r})\,d\mathrm{r} + D_{XC}[\Psi,\rho] - E_{Hartree}[\Psi,\rho]\end{aligned} \tag{6.121}$$

Now, remembering the expression (6.98) from the perturbation theory with the interaction Hamiltonian (6.115) one can formulate the general expression for the difference (6.120)

$$\begin{aligned}\Delta\Omega &= -\frac{1}{\beta}Tr\left(\rho_{Z}^{KS}\hat{\mathrm{T}}\left[-\frac{1}{\hbar}\int_0^{\hbar\beta} d\tau H_1(\tau)\right]\right) = -\frac{1}{\beta}\left\langle \hat{\mathrm{T}}\left[-\frac{1}{\hbar}\int_0^{\hbar\beta} d\tau H_1(\tau)\right]\right\rangle_T^{KS}\\&= -\frac{1}{\beta}\sum_{n=1}^{\infty}\left(-\frac{1}{\hbar}\right)^n\int_0^{\hbar\beta} d\tau_1 ... \int_0^{\hbar\beta} d\tau_n \left\langle \hat{\mathrm{T}}[H_1(\tau_1)...H_1(\tau_n)]\right\rangle_T^{KS}\end{aligned} \tag{6.122}$$

and finally for the XC Bose–Einstein functional:

$$
\begin{aligned}
&E_{XC}[\Psi,\rho] \\
&= -\frac{1}{\beta}\sum_{n=1}^{\infty}\left(-\frac{1}{\hbar}\right)^{n}\int_{0}^{\hbar\beta} d\tau_1 ... \int_{0}^{\hbar\beta} d\tau_n \\
&\quad \left\langle \hat{\mathrm{T}}\left[\left(W - V_{Hartree} - V_{XC} - D_{XC}\right)(\tau_1)...\left(W - V_{Hartree} - V_{XC} - D_{XC}\right)(\tau_n)\right]\right\rangle_T^{KS} \\
&+\int \left(V_{Hartree}(\mathrm{r}) + V_{XC}(\mathrm{r})\right)\rho(\mathrm{r})d\mathrm{r} + D_{XC}[\Psi,\rho] - E_{Hartree}[\Psi,\rho]
\end{aligned}
\tag{6.123}
$$

while being susceptible to be specialized in the first order for getting the exchange-functional only. This will be subjected to the next section.

6.4.3 Ψ-Exchange Functional

Since, the present purpose of dealing with dilute–gas states the first order approximation to DFT-BEC under the first order parameter expansion, the g-DFT form with

$$
g = \left|\Psi(\mathrm{r})\right|^2 \equiv \left\langle \psi(\mathrm{r})\right\rangle_T^2 \tag{6.124}
$$

is wholly sufficient and that will be the case also with the XC energy above, equation (6.123), that this way will restraint to *exchange* energy only. Therefore one writes the exchange BEC-density functional explicitly as:

$$
\begin{aligned}
E_X[\Psi,\rho] &= \frac{1}{\beta\hbar}\int_0^{\hbar\beta} d\tau \left\langle \hat{\mathrm{T}}\left[\left(W - V_{Hartree} - V_{XC} - D_{XC}\right)(\tau)\right]\right\rangle_T^{KS} \\
&+\int \left(V_{Hartree}(\mathrm{r}) + V_{XC}(\mathrm{r})\right)\rho(\mathrm{r})d\mathrm{r} + \int D_{XC}[\Psi,\rho]d\mathrm{r} - E_{Hartree}[\Psi,\rho] \\
&= \frac{1}{\beta\hbar}\int_0^{\hbar\beta} d\tau \left\langle \hat{\mathrm{T}}\left[W(\tau)\right]\right\rangle_T^{KS} - E_{Hartree}[\Psi,\rho] \\
&\underbrace{-\frac{1}{\beta\hbar}\int_0^{\hbar\beta} d\tau \left\langle \hat{\mathrm{T}}\left[V_{Hartree}(\tau)\right]\right\rangle_T^{KS}}_{\int V_{Hartree}(\mathrm{r})\rho(\mathrm{r})d\mathrm{r}}
\underbrace{-\frac{1}{\beta\hbar}\int_0^{\hbar\beta} d\tau \left\langle \hat{\mathrm{T}}\left[V_{XC}(\tau)\right]\right\rangle_T^{KS}}_{\int V_{XC}(\mathrm{r})\rho(\mathrm{r})d\mathrm{r}}
\underbrace{-\frac{1}{\beta\hbar}\int_0^{\hbar\beta} d\tau \left\langle \hat{\mathrm{T}}\left[D_{XC}(\tau)\right]\right\rangle_T^{KS}}_{\int D_{XC}(\mathrm{r})\rho(\mathrm{r})d\mathrm{r}} \\
&+\int V_{Hartree}(\mathrm{r})\rho(\mathrm{r})d\mathrm{r} + \int V_{XC}(\mathrm{r})\rho(\mathrm{r})d\mathrm{r} + \int D_{XC}[\Psi,\rho]d\mathrm{r} \\
&= \frac{1}{\beta\hbar}\int_0^{\hbar\beta} d\tau \left\langle \hat{\mathrm{T}}\left[W(\tau)\right]\right\rangle_T^{KS} - E_{Hartree}[\Psi,\rho]
\end{aligned}
\tag{6.125}
$$

or in the practical form

$$
\begin{aligned}
E_X[\Psi,\rho] &= \frac{1}{2}\int W(\mathrm{r},\mathrm{r}')\rho(\mathrm{r})\rho(\mathrm{r}')d\mathrm{r}d\mathrm{r}' - E_{Hartree}[\Psi,\rho] \\
&= \frac{1}{2}\int W(\mathrm{r},\mathrm{r}')\left\langle \psi^{+}(\mathrm{r})\psi^{+}(\mathrm{r}')\psi(\mathrm{r})\psi(\mathrm{r}')\right\rangle_T^{KS} d\mathrm{r}d\mathrm{r}' - E_{Hartree}[\Psi,\rho]
\end{aligned}
\tag{6.126}
$$

while remembering the main Y-DFT ingredients, especially the superfluidic density of (6.100).

Next, one may evaluate the average contained within the inter-particle interaction in (6.126) through unfolding it through the Bogoliubov transformations

$$\psi(\mathrm{r}) = \Psi(\mathrm{r}) + \tilde{\psi}(\mathrm{r}) \tag{6.127}$$

$$\psi^{+}(\mathrm{r}) = \Psi^{+}(\mathrm{r}) + \tilde{\psi}^{+}(\mathrm{r}) \tag{6.128}$$

and firstly yields

$$\begin{aligned}
&\psi^{+}(\mathrm{r})\psi^{+}(\mathrm{r'})\psi(\mathrm{r})\psi(\mathrm{r'}) \\
&= |\Psi(\mathrm{r})|^2 |\Psi(\mathrm{r'})|^2 + \tilde{\psi}^{+}(\mathrm{r})\tilde{\psi}^{+}(\mathrm{r'})\tilde{\psi}(\mathrm{r'})\tilde{\psi}(\mathrm{r}) \\
&+2|\Psi(\mathrm{r})|^2 \tilde{\psi}^{+}(\mathrm{r'})\tilde{\psi}(\mathrm{r'}) + 2\Psi(\mathrm{r})\Psi^{+}(\mathrm{r'})\tilde{\psi}^{+}(\mathrm{r})\tilde{\psi}(\mathrm{r'}) \\
&+\Psi(\mathrm{r})\Psi(\mathrm{r'})\tilde{\psi}^{+}(\mathrm{r})\tilde{\psi}^{+}(\mathrm{r'}) + \Psi^{+}(\mathrm{r})\Psi^{+}(\mathrm{r'})\tilde{\psi}(\mathrm{r})\tilde{\psi}(\mathrm{r'}) \\
&+2|\Psi(\mathrm{r})|^2 \Psi(\mathrm{r'})\underbrace{\tilde{\psi}^{+}(\mathrm{r'})}_{!} + 2|\Psi(\mathrm{r})|^2 \Psi^{+}(\mathrm{r'})\underbrace{\tilde{\psi}(\mathrm{r'})}_{!} \\
&+2\Psi^{+}(\mathrm{r})\underbrace{\tilde{\psi}^{+}(\mathrm{r'})\tilde{\psi}(\mathrm{r'})\tilde{\psi}(\mathrm{r})}_{!} + 2\Psi(\mathrm{r})\underbrace{\tilde{\psi}(\mathrm{r})\tilde{\psi}^{+}(\mathrm{r'})\tilde{\psi}(\mathrm{r'})}_{!}
\end{aligned} \tag{6.129}$$

When the grand statistical average is taken over expression (6.129) one individually gets:

- The order parameter containing term behaves like a scalar and is let untouched

$$\left\langle |\Psi(\mathrm{r})|^2 |\Psi(\mathrm{r'})|^2 \right\rangle_T^{KS} = |\Psi(\mathrm{r})|^2 |\Psi(\mathrm{r'})|^2 \tag{6.130}$$

- The terms with conjugated product of the fluctuation (or thermal) field turns out the associate density

$$\left\langle 2|\Psi(\mathrm{r})|^2 \tilde{\psi}^{+}(\mathrm{r'})\tilde{\psi}(\mathrm{r'}) \right\rangle_T^{KS} = 2|\Psi(\mathrm{r})|^2 \tilde{\rho}(\mathrm{r'}) \tag{6.131}$$

- At different space-points the conjugate products of the fluctuating fields turn out the Green function (at the same time/temperature) according with its basic definition

$$G(\mathrm{r}\tau;\mathrm{r'}\tau) = -\left\langle \tilde{\psi}^{+}(\mathrm{r})\tilde{\psi}(\mathrm{r'}) \right\rangle_T^{KS} \tag{6.132}$$

- so that one has for the two specific terms in (6.129):

$$\left\langle 2\Psi(\mathrm{r})\Psi^{+}(\mathrm{r'})\tilde{\psi}^{+}(\mathrm{r})\tilde{\psi}(\mathrm{r'}) \right\rangle_T^{KS} = -2\Psi(\mathrm{r})\Psi^{+}(\mathrm{r'})G(\mathrm{r}\tau;\mathrm{r'}\tau) \tag{6.133}$$

$$\left\langle \tilde{\psi}^{+}(\mathrm{r})\tilde{\psi}^{+}(\mathrm{r'})\tilde{\psi}(\mathrm{r'})\tilde{\psi}(\mathrm{r}) \right\rangle_T^{KS} = G(\mathrm{r}\tau;\mathrm{r'}\tau)G(\mathrm{r'}\tau;\mathrm{r}\tau) \tag{6.134}$$

- The sum of the conjugated terms that cannot turn out through averaging (trace operation) neither Green function nor the thermal density it is considered as giving a double real part of the associate product of thermal/fluctuation densities that by averaging is divided by factor 2 and turns out in the simple form

$$\left\langle \Psi(\mathrm{r})\Psi(\mathrm{r'})\tilde{\psi}^{+}(\mathrm{r})\tilde{\psi}^{+}(\mathrm{r'}) + \Psi^{+}(\mathrm{r})\Psi^{+}(\mathrm{r'})\tilde{\psi}(\mathrm{r})\tilde{\psi}(\mathrm{r'}) \right\rangle_T^{KS} = \tilde{\rho}(\mathrm{r})\tilde{\rho}(\mathrm{r'}) \tag{6.135}$$

- The terms containing odd powers of fluctuating/thermal field in (6.129) averages to zero since by construction it originates into the statistical average of the fluctuating/thermal field to zero,

$$\left\langle \tilde{\psi}(\mathrm{r}) \right\rangle_T = 0 = \left\langle \tilde{\psi}^+(\mathrm{r}) \right\rangle_T \tag{6.136}$$

- that is it does not produces any Bose–Einstein condensate that is characterized by the order parameter (6.100); thus, one has

$$\begin{aligned} 0 &= \left\langle 2\left|\Psi(\mathrm{r})\right|^2 \Psi(\mathrm{r'})\tilde{\psi}^+(\mathrm{r'}) \right\rangle_T^{KS} = \left\langle 2\left|\Psi(\mathrm{r})\right|^2 \Psi^+(\mathrm{r'})\tilde{\psi}(\mathrm{r'}) \right\rangle_T^{KS} \\ &= \left\langle 2\Psi^+(\mathrm{r})\tilde{\psi}^+(\mathrm{r'})\tilde{\psi}(\mathrm{r'})\tilde{\psi}(\mathrm{r}) \right\rangle_T^{KS} = \left\langle 2\Psi(\mathrm{r})\tilde{\psi}(\mathrm{r})\tilde{\psi}^+(\mathrm{r'})\tilde{\psi}(\mathrm{r'}) \right\rangle_T^{KS} \end{aligned} \tag{6.137}$$

All in all, with equations (6.129–6.137) the exchange expression (6.126) now specializes to

$$\begin{aligned} E_X[\Psi,\rho] = \frac{1}{2}\int d\mathrm{r}d\mathrm{r'}W(\mathrm{r},\mathrm{r'})\Big\{&\left|\Psi(\mathrm{r})\right|^2 \left|\Psi(\mathrm{r'})\right|^2 + 2\left|\Psi(\mathrm{r})\right|^2 \tilde{\rho}(\mathrm{r'}) + \tilde{\rho}(\mathrm{r})\tilde{\rho}(\mathrm{r'}) \\ -2\Psi(\mathrm{r})\Psi^+(\mathrm{r'})G(\mathrm{r}\tau;\mathrm{r'}\tau) + G(\mathrm{r}\tau;\mathrm{r'}\tau)G(\mathrm{r'}\tau;\mathrm{r}\tau)\Big\}& - E_{Hartree}[\Psi,\rho] \end{aligned} \tag{6.138}$$

Now, while considering the relationships between the superfluid, condensate and fluctuating densities in the *main field approximation of the first order*, one has, from (6.127) for instance,

$$\rho(\mathrm{r}) = \left|\Psi(\mathrm{r})\right|^2 + \tilde{\rho}(\mathrm{r}) \tag{6.139}$$

so that recognizing the vanishing of the first order for the fluctuation/thermal field

$$\tilde{\psi}(\mathrm{r}) \to 0 \tag{6.140}$$

With the condition (6.139) the first three integrand terms of (6.138) become

$$\begin{aligned} &\left|\Psi(\mathrm{r})\right|^2 \left|\Psi(\mathrm{r'})\right|^2 + 2\left|\Psi(\mathrm{r})\right|^2 \tilde{\rho}(\mathrm{r'}) + \tilde{\rho}(\mathrm{r})\tilde{\rho}(\mathrm{r'}) \\ &= \left|\Psi(\mathrm{r})\right|^2 \left|\Psi(\mathrm{r'})\right|^2 + 2\left|\Psi(\mathrm{r})\right|^2 \left(\rho(\mathrm{r'}) - \left|\Psi(\mathrm{r'})\right|^2\right) + \left(\rho(\mathrm{r}) - \left|\Psi(\mathrm{r})\right|^2\right)\left(\rho(\mathrm{r'}) - \left|\Psi(\mathrm{r'})\right|^2\right) \\ &= \rho(\mathrm{r})\rho(\mathrm{r'}) \end{aligned} \tag{6.141}$$

and therefore providing the Hartree integral (6.118) that back in (6.138) leads with the final result for exchange functional (Vetter, 1997)

$$E_X[\Psi,\rho] = \frac{1}{2}\int d\mathrm{r}d\mathrm{r'}W(\mathrm{r},\mathrm{r'})\left\{-2\Psi(\mathrm{r})\Psi^+(\mathrm{r'})G(\mathrm{r}\tau;\mathrm{r'}\tau) + \left|G(\mathrm{r}\tau;\mathrm{r'}\tau)\right|^2\right\} \tag{6.142}$$

Note that if one like to visualize (6.142) in Feynman-like diagrams it worth being firstly rewritten as the Green function series expansion at the equal time

$$E_X[\Psi,\rho] = \int d\mathrm{r}d\mathrm{r'}W(\mathrm{r},\mathrm{r'})\left\{\Psi(\mathrm{r})\Psi^+(\mathrm{r'})g(\mathrm{r}\tau;\mathrm{r'}\tau) + \frac{1}{2}\left|g(\mathrm{r}\tau;\mathrm{r'}\tau)\right|^2\right\} \tag{6.143}$$

where we considered the Green-function re-notation

$$G(\mathrm{r}\tau;\mathrm{r}'\tau) = -g(\mathrm{r}\tau;\mathrm{r}'\tau) \tag{6.144}$$

so that having the final Feynman diagrammatical result:

$$E_X[\Psi,\rho] =$$

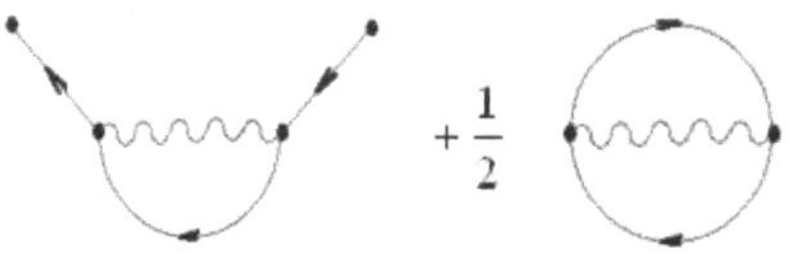

(6.145)

One may found this result instructive when like to go to the higher orders based on systematical Feynman diagrams' expansion. For instance, in the second order expansion one would expect also correlation contribution to be turning out, although one could claim that they are of no use for bosonic systems, unless it is very inhomogeneous.

6.5 FROM Ψ-KOHN–SHAM TO GROSS–PITAEVSKY EQUATION

Here one likes to check whether the Y-DFT in general and the KS equation in special provides contact with the specific BEC equation, in the mean field approximation, as the Gross–Pitaevsky is. To this end one may start from noticing the difference between *perturbation* and *interaction* when we speak about the two DFT systems:

- the natural or HK system with external $V_{ext}(\mathbf{r})$ and interacting $W(\mathbf{r},\mathbf{r}')$ potentials, the last one being included in the HK-functional F_{HK}; the specific auxiliary order parameter related potential D_{HK} is just considered formally since at the end it has to be set to zero due to the fact its coupling currents are vanishing ($J_{HK}\to 0$) for the macroscopic point of view (the equation of motion) it has the grand potential, see (6.103), here rewritten in more convenient form with unifying the HK notations for the external, that is "*ext* = *HK*", as

$$\Omega_{HK}[\Psi,\rho] = F_{HK}[\Psi,\rho] + \int d\mathrm{r}\left[V_{HK}(\mathrm{r}) - \mu\right]\rho(\mathrm{r}) + D_{HK}[\Psi,\rho] \tag{6.146}$$

  and
- the non-interacting or KS (auxiliary) construction with artificial KS potential V_{KS} that has the effective potential role, which includes also the external or HK potential already, but having no interaction potential anymore, yet with the same density (and order parameter) as that characterizing the HK system, while here, being about of a non-interacting state, specifically for order parameter existence the associate potential D_{HK} does not vanish anymore, while being characterized by the currents J_{KS} with an important role in KS equation itself, as it will be soon evident; for convenience, also for KS one will unify all notation in the grand canonical potential that reads as

$$\Omega_{KS}[\Psi,\rho] = F_{KS}[\Psi,\rho] + \int d\mathrm{r}\left[V_{KS}(\mathrm{r}) - \mu\right]\rho(\mathrm{r}) + D_{KS}[\Psi,\rho] \tag{6.147}$$

In particular, the grand canonical potential (6.147) help in producing the Y-KS equation throughout applying the first equation of the stationary system

$$\begin{cases} \dfrac{\delta}{\delta\Psi^{+}}\Omega_{KS}=0 \\ \dfrac{\delta}{\delta\rho}\Omega_{KS}=0 \end{cases} \tag{6.148}$$

This means that one should rewrite first equation (6.147) in terms of the order parameter with the help of the Bogoliubov transformation (6.127), while taking into account for the vanishing rule for the integrals containing odd powers of the fluctuating (or thermal) field. Therefore in the zero-temperature limit $T\rightarrow 0;\beta\rightarrow\infty$ the grand canonical potential (6.147) looks like this:

$$\begin{aligned} \Omega_{KS}[\Psi,\tilde{\psi}] &= \int dr\Psi^{+}(\mathrm{r})\hat{T}\Psi(\mathrm{r})+\int dr\tilde{\psi}^{+}(\mathrm{r})\hat{T}\tilde{\psi}(\mathrm{r}) \\ &+\int dr\Psi^{+}(\mathrm{r})\left[V_{KS}(\mathrm{r})-\mu\right]\Psi(\mathrm{r})+\int dr\tilde{\psi}^{+}(\mathrm{r})\left[V_{KS}(\mathrm{r})-\mu\right]\tilde{\psi}(\mathrm{r}) \\ &+\int dr\left[\Psi^{+}(\mathrm{r})J_{KS}(\mathrm{r})+J_{KS}^{*}(\mathrm{r})\Psi(\mathrm{r})\right] \end{aligned} \tag{6.149}$$

and its functional derivative respecting the order parameter, the first equation of (6.148), brings the Y-KS equation under the form

$$\left(-\frac{\hbar^{2}\nabla^{2}}{2m}+V_{KS}(\mathrm{r})-\mu\right)\Psi(\mathrm{r})=-J_{KS}(\mathrm{r}) \tag{6.150}$$

To be next employed for the real system characterized by the HK grand-potential (6.146). However, before proceeding further one should note the second equation of the system (6.148) will produce the Schrödinger equation associated with the thermal part ($\tilde{\psi}$) of the (superfluid) system, as was earlier proofed, see equation (6.59). Instead new physics information is expected from the order parameter equation that is the Y-KS equation (6.150).

Returning to the natural or HK system, its great potential (6.146) has to fulfill the same optimization conditions as the KS one, namely:

$$\begin{aligned} \rho(\mathrm{r}) &= \rho_{KS}(\mathrm{r})=\left\langle\psi^{+}(\mathrm{r})\psi(\mathrm{r})\right\rangle_{T}^{KS} \\ &=\left\langle\left[\Psi^{+}(\mathrm{r})+\tilde{\psi}^{+}(\mathrm{r})\right]\left[\Psi(\mathrm{r})+\tilde{\psi}(\mathrm{r})\right]\right\rangle_{T}^{KS} \\ &=\left\langle\Psi^{+}(\mathrm{r})\Psi(\mathrm{r})\right\rangle_{T}^{KS}+\left\langle\Psi^{+}(\mathrm{r})\tilde{\psi}(\mathrm{r})\right\rangle_{T}^{KS}+\left\langle\tilde{\psi}^{+}(\mathrm{r})\Psi(\mathrm{r})\right\rangle_{T}^{KS}+\left\langle\tilde{\psi}^{+}(\mathrm{r})\tilde{\psi}(\mathrm{r})\right\rangle_{T}^{KS} \\ &=\left|\Psi(\mathrm{r})\right|^{2}+\Psi^{+}(\mathrm{r})\underbrace{\left\langle\tilde{\psi}(\mathrm{r})\right\rangle_{T}^{KS}}_{0}+\underbrace{\left\langle\tilde{\psi}^{+}(\mathrm{r})\right\rangle_{T}^{KS}}_{0}\Psi(\mathrm{r})+\left\langle\tilde{\psi}^{+}(\mathrm{r})\tilde{\psi}(\mathrm{r})\right\rangle_{T}^{KS} \\ &=\left|\Psi(\mathrm{r})\right|^{2}+\left\langle\tilde{\psi}^{+}(\mathrm{r})\tilde{\psi}(\mathrm{r})\right\rangle_{T}^{KS} \\ &=\left|\Psi(\mathrm{r})\right|^{2}+\sum_{i,j}\varphi_{i}^{*}(\mathrm{r})\varphi_{j}(\mathrm{r})\left\langle\hat{a}_{i}^{+}\hat{a}_{j}\right\rangle_{T}^{KS} \end{aligned} \tag{6.151}$$

Now, in order to make use of the KS variational equations (6.148) one will employ (6.146) to the form

$$\begin{aligned}\Omega_{HK}[\Psi,\rho]=\Omega_{KS}[\Psi,\rho]+\int f_{\text{int}}(\Psi,\rho)d\text{r}\\-\int d\text{r}\left[V_{KS}(\text{r})-V_{HK}(\text{r})\right]\rho(\text{r})-\int d\text{r}\left[\Psi^{+}(\text{r})J_{KS}(\text{r})+J_{KS}^{*}(\text{r})\Psi(\text{r})\right]\end{aligned} \tag{6.152}$$

where the LDA for the interaction term was introduce, namely

$$F_{\text{int}}[\Psi,\rho]=\int f_{\text{int}}(\Psi,\rho)d\text{r} \tag{6.153}$$

while remembering that here the inter-particle energy W stays for the identified interaction; the way to put its double coordinate dependency into LDA framework will be in an instance revealed. For the moment, let's apply the equations (6.151–6.152) while counting for (6.148) as well; then we will have respectively the

$$\begin{cases}\dfrac{\delta}{\delta\Psi^{+}}f_{\text{int}}(\Psi,\rho)=J_{KS}\\[2mm]\dfrac{\delta}{\delta\rho}f_{\text{int}}(\Psi,\rho)=V_{KS}(\text{r})-V_{HK}(\text{r})\end{cases} \tag{6.154}$$

Now, one has to replace the KS current and potential from equation (6.154) to (6.150) and will get the working form for Y-KS

$$\left(-\frac{\hbar^{2}\nabla^{2}}{2m}+V_{ext}(\text{r})-\mu+\frac{\delta}{\delta\rho}f_{\text{int}}(\Psi,\rho)\right)\Psi(\text{r})=-\frac{\delta}{\delta\Psi^{+}}f_{\text{int}}(\Psi,\rho) \tag{6.155}$$

where, eventually, we replaced back the "HK = ext." notation for acquiring much preciseness. Equation (6.155) definitely stays for a modified Schrödinger equation, being specific to condensate (order parameter) or to superfluid in its mean field approximation.

The next point is to further specialize equation (6.155) to the present case of interaction. One can identify it based on previous exchange relation (6.138) the actual interaction functional looks like

$$\begin{aligned}\left[\left(\hat{1}E_{KS}\right),\psi(\text{r})\right]&=\left[\left(\hat{1}E_{KS}\right),\Psi(\text{r})+\sum_{i}\varphi_{i}(\text{r})\hat{a}_{i}\right]=\sum_{i}\varphi_{i}(\text{r})\left[\left(\hat{1}E_{KS}\right),\hat{a}_{i}\right]\\\left[\left(\hat{1}E_{KS}\right),\psi(\text{r})\right]&=-\sum_{i}E_{i}\varphi_{i}(\text{r})\hat{a}_{i}\end{aligned} \tag{6.156}$$

Eq. (6.156) becomes of the LDA type (6.153) in the case of bilocal-to-local transformation of the interaction upon the order parameter's delta function dependency

$$W(\text{r},\text{r}')\rightarrow g\delta(\text{r}-\text{r}') \tag{6.157}$$

in which case its coupling with Green function turns to be

$$\int dr'\delta(r-r')G(r\tau;r'\tau) = -\int dr'\delta(r-r')\left\langle\tilde{\psi}^{+}(r)\tilde{\psi}(r')\right\rangle_T^{KS} = -\left\langle\tilde{\rho}(r)\right\rangle_T^{KS} \quad (6.158)$$

and the interaction functional will look like

$$F_{\text{int}}[\Psi,\rho] = \frac{g}{2}\int dr\left\{\begin{array}{l}\Psi^{+2}(r)\Psi^{2}(r) + 2\left|\Psi(r)\right|^{2}\tilde{\rho}(r) + \tilde{\rho}^{2}(r)\\ +2\Psi(r)\Psi^{+}(r)\left\langle\tilde{\rho}(r)\right\rangle_T^{KS} + \left\langle\tilde{\rho}(r)\right\rangle_T^{KS}\left\langle\tilde{\rho}(r)\right\rangle_T^{KS}\end{array}\right\} \quad (6.159)$$

Next, since we like to arrive in order parameter and density dependence one will replace the fluctuating density from (6.139) to (6.159) to successively obtain

$$\begin{aligned}F_{\text{int}}[\Psi,\rho] &= \frac{g}{2}\int dr\left\{\begin{array}{l}\rho^{2}(r) + 2\Psi(r)\Psi^{+}(r)\left\langle\left[\rho(r)-\left|\Psi(r)\right|^{2}\right]\right\rangle_T^{KS}\\ +\left\langle\left[\rho(r)-\left|\Psi(r)\right|^{2}\right]\right\rangle_T^{KS}\left\langle\left[\rho(r)-\left|\Psi(r)\right|^{2}\right]\right\rangle_T^{KS}\end{array}\right\}\\ &= \frac{g}{2}\int dr\left\{\begin{array}{l}\rho(r)^{2} + 2\Psi(r)\Psi^{+}(r)\left\langle\rho(r)\right\rangle_T^{KS} - 2\Psi(r)\Psi^{+}(r)\left\langle\left|\Psi(r)\right|^{2}\right\rangle_T^{KS}\\ +\left\langle\rho(r)\right\rangle_T^{KS}\left\langle\rho(r)\right\rangle_T^{KS} - 2\left\langle\rho(r)\right\rangle_T^{KS}\left\langle\left|\Psi(r)\right|^{2}\right\rangle_T^{KS} + \left\langle\left|\Psi(r)\right|^{2}\right\rangle_T^{KS}\left\langle\left|\Psi(r)\right|^{2}\right\rangle_T^{KS}\end{array}\right\}\end{aligned} \quad (6.160)$$

At this point one will further employ the terms in equation (6.160) as following:

- the thermodynamic average over the order parameter may leave it unaffected since it characterizes the condensate order anyway; that is the case of the third term of (6.160)

$$\left\langle\left|\Psi(r)\right|^{2}\right\rangle_T^{KS} = \left|\Psi(r)\right|^{2} \quad (6.161)$$

- the product of the two thermodynamic average over the superfluid density reduces to that of the order parameter in the mean field approximation

$$\left\langle\rho(r)\right\rangle_T^{KS}\left\langle\rho(r)\right\rangle_T^{KS} \cong \left\langle\left|\Psi(r)\right|^{2}\right\rangle_T^{KS}\left\langle\left|\Psi(r)\right|^{2}\right\rangle_T^{KS} = \left[\Psi^{+}(r)\Psi(r)\right]^{2} \quad (6.162)$$

- in the same framework of main field approximation, for the last term of equation (6.160), the fourth power of the mean field is considered as giving the effect of the squared superfluid density, respectively

$$\left|\Psi(r)\right|^{4} \cong \rho^{2}(r) \quad (6.163)$$

This way, the expression (6.160) simplifies as

$$F_{\text{int}}[\Psi,\rho] = \frac{g}{2}\int dr\left\{2\rho^{2}(r) - \left[\Psi^{+}(r)\Psi(r)\right]^{2}\right\} \quad (6.164)$$

from where through comparison with (6.153) the interaction density is extracted with the working expression

$$f_{\text{int}}(\Psi,\rho)=g\left\{\rho^{2}(\mathrm{r})-\frac{1}{2}\left[\Psi^{+}(\mathrm{r})\Psi(\mathrm{r})\right]^{2}\right\} \tag{6.165}$$

Immediately, the derivatives needed in (6.155) are formed:

$$\begin{cases}\dfrac{\delta}{\delta\rho}f_{\text{int}}(\Psi,\rho)=2g\rho(\mathrm{r})\\ \dfrac{\delta}{\delta\Psi^{+}}f_{\text{int}}(\Psi,\rho)=-g\left[\Psi^{+}(\mathrm{r})\Psi(\mathrm{r})\right]\Psi(\mathrm{r})\end{cases} \tag{6.166}$$

that when plugged into Ψ-KS equation (6.155) produces the mean field equation

$$\left(-\frac{\hbar^{2}\nabla^{2}}{2m}+V_{ext}(\mathrm{r})-\mu+g\left[2\rho(\mathrm{r})-\Psi^{+}(\mathrm{r})\Psi(\mathrm{r})\right]\right)\Psi(\mathrm{r})=0 \tag{6.167}$$

Finally, one may once more apply the mean field expansion (6.139) in order to evaluate the g-coupling in (6.167):

$$2\rho(\mathrm{r})-\Psi^{+}(\mathrm{r})\Psi(\mathrm{r})=\Psi^{+}(\mathrm{r})\Psi(\mathrm{r})+2\tilde{\psi}^{+}(\mathrm{r})\tilde{\psi}(\mathrm{r})+2\left[\Psi^{+}(\mathrm{r})\tilde{\psi}(\mathrm{r})+\Psi(\mathrm{r})\tilde{\psi}^{+}(\mathrm{r})\right] \tag{6.168}$$

The last term of (6.168) vanishes through the left multiplication of (6.167) with $\Psi^{+}(\mathrm{r})$ followed by integration due to the appearance of the first order fluctuating/thermal field, so leaving the equation (6.167) with the so called modified Gross–Pitaevsky equation (or Hartree-Fock-Bogoliubov) equation

$$\left(-\frac{\hbar^{2}\nabla^{2}}{2m}+V_{ext}(\mathrm{r})-\mu+2g\tilde{\rho}(\mathrm{r})+g\Psi^{+}(\mathrm{r})\Psi(\mathrm{r})\right)\Psi(\mathrm{r})=0 \tag{6.169}$$

Eq. (6.169) formally differs from the Landau version (5.100) by the thermal density presence, $\tilde{\rho}(\mathrm{r})$, yet noting that the original Gross–Pitaevsky equation was given in general terms of the superfluid wave-function

$$\left(-\frac{\hbar^{2}\nabla^{2}}{2m}+V_{ext}(\mathrm{r})+g\left|\psi(\mathrm{r})\right|^{2}\right)\psi(\mathrm{r})=\mu\psi(\mathrm{r}) \tag{6.170}$$

so viewed as the non-linear (or generalized) Schrödinger equation for superfluid systems. Nevertheless, the main point here is that the DFT in general and its order parameter version Y-KS fully agrees with the conventional treatment of the Bose–Einstein condensate, being even more general, since recovering the fashioned Gross–Pitaevsky equation in LDA when the inter-particle potential reduced to a point interaction, that is for a system with the interaction Hamiltonian

$$\hat{H}_{\text{int}}=\frac{g}{2}\int d\mathrm{r}\psi^{+}(\mathrm{r})\psi^{+}(\mathrm{r})\psi(\mathrm{r})\psi(\mathrm{r})\,. \tag{6.171}$$

It is just the LDA that is the next topic, to see how the exchange functional in Y-KS analytically looks like for an homogeneous bosonic system.

6.6 DENSITY FUNCTIONALS OF HOMOGENEOUS SYSTEMS—FERMIONIC CASE

We are going to expose now the homogeneous case of the exchange energy functional, for both fermions and bosons; note that homogeneous system is essential to understand it since staying as the weak-interaction limit of the inhomogeneous case.

For fermions, electrons for instance, before pursuit to exchange energy deduction, one should learn the main working tools for the free electronic case, usually known as the Thomas–Fermi approximation to the kinetic energy.

The present sections detail in elementary manner the results presented in Sections 3.2 and 3.3.

6.6.1 Thomas–Fermi Approximation of Kinetic Energy

One elegant way of introducing Thomas–Fermi approximation is starting with the picture of an N-electronic gas within an infinite potential of a 3D-box cavity with L-width (and volume $V = L^3$). Then the Schrodinger solution for a single particle with mass m from the electronic gas having to such fundamental motion gives the paradigmatic eigen-wave-function and eigen-energy

$$\begin{cases} \psi_{\mathrm{k}}(\mathrm{r}) = \dfrac{1}{\sqrt{V}} e^{i\mathrm{kr}} \\ \varepsilon = \dfrac{h^2}{8mL^2}\left(n_x^2 + n_y^2 + n_z^2\right) \end{cases} \tag{6.172}$$

From the energy one can obtain the so called 3D-quantum number

$$\zeta = \sqrt{n_x^2 + n_y^2 + n_z^2} = \left(\frac{8mL^2}{h^2}\varepsilon\right)^{1/2} \tag{6.173}$$

With the help of which one can construct the sphere of the quantum states containing energies less or equal ε; from this sphere the number of distinct levels, say $\gamma(\varepsilon)$, belonging to the initial box-problem is identified as 1/8 from this sphere as the associate cube inscribes itself on just one quarter of one hemisphere:

$$\gamma(\varepsilon) = \frac{1}{8}\left(\frac{4\pi\zeta^3}{3}\right) = \frac{\pi}{6}\left(\frac{8mL^2}{h^2}\varepsilon\right)^{3/2} \tag{6.174}$$

One may notice that, phenomenological speaking, we have combined the conceptual cubic potential (trap) with the spherical representation of states, so that involving the most general symmetrical structures in nature; this guaranties on the generality of the analytical results. Now, the infinitesimal number of states belonging to the interval $(\varepsilon, \varepsilon + d\varepsilon)$ gives the successive expressions

$$\begin{aligned} g(\varepsilon)d\varepsilon &= \gamma(\varepsilon + d\varepsilon) - \gamma(\varepsilon) \\ &= \frac{\pi}{6}\left(\frac{8mL^2}{h^2}\right)^{3/2}\left[(\varepsilon + d\varepsilon)^{3/2} - \varepsilon^{3/2}\right] \\ &= \frac{\pi}{6}\left(\frac{8mL^2}{h^2}\right)^{3/2}\varepsilon^{3/2}\left[\left(1+\frac{d\varepsilon}{\varepsilon}\right)^{3/2} - 1\right] \\ &\cong \frac{\pi}{4}\left(\frac{8mL^2}{h^2}\right)^{3/2}\varepsilon^{1/2}d\varepsilon \end{aligned} \tag{6.175}$$

Going now from the single particle to the sample of N-particles the Fermi–Dirac statistics is taken into account while noticing its low-temperature limit

$$f_D(\varepsilon) = \frac{1}{1+e^{\beta(\varepsilon-\mu)}} \underset{\beta\to\infty}{\overset{T\to 0K}{\longrightarrow}} \begin{cases} 1 \ldots \varepsilon \le \varepsilon_F \\ 0 \ldots \varepsilon > \varepsilon_F \end{cases} \tag{6.176}$$

With the important mention that the so called Fermi energy corresponds to the 0K limit of the chemical potential:

$$\varepsilon_F = \lim_{T\to 0}\mu \tag{6.177}$$

So, practically, in low-temperature limit we are interested for the Fermi–Dirac statistical contribution reduces to identity; in these conditions, while considered also the spin contribution by the multiplication factor "2" in front, the total (kinetic) energy and the number of particles in the system are found, respectively, in relation with Fermi energy, with the expressions:

$$T = 2\int_{\varepsilon<\varepsilon_F} \varepsilon f_D(\varepsilon)g(\varepsilon)d\varepsilon = 2\frac{\pi}{4}\left(\frac{8mL^2}{h^2}\right)^{3/2}\int_0^{\varepsilon_F}\varepsilon^{3/2}d\varepsilon = \frac{8\pi}{5}\left(\frac{2m}{h^2}\right)^{3/2}V\varepsilon_F^{5/2} \tag{6.178}$$

$$N = 2\int_{\varepsilon<\varepsilon_F} f_D(\varepsilon)g(\varepsilon)d\varepsilon = 2\frac{\pi}{4}\left(\frac{8mL^2}{h^2}\right)^{3/2}\int_0^{\varepsilon_F}\varepsilon^{1/2}d\varepsilon = \frac{8\pi}{3}\left(\frac{2m}{h^2}\right)^{3/2}V\varepsilon_F^{3/2} \tag{6.179}$$

From the last relation one can form the electronic density

$$\rho = \frac{N}{V} = \frac{8\pi}{3}\left(\frac{2m}{h^2}\right)^{3/2}\varepsilon_F^{3/2} = \frac{8\pi}{3}\left(\frac{2m}{h^2}\right)^{3/2}\left(\frac{\hbar^2 k_F^2}{2m}\right)^{3/2} = \frac{k_F^3}{3\pi^2} \tag{6.180}$$

allowing the density expression for the Fermi-wave number

$$k_F = \left(3\pi^2\rho\right)^{1/3} \tag{6.181}$$

On the other side, by comparing the relations (6.178) and (6.179) one can learn their inter-connection successively transcribed as

$$\begin{aligned} T_{TF} &= \frac{3}{5}N\varepsilon_F = \frac{3}{5}\rho V\frac{\hbar^2 k_F^2}{2m} = \frac{3}{10}\frac{\hbar^2}{m}\rho V\left(3\pi^2\rho\right)^{2/3} \\ &= \frac{3}{10}\left(3\pi^2\right)^{2/3}\frac{\hbar^2}{m}V\rho^{5/3} = \frac{3}{10}\left(\frac{3}{8\pi}\right)^{2/3}\frac{h^2}{m}V\rho^{5/3} \end{aligned} \tag{6.182}$$

Now a fundamental transformation is performed that is aimed to correctly write the expression (6.182) that in its actual form is inconsistent since having energy (in left side) expressed as a function of density (in right side), whereas their relationship should be of the functional nature; this can be immediately rewritten when recognizing the equation (6.182) correctness at the energy density level, that is dividing it to the space volume

$$t = \frac{T_{TF}}{V} = \frac{3}{10}\left(3\pi^2\right)^{2/3}\frac{\hbar^2}{m}\rho^{5/3} \tag{6.183}$$

that allows writing the kinetic energy as a density functional within the LDA, specific to homogeneous systems, by integrating the *local* expression (6.183), see also (3.14)

$$T_{TF}^{LDA}[\rho] = \frac{\hbar^2}{m}C_F\int\rho^{5/3}(\mathrm{r})d\mathrm{r},\; C_F = \frac{3}{10}\left(3\pi^2\right)^{2/3} = 2.871 \tag{6.184}$$

Next, the LDA framework will be applied to evaluate other important energetic quantity, here very much concerned: the exchange functional.

6.6.2 Hartree–Fock–Slater Approximation of Exchange Energy and Potential

There is quite instructive presenting the exchange energy analytics for fermionic case, after having unfolded the Thomas–Fermi approximation for kinetic energy, and having previously in depth detailed the way the exchange functional is formulated within DFT.

Therefore, as in the latter case, one has to deal in fact with computing the *double summed double integral*

$$E_X = -\sum_{\mathrm{k},k<k_F}\sum_{\mathrm{k'},k'<k_F}\int d\mathrm{r}\int d\mathrm{r'}\psi_{\mathrm{k}}^*(\mathrm{r})\psi_{\mathrm{k'}}^*(\mathrm{r'})\frac{e^2}{|\mathrm{r}-\mathrm{r'}|}\psi_{\mathrm{k}}(\mathrm{r'})\psi_{\mathrm{k'}}(\mathrm{r}) \tag{6.185}$$

that encompasses the main features of the exchange phenomenon at the fermionic level:

- it bears the minus sign since the energy for exchanging the electronic spins is "consuming" from the total or kinetic energy of the system so it has to be subtracted from it;
- since the two-interacting electrons are in different positions and with different momenta then the exchanging spins is also equivalent with exchanging their places, a matter reflected in the ordering of the last two wave-function product of (6.185); all such position-momentum combinations are taking into account by respective double summation and double integration;

- the homogeneous exchange energy will be obtained when the wave-functions in (6.185) are considered with the plane-wave expressions of equation (6.172)

We start with performing the double integration first. So one has the homogeneous unfolding:

$$\begin{aligned}
&\int dr \int dr' \psi_k^*(r)\psi_{k'}^*(r') \frac{e^2}{|r-r'|} \psi_k(r')\psi_{k'}(r) \\
&= \frac{1}{V^2}\int dr \int dr' e^{-ikr} e^{-ik'r'} \frac{e^2}{|r-r'|} e^{ikr'} e^{ik'r} \\
&= \frac{1}{V}\int dr \frac{1}{V}\int dr'' e^{i(k'-k)r''} \frac{e^2}{|r''|}
\end{aligned} \tag{6.186}$$

Note that the last integral of (6.186) was arranged in such as the so called *Fourier-transformation of the Coulomb potential* can be applied, namely:

$$\begin{aligned}
&\int dr'' e^{i(k'-k)r''} \frac{e^2}{|r''|} \equiv \lim_{\alpha\to 0}\int ds \frac{e^{iqs} e^{-\alpha s}}{s} \\
&= \lim_{\alpha\to 0}\int_0^\infty ds 2\pi s^2 \int_{-1}^{+1} du e^{iqsu} \frac{e^{-\alpha s}}{s} \\
&= \lim_{\alpha\to 0}\int_0^\infty ds \frac{2\pi}{iq}\left(e^{(iq-\alpha)s} - e^{-(iq+\alpha)s}\right) \\
&= \lim_{\alpha\to 0}\frac{2\pi}{iq}\left(-\frac{1}{iq-\alpha} - \frac{1}{iq+\alpha}\right) \\
&= \lim_{\alpha\to 0}\frac{2\pi 2iq}{iq\left(q^2+\alpha^2\right)} \\
&= \frac{4\pi}{q^2} = \frac{4\pi}{|k'-k|^2}
\end{aligned} \tag{6.187}$$

With (6.187) back in (6.186) one gets for the double integration the immediate result

$$\int dr \int dr' \psi_k^*(r)\psi_{k'}^*(r') \frac{e^2}{|r-r'|} \psi_k(r')\psi_{k'}(r) = \frac{1}{V}\frac{4\pi e^2}{|k'-k|^2} \tag{6.188}$$

Next we are going to undertake the double momentum summation over (6.188); they will be nevertheless done by transforming into the *thermodynamic limit* of the summation into integral according with the simple rule, see also (5.69):

$$\sum_{\mathrm{k}}(\bullet) \overset{\substack{V\to\infty\\N\to\infty\\N/V\to\rho}}{\longrightarrow} \frac{V}{(2\pi)^3}\int d\mathrm{k} \tag{6.189}$$

So that the first inner summation of (6.185) successively yields:

$$\begin{aligned}
&\sum_{\mathrm{k}',k'<k_F}\int d\mathrm{r}\int d\mathrm{r}'\psi_{\mathrm{k}}^*(\mathrm{r})\psi_{\mathrm{k}'}^*(\mathrm{r}')\frac{e^2}{|\mathrm{r}-\mathrm{r}'|}\psi_{\mathrm{k}}(\mathrm{r}')\psi_{\mathrm{k}'}(\mathrm{r})\\
&=\sum_{\mathrm{k}',k'<k_F}\frac{1}{V}\frac{4\pi e^2}{|\mathrm{k}'\text{-}\mathrm{k}|^2}\\
&=\frac{4\pi e^2}{(2\pi)^3}\int\frac{1}{|\mathrm{k}'\text{-}\mathrm{k}|^2}d\mathrm{k}\\
&=\frac{4\pi e^2}{(2\pi)^2}\int_0^{k_F}dk'k'^2\int_{-1}^{+1}\frac{du}{k^2+k'^2-2kk'u}\\
&=\frac{e^2}{2\pi}k_F\left[2+\frac{k_F^2-k^2}{kk_F}\ln\left|\frac{k_F+k}{k_F-k}\right|\right]
\end{aligned} \tag{6.190}$$

that can be finally completed with the final outer-summation in (6.185) to have

$$\begin{aligned}
&\sum_{\mathrm{k},k<k_F}\sum_{\mathrm{k}',k'<k_F}\int d\mathrm{r}\int d\mathrm{r}'\psi_{\mathrm{k}}^*(\mathrm{r})\psi_{\mathrm{k}'}^*(\mathrm{r}')\frac{e^2}{|\mathrm{r}-\mathrm{r}'|}\psi_{\mathrm{k}}(\mathrm{r}')\psi_{\mathrm{k}'}(\mathrm{r})\\
&=\frac{e^2}{2\pi}k_F\sum_{\mathrm{k},k<k_F}\left[2+\frac{k_F^2-k^2}{kk_F}\ln\left|\frac{k_F+k}{k_F-k}\right|\right]
\end{aligned} \tag{6.191}$$

At this point one notes that the summation over "2" will produce two times the result of summation over the second term in the bracket of (6.191). For the last one can perform the summation over the momentum by applying the same recipe as performed for the first summation above and get:

$$\begin{aligned}
&\sum_{\mathrm{k},k<k_F}\frac{k_F^2-k^2}{kk_F}\ln\left|\frac{k_F+k}{k_F-k}\right|\\
&=\frac{V}{(2\pi)^3}\int_0^{k_F}dk2\pi k^2\int_{-1}^{+1}\frac{k_F^2-k^2}{kk_F}\ln\left|\frac{k_F+k}{k_F-k}\right|dua\\
&=\frac{V}{2\pi^2k_F}\underbrace{\int_{-k_F}^{k_F}dk\left(k_F^2k-k^3\right)\ln(k_F+k)}_{k_F^4/3}\\
&=\frac{Vk_F^3}{6\pi^2}
\end{aligned} \tag{6.192}$$

Considering (6.192) as the "unity" benchmark result for the summation on members of bracket of (6.191) one eventually obtains for the exchange energy (6.185) the result (counting also for its physical minus sign)

$$\begin{aligned} E_X &= -\sum_{\mathrm{k},k<k_F}\sum_{\mathrm{k}',k'<k_F}\int d\mathrm{r}\int d\mathrm{r}'\psi_{\mathrm{k}}^*(\mathrm{r})\psi_{\mathrm{k}'}^*(\mathrm{r}')\frac{e^2}{|\mathrm{r}-\mathrm{r}'|}\psi_{\mathrm{k}}(\mathrm{r}')\psi_{\mathrm{k}'}(\mathrm{r}) \\ &= \frac{e^2}{2\pi}k_F(2+1)\left(\frac{Vk_F^3}{6\pi^2}\right) = -\frac{e^2V}{4\pi^3}k_F^4 \end{aligned} \tag{6.193}$$

In the same manner like in Thomas–Fermi approximation for kinetic energy, now one can replace the Fermi-wave-number with the correspondent equilibrium density (6.181) so that to have

$$E_X = -\frac{e^2V}{4\pi^3}\left(3\pi^2\rho\right)\left(3\pi^2\rho\right)^{1/3} = -\frac{3e^2}{4\pi}V\rho\left(3\pi^2\rho\right)^{1/3} \tag{6.194}$$

Now we have the same density functional problem as earlier signalized with the density functional of kinetic energy, see equation (6.182); therefore we maintain the same LDA procedure to overcome the conceptual problem having in one part the density functional (that is a number) and in the other side we have a function. So that we firstly take the energy density

$$e_X = \frac{E_X}{V} = -\frac{3e^2}{4\pi}\left(3\pi^2\right)^{1/3}\rho^{4/3} \tag{6.195}$$

and then spatially integrating it to obtain the density functional exchange energy within LDA approximation, see also (3.25):

$$E_X^{LDA}[\rho] = -\frac{3e^2}{4\pi}\left(3\pi^2\right)^{1/3}\int d\mathrm{r}\rho(\mathrm{r})^{4/3} \tag{6.196}$$

or in atomic units

$$E_X^{LDA}[\rho] = -C_X\int d\mathrm{r}\rho(\mathrm{r})^{4/3},\ C_X = \frac{3}{4}\left(\frac{3}{\pi}\right)^{1/3} = 0.7386 \tag{6.197}$$

However, having the exchange energy as the density functional, also the exchange potential can be immediately deduced since taking the first density functional derivative that means

$$V_X(\mathrm{r}) = \frac{\delta E_X[\rho(\mathrm{r})]}{\delta\rho(\mathrm{r})} = -e^2\left(\frac{3}{\pi}\right)^{1/3}\rho(\mathrm{r})^{1/3} \tag{6.198}$$

Even more interestingly, due to the present derivation one can even more analytical express the exchange potential as an explicit function of coordinate, Fermi momentum and the total number of electrons in the system. The procedure unfolds as following:

take the exchange-wave-function expression (6.185) and performing the functional derivation respecting the product $\{\psi_{k}^{*}(r)\psi_{k'}(r)\}$ and obtain the wave-function version of the exchange potential

$$\begin{aligned}
V_X(\mathrm{r},\mathrm{k}) &= \frac{\delta E_X}{\delta\left\{\psi_{\mathrm{k}}^{*}(\mathrm{r})\psi_{\mathrm{k'}}(\mathrm{r})\right\}} \\
&= -\sum_{\mathrm{k'},k'<k_F}\int d\mathrm{r'}\psi_{\mathrm{k'}}^{*}(\mathrm{r'})\frac{e^2}{|\mathrm{r}-\mathrm{r'}|}\psi_{\mathrm{k}}(\mathrm{r'}) \\
&= -e^2\sum_{\mathrm{k'},k'<k_F}\frac{e^{i(\mathrm{k}-\mathrm{k'})\mathrm{s}}}{s} \\
&= -e^2\frac{e^{i\mathrm{ks}}}{s}\sum_{\mathrm{k'},k'<k_F}e^{-i\mathrm{k's}}
\end{aligned} \tag{6.199}$$

Then, the k'-sum of (6.199) is performed like above in the statistical limit:

$$\begin{aligned}
\sum_{\mathrm{k'},k'<k_F}e^{-i\mathrm{k's}} &= \frac{V}{(2\pi)^3}2\pi\int_0^{k_F}dk'k'^2\int_{-1}^{+1}e^{-ik'su}du \\
&= \frac{V}{2\pi^2 s}\int_0^{k_F}dk'k'\sin(k's) \\
&= \frac{V}{2\pi^2 s}\left\{-\frac{d}{ds}\int_0^{k_F}dk'\cos(k's)\right\} \\
&= -\frac{V}{2\pi^2 s}\frac{d}{ds}\frac{\sin(k_F s)}{s} \\
&= -\frac{3}{2}N\frac{1}{k_F^3}\left\{\frac{k_F\cos(k_F s)}{s^2}-\frac{\sin(k_F s)}{s^3}\right\}
\end{aligned} \tag{6.200}$$

being the last line due to the replacement of the volume with its expression in terms of Fermi-momentum and the total number of electrons, according with equation (6.181) here employed as

$$V = N\frac{3\pi^2}{k_F^3} \tag{6.201}$$

With (6.200) back in (6.199) one has the space-momentum dependence of the exchange potential with "*s*" going back to the "*r*" signature

$$V_X(\mathrm{r},\mathrm{k}) = \frac{3}{2}Ne^2\frac{e^{i\mathrm{kr}}}{r}\left\{\frac{k_F r\cos(k_F r)}{k_F^3 r^3}-\frac{\sin(k_F r)}{k_F^3 r^3}\right\} \tag{6.202}$$

that can be shortened under the already consecrated notation:

$$V_X(\mathrm{r},\mathrm{k}) = \frac{3}{2} Ne^2 \frac{e^{i\mathrm{kr}}}{r} g(k_F r) \tag{6.203}$$

with

$$g(x) = \frac{x\cos(x) - \sin(x)}{x^3} \tag{6.204}$$

In accordance with the early formulation – see equation (1.68). Nevertheless, this expression has the advantage as being single coordinate dependent, and therefore may be found useful for some practical and analytical implementation of some of the chemical reactivity related indices of chemical hardness and softness.

However, having now both kinetic and exchange energetic terms worth combining them into an expression containing universal constants and electronic parameter r_s, see also equation (3.69),

$$r_s = \frac{r_0}{a_0} \tag{6.205}$$

regarded as the ratio between the so called Wigner–Seitz radius specific to solid state representation of an electron placed in *a center* of sphere with a calibrated radius such that the resulting density to fit with that of the resulting many-body ensemble

$$r_0 = \left(\frac{3}{4\pi\rho}\right)^{1/3} \tag{6.206}$$

and the classical (first) Bohr radius having the electron at the frontier of the "atomic" sphere

$$a_0 = \frac{\hbar^2}{me^2} \tag{6.207}$$

Therefore, the parameter r_s practically measures, in Bohr radii, the "strength" between two electrons (one in a center and other on the frontier of the sample density sphere). With its help the Fermi-wave-number (and momentum) may be rewritten through combining (6.181) with (6.205) and (6.206); it gets the actual form:

$$\begin{cases} k_F = \left(3\pi^2\rho\right)^{1/3} = \left(\frac{9\pi}{4}\right)^{1/3} \frac{1}{r_0} = \left(\frac{9\pi}{4}\right)^{1/3} \frac{1}{a_0 r_s} = \frac{1.92}{a_0 r_s} \\ N = \frac{V}{3\pi^2} k_F^3 \end{cases} \tag{6.208}$$

With (6.208) and by using (6.207) one can collect the k_F-kinetic form (6.182) and k_F-exchange form (6.193), to form the Hartree–Fock (HF) total energy:

$$\begin{aligned} E_{HF} &= T_{TF} + E_X \\ &= \frac{3}{5}N\frac{\hbar^2 k_F^2}{2m} - \frac{3}{4\pi}e^2 N k_F \\ &= N\frac{\hbar^2}{2ma_0^2}\left(\frac{3}{5}\frac{(1.92)^2}{r_s^2} - \frac{3}{2\pi}\frac{1.92}{r_s}\right) \end{aligned} \tag{6.209}$$

from where the reduced HF-total energy looks like

$$\varepsilon_{HF} = \frac{E_{HF}}{N} = \frac{\hbar^2}{2ma_0^2}\left(\frac{2.21}{r_s^2} - \frac{0.917}{r_s}\right) \tag{6.210}$$

or even simpler in atomic units

$$\varepsilon_{HF} = \frac{1.11}{r_s^2} - \frac{0.458}{r_s} \tag{6.211}$$

Eq. (6.211) brings a beautiful physical interpretation; if one assumes that energy (6.211) is the exact energy of the N-fermionic system, then, "playing" with the exchange term under force or potential forms one can distinguish the for the correlation effects; they are arising from the fact that each electrons moves in the field of other (N-1) and the same happens with all electrons, so that the correlation is a subtle effects relaying on both kinetic energy (the movement of electrons) as well as on the exchange energy (the interchange of electrons). Consequently in identifying the correlation part from (6.211) one has to combine the information as coming from kinetic and exchange energies, simultaneously. This can be done in two ways, depending which picture is assumed the force or the potential one.

For instance, if one considers the so called exchange density force having the form:

$$f_X = \frac{0.458}{r_s^2} \tag{6.212}$$

then the result (6.211) may be transcribed equivalently as

$$\begin{aligned} H_{HK} &= \left(T + W + \frac{\ln \rho_{\not Z}}{\beta}\right) + (V - \mu N) + D_{ext} \\ &= \left(T + \frac{\ln \rho_{\not Z}}{\beta}\right) + \underbrace{V + V_{Hartree} + V_{XC}}_{V_{KS}} - \mu N + D_{KS} + W - V_{Hartree} - V_{XC} - D_{XC} \\ &= H_{KS} + H_1 \end{aligned} \tag{6.213}$$

from where there is apparent the total energy may be expressed in combination of exchange effects only, under an integral equation, from where appears naturally also the necessity the correlation term to come out naturally, here with the *force like-correlation expression*

$$\varepsilon^{f}_{correlation} = \frac{0.193}{r_s^2} \tag{6.214}$$

that differs somehow from the earlier LDA formulation (3.71) in both coefficient and the power order of the Wigner–Seitz radius. On the other side, if one likes to have the *potential-like correlation expression*, introduces the *exchange density potential*

$$v_X = \frac{0.55}{r_s} \tag{6.215}$$

then the potential-like variant of (6.214) will be looking like (6.216).

$$\varepsilon_{HF} = -2\nabla v_X - v_X + \varepsilon^{v}_{correlation} \tag{6.216}$$

Now, the potential-like correlation energy takes the form:

$$\varepsilon^{v}_{correlation} = -\frac{0.097}{r_s} \tag{6.217}$$

which is more on the size of earlier result, see equation (3.71).

In any case, the equation (6.211) tells us that the exchange potential, or force, or energy in general may constitute as the main ingredient to be worked up in modeling the total energy of an *N*-body interacting system, while providing close relationship with the correlation effects also.

However, one can go beyond the Hartree–Fock–Slater approximation in geenal, and beyond LDA in special following the guidance in deriving the popular density functionals, as described in Chapter 3. Nevertheless, in what regards the bosonic case, by their inner quantum nature as being not restraining to the exclusion Pauli principle are quite appropriate for being treated within the LDA and for dilute systems, or homogenous systems with week interaction, it is often the only case it matters. As a consequence in what follows only the LDA approximation will be unformded for developing Y-density functionals of BEC, and especially that of exchange energy, since, again peculiar to bosonic systems, the kinetic energy eventually vanishes at condensate regime since the zero wave-vector conditions applies, see equations (5.75–5.77) and equation (5.88).

6.7 EXCHANGE DENSITY FUNCTIONAL OF HOMOGENEOUS SYSTEMS—BOSONIC CASE

In bosonic homogeneous systems one considers the KS equations (6.59) as being specialized to constant external potential and stationary square root of the solution of the associate Schrödinger equation; note that such constant situation is also reflected in KS potential of equation (6.16)

$$V_{ext}(\mathrm{r}) = \varepsilon_0 \tag{6.218}$$

that gives for the particle thermal wave-spectrum $\varphi_k(\mathrm{r})$ the working equation

$$\left(-\frac{\hbar^2\nabla^2}{2m}+\varepsilon_0-\mu\right)\varphi_k(\mathrm{r})=E_k\varphi_k(\mathrm{r}) \tag{6.219}$$

with the immediate plane-wave solution:

$$\begin{cases}\varphi_k(\mathrm{r})=\dfrac{1}{\sqrt{V}}e^{ik\mathrm{r}}\\ E_k=\dfrac{\hbar^2k^2}{2m}+\varepsilon_0-\mu\end{cases} \tag{6.220}$$

One should nevertheless observe that since the associate statistical bosonic occupation number at finite temperature

$$f_B(E_k)=\frac{1}{e^{\beta E_k}-1} \tag{6.221}$$

has the non-negative form also in the limiting case of zero-momentum state (condensate state)

$$E_k\xrightarrow{k\to 0}\varepsilon_0-\mu \tag{6.222}$$

the chemical potential cannot overpass the external applied or the present KS potential (6.218), thus laying on the interval

$$\mu\in(-\infty,\varepsilon_0] \tag{6.223}$$

In the same condensate or zero-mode limit (6.222) one has for the condensate equation of (6.59) or (6.150) here with the form

$$\left.\left(\frac{\hbar^2k^2}{2m}+\varepsilon_0-\mu\right)\right|_{k\to 0}\Psi(\mathrm{r})=-J_{KS}(\mathrm{r}) \tag{6.224}$$

resulting in the constancy of the KS-auxiliary or reference current field: for homogeneous bosonic systems

$$J_{KS}(\mathrm{r},k=0)=(\mu-\varepsilon_0)\Psi(\mathrm{r}) \tag{6.225}$$

thus contributing to the condensation phase in a non-zero manner, yet depending on the chemical potential.

Nevertheless, for the thermal or fluctuating part of the superfluid one should sum up to the entire momentum spectra in (6.53) with the actual plane wave solution of (6.220)

$$\tilde{\rho}=\sum_k{}'\left|\varphi_k(\mathrm{r})\right|^2\frac{1}{\exp(\beta E_k)-1}==\frac{1}{V}\sum_{k\neq 0}\frac{1}{\exp(\beta E_k)-1} \tag{6.226}$$

Now, the thermodynamic limit can be applied to transform the spectrum summation into the continuum integral upon momentum, as introduced in (5.69) and (6.189), to successively become:

$$\begin{aligned}\tilde{\rho} &= \frac{1}{(2\pi)^3}\int d\mathbf{k}\frac{1}{\exp(\beta E_k)-1} \\ &= \frac{1}{(2\pi)^3}\int d\mathbf{k}\frac{1}{\exp\left(\beta\left[\frac{\hbar^2 k^2}{2m}+\varepsilon_0-\mu\right]\right)-1} \\ &= \frac{4\pi}{(2\pi)^3}\int_0^\infty dk\frac{k^2}{\exp\left(\beta\frac{\hbar^2 k^2}{2m}-\beta[\mu-\varepsilon_0]\right)-1}\end{aligned} \tag{6.227}$$

The integrant of (6.227) suggests that one may do the notations

$$\begin{cases} x = \beta\dfrac{\hbar^2 k^2}{2m} \\ \alpha = \beta(\mu-\varepsilon_0) \end{cases} \tag{6.228}$$

with the help of which the integration variable changes in (6.227) by the transformation:

$$k^2 dk = 2^{1/2}\left(\frac{m}{\beta\hbar^2}\right)^{3/2}\sqrt{x}dx \tag{6.229}$$

to provide the polylogarithm form for of the thermal density (6.227)

$$\tilde{\rho} = \left(\frac{m}{2\pi\beta\hbar^2}\right)^{3/2}\frac{1}{\Gamma\left(\frac{3}{2}\right)}\int_0^\infty dx\frac{x^{1/2}}{\exp(x-\alpha)-1} = \lambda_{dB}^{-3} g_{\frac{3}{2}}(e^\alpha) \tag{6.230}$$

so that being in full accordance with previous fundamental result of (5.5) in the case of critical regime ($e^\alpha = 1; \alpha = 0$), while recognizing the de Broglie thermal wave-length (5.11), the Gamma function

$$\Gamma\left(\frac{3}{2}\right) = \frac{1}{2}\Gamma\left(\frac{1}{2}\right) = \frac{1}{2}\sqrt{\pi} \tag{6.231}$$

and the basic polylogarithm function in integral and series representations:

$$g_s(z) = \frac{z}{\Gamma(s)}\int_0^\infty dx\frac{x^{s-1}}{\exp(x)-z} = \sum_{k=1}^\infty \frac{z^k}{k^s} \tag{6.232}$$

This way, in fact, the whole Bose–Einstein phenomenology was re-derived from the KS case of homogeneous system.

Very interesting, when the first derivative of the thermal density (6.230) is taken respecting the α parameter of (6.228) one has the direct result:

$$\frac{\partial\tilde{\rho}}{\partial\alpha}=\frac{\partial}{\partial\alpha}\left(\lambda_{dB}^{-3}g_{\frac{3}{2}}\left(e^{\alpha}\right)\right)=\lambda_{dB}^{-3}g_{\frac{1}{2}}\left(e^{\alpha}\right) \tag{6.233}$$

On the other side, through relating with the chemical potential derivative one finds the interesting connection with the local softness as was defined within DFT by equation (4.31):

$$\frac{\partial\tilde{\rho}}{\partial\alpha}=\frac{1}{\beta}\left(\frac{\partial\tilde{\rho}}{\partial\mu}\right)_{\varepsilon_0}=\frac{1}{\beta}\tilde{s}(\mathrm{r}) \tag{6.234}$$

By comparing equations (6.233) and (6.234) an important DFT-BEC reactivity information may be formulated, namely that the local chemical softness acquires uniform (i.e., global) value for a homogeneous condensate, with the form:

$$\tilde{s}(\mathrm{r})=\beta\frac{\partial\tilde{\rho}}{\partial\alpha}=\beta\lambda_{dB}^{-3}g_{\frac{1}{2}}\left(e^{\beta(\mu-\varepsilon_0)}\right) \tag{6.235}$$

That eventually leads also with the BEC global softness formulation

$$\tilde{S}=\int d\mathrm{r}\tilde{s}(\mathrm{r})=V\beta\lambda_{dB}^{-3}g_{\frac{1}{2}}\left(e^{\beta(\mu-\varepsilon_0)}\right) \tag{6.236}$$

so offering also a DFT-BEC chemical hardness formulation for homogeneous systems, according with equation (4.13):

$$\tilde{\eta}=\frac{\lambda_{dB}^{3}}{2V\beta}g_{\frac{1}{2}}^{-1}\left(e^{\beta(\mu-\varepsilon_0)}\right) \tag{6.237}$$

One definitely notes for the $\tilde{\eta}=\tilde{\eta}(\mu)$ dependency in (6.237) that is now by means of an integral formulation than in DFT-chemical reactivity derivative expression (4.53); it can provide further insight into the fermionic-bosonic mixtures in molecules and clusters and of their reactivity.

Next, following the same basic procedure the bilocal (non-diagonal) or generalized version of thermal density (6.226) may be advanced with a direct impact on the BEC formulation of the LDA exchange functional (6.143) through the negative Green function, see (6.144) and (6.132) further combined with (6.41) and (6.42) to successively become for the actual homogeneous case with the plane waves (6.220)

$$
\begin{aligned}
g^{\text{hom}}(\mathbf{r},\mathbf{r}') &= \sum_k \phi_k^*(\mathbf{r})\phi_k(\mathbf{r}')\frac{1}{\exp(\beta E_k)-1} \\
&= \frac{1}{V}\sum_k \frac{e^{i\mathbf{k}(\mathbf{r}-\mathbf{r}')}}{\exp\left(\beta\frac{\hbar^2 k^2}{2m}-\alpha\right)-1} \\
&= \frac{1}{(2\pi)^3}\int d\mathbf{k}\frac{e^{i\mathbf{k}(\mathbf{r}-\mathbf{r}')}}{\exp\left(\beta\frac{\hbar^2 k^2}{2m}-\alpha\right)-1} \\
&= g^{\text{hom}}(\mathbf{r}-\mathbf{r}') \\
&\overset{\mathbf{s}=\mathbf{r}-\mathbf{r}'}{=} \frac{2\pi}{(2\pi)^3}\int_0^\infty dk k^2\left\{\int_{-1}^{+1} e^{iks\cos\theta}d(\cos\theta)\right\}\frac{1}{\exp\left(\beta\frac{\hbar^2 k^2}{2m}-\alpha\right)-1} \\
&= \frac{1}{(2\pi)^2}\int_0^\infty dk k^2\frac{e^{iks}-e^{-iks}}{iks}\frac{1}{\exp\left(\beta\frac{\hbar^2 k^2}{2m}-\alpha\right)-1} \\
&= \frac{1}{2\pi^2}\int_0^\infty dk\frac{k}{s}\frac{\sin(ks)}{\exp\left(\beta\frac{\hbar^2 k^2}{2m}-\alpha\right)-1} \\
&= \frac{1}{(2\pi)^2}\int_0^\infty dk k^2\frac{2\sin(ks)}{ks}\frac{1}{\exp\left(\beta\frac{\hbar^2 k^2}{2m}-\alpha\right)-1} \\
&\overset{\frac{\beta\hbar^2}{2m}=\frac{\lambda_{dB}^2}{2\pi}}{=} \frac{1}{2\pi^2}\frac{2\pi}{s\lambda_{dB}^2}\int_0^\infty dy\sin\left(\sqrt{2\pi}\frac{sy}{\lambda_{dB}}\right)\frac{ydy}{\exp(y^2-\alpha)-1}
\end{aligned}
\tag{6.238}
$$

so obtaining the resumed result (Vetter, 1997):

$$
g^{\text{hom}}(s) = \frac{1}{\pi}\frac{1}{s\lambda_{dB}^2}h_\alpha\left(\frac{s}{\lambda_{dB}}\right) \tag{6.239}
$$

with the new integral-function

$$
h_\alpha(z) = \int_0^\infty dy\sin\left(\sqrt{2\pi}yz\right)\frac{ydy}{\exp(y^2-\alpha)-1} \tag{6.240}
$$

Expression (6.239) may be now utilized in formulating the LDA version of the exchange DFT-BEC functional (6.143). Likewise in LDA approach to fermionic case, see equations (6.183) and (6.195), firstly, one deals with the energy density that is equation (6.143) is divided by the available volume space, as

$$\begin{aligned} e_X^{Bose-LDA} &= \frac{E_X[\Psi,\rho]}{V} \\ &= \frac{1}{V}\int drdr' W(\mathrm{r},\mathrm{r}')\left\{|\Psi|^2 g^{\mathrm{hom}}(\mathrm{r}-\mathrm{r}') + \frac{1}{2}\left|g^{\mathrm{hom}}(\mathrm{r}-\mathrm{r}')\right|^2\right\} \\ &= \frac{1}{V}\int d\mathrm{R}ds W(s)\left\{|\Psi|^2 g^{\mathrm{hom}}(s) + \frac{1}{2}\left|g^{\mathrm{hom}}(s)\right|^2\right\} \\ &= 4\pi\int_0^\infty dss^2 W(s)\left\{|\Psi|^2 g^{\mathrm{hom}}(s) + \frac{1}{2}\left|g^{\mathrm{hom}}(s)\right|^2\right\} \end{aligned} \tag{6.241}$$

Expression (6.241) can be seen as composed by two contributions: one from the condensate's amplitude and one from the thermal or fluctuating basin of the condensate, yet both depending on the temperature and the chemical potential of the system, namely:

$$e_X^{Bose-LDA} = |\Psi|^2 e_X^{\Psi-LDA}(\alpha,\beta) + e_X^{\tilde{\psi}-LDA}(\alpha,\beta) \tag{6.242}$$

with the condensate and thermal terms, unfolded by plugging the above equation (6.239) into the corresponded terms, respectively:

$$\begin{aligned} e_X^{\Psi-LDA}(\alpha,\beta) &= 4\pi\int_0^\infty dss^2 W(s) g^{\mathrm{hom}}(s) \\ &= \frac{4}{\lambda_{dB}^2}\int_0^\infty dss W(s) h_\alpha\left(\frac{s}{\lambda_{dB}}\right) \end{aligned} \tag{6.243}$$

$$\begin{aligned} e_X^{\tilde{\psi}-LDA}(\alpha,\beta) &= 4\pi\int_0^\infty dss^2 W(s)\left|g^{\mathrm{hom}}(s)\right|^2 \\ &= \frac{4}{\pi\lambda_{dB}^4}\int_0^\infty ds W(s) h_s^2\left(\frac{s}{\lambda_{dB}}\right) \end{aligned} \tag{6.244}$$

The full LDA-DFT-BEC formulation for exchange functional then will be formed integrating the energetic density (6.242), in the same manner previously done for the fermionic case of kinetic energy (6.183) and exchange functional (6.195):

$$E_X^{Bose-LDA}[\Psi,\rho] = \int d\mathrm{r}\, e_X^{Bose-LDA}[\Psi,\rho]\Big|_{\substack{\Psi=\Psi(\mathrm{r}) \\ \rho=\rho(\mathrm{r})}} \tag{6.245}$$

However, as it was the case with the fermionic systems, see equation (6.198), for further self-consistent KS computational implementations, worth formulating also the exchange potential, viewed as the derivation of the exchange functional respecting its density variable; however, since the BEC variables $[\Psi,\rho]$ and the LDA framework one has for the exchange potential and exchange current the working expressions:

$$\begin{cases} V_X^{Bose-LDA} = \dfrac{\partial}{\partial \rho} e_X^{Bose-LDA}[\Psi, \rho] \\ J_X^{Bose-LDA} = \dfrac{\partial}{\partial \Psi^+} e_X^{Bose-LDA}[\Psi, \Psi^+, \rho] \end{cases} \tag{6.246}$$

While noting that we have the obvious derivative equivalence

$$\frac{\partial}{\partial \rho} = \frac{\partial}{\partial \tilde{\rho}} \tag{6.247}$$

the exchange potential of (6.246) successively writes:

$$\begin{aligned} V_X^{Bose-LDA} &= \frac{\partial}{\partial \tilde{\rho}} e_X^{Bose-LDA}[\Psi, \rho] \\ &= \frac{\partial e_X^{Bose-LDA}[\Psi, \rho]}{\partial \alpha} \cdot \frac{\partial \alpha}{\partial \tilde{\rho}} \\ &= \left[|\Psi|^2 \frac{\partial e_X^{\Psi-LDA}(\alpha, \beta)}{\partial \alpha} + \frac{\partial e_X^{\tilde{\psi}-LDA}(\alpha, \beta)}{\partial \alpha} \right] \left[\frac{\partial \tilde{\rho}}{\partial \alpha} \right]^{-1} \\ &= \frac{\beta}{\tilde{s}(\mathrm{r})} \left[|\Psi|^2 \frac{\partial e_X^{\Psi-LDA}(\alpha, \beta)}{\partial \alpha} + \frac{\partial e_X^{\tilde{\psi}-LDA}(\alpha, \beta)}{\partial \alpha} \right] \end{aligned} \tag{6.248}$$

where the BEC-local softness (6.235) was formally considered; this way equation (6.248) provides the chemical reactivity coupling with physical formulation of the DFT-BEC; in other words it shows that chemical reactivity may have an important role, at least for scaling as is apparent from (6.248); the reverse implications, that is of DFT-BEC in chemical reactivity and bonding will be later revealed, see Chapter 9. However, the analytical version of the expression (6.248) is obtained when equations (6.235), (6.240), (6.243), and (6.244) are implemented to yield (Vetter, 1997)

$$\begin{aligned} V_X^{Bose-LDA} &== \frac{\lambda_{dB}^3}{g_{\frac{1}{2}}\left(e^{\beta(\mu-\varepsilon_0)}\right)} \left[\begin{array}{l} \dfrac{4|\Psi|^2}{\lambda_{dB}^2} \displaystyle\int_0^\infty dssW(s) \frac{\partial}{\partial \alpha} h_\alpha \left(\frac{s}{\lambda_{dB}} \right) \\ + \dfrac{4}{\pi \lambda_{dB}^4} \displaystyle\int_0^\infty dsW(s) \frac{\partial}{\partial \alpha} h_s^2 \left(\frac{s}{\lambda_{dB}} \right) \end{array} \right] \\ &= \frac{1}{g_{\frac{1}{2}}\left(e^{\beta(\mu-\varepsilon_0)}\right)} \left[\begin{array}{l} 4|\Psi|^2 \lambda_{dB} \displaystyle\int_0^\infty dssW(s) \frac{\partial}{\partial \alpha} h_\alpha \left(\frac{s}{\lambda_{dB}} \right) \\ + \dfrac{8}{\pi \lambda_{dB}} \displaystyle\int_0^\infty dsW(s) h_\alpha \left(\frac{s}{\lambda_{dB}} \right) \frac{\partial}{\partial \alpha} h_\alpha \left(\frac{s}{\lambda_{dB}} \right) \end{array} \right] \end{aligned} \tag{6.249}$$

with the derivative of the function (6.240) looking like

$$\frac{\partial}{\partial\alpha}h_{\alpha}(z)=\int_{0}^{\infty}dy\sin\left(\sqrt{2\pi}yz\right)\frac{ydy}{\left[\exp\left(y^{2}-\alpha\right)-1\right]^{2}} \tag{6.250}$$

In the same manner the exchange current from (6.246) takes the successive forms:

$$\begin{aligned} J_{X}^{Bose-LDA} &= \frac{\partial}{\partial\Psi^{*}}e_{X}^{Bose-LDA}[\Psi,\rho] \\ &= \Psi e_{X}^{\Psi-LDA}(\alpha,\beta)+V_{X}^{Bose-LDA}\cdot[-\Psi] \\ &= \Psi\left[e_{X}^{\Psi-LDA}(\alpha,\beta)-V_{X}^{Bose-LDA}\right] \end{aligned} \tag{6.251}$$

so that having also the formal confirmation the exchange current is directly related with the Bosonic condensation via the order parameter amplitude, while modulated by the difference between the condensate exchange energy and the global exchange potential; this way we have a better picture also in the condensate formation through the breaking symmetry between the global (LDA) exchange potential and the exchange energy part flowing into the condensate. Finally, one would stress that the exchange discussion is as important as furnishing the leading contribution in the XC KS system, the actual Bose-condensate; in fact, there is immediate from (6.33) with the external vanishing current condition (6.38) that the exchange current models the KS-current of the condensate.

Finally, worth noting that further developments of the actual DFT-BEC may be obtained when even another current is considered, the so called anomalous current, producing the associate anomalous density, and practically splitting the fluctuation contribution in the thermal and athermal parts, thus accounting for non-homogeneous effects as well (Argaman and Band, 2011).

KEYWORDS

- **Bose–Einstein Condensation**
- **Density Functional Theory**
- **Exchange-correlation**
- **Gross–Pitaevsky equation**
- **Hartree–Fock–Slater approximation**
- **Hohenberg–Kohn functional**
- **Kohn–Sham functional**
- **Superfluidic density**
- **Thomas–Fermi approximation**

PART III

MODERN QUANTUM THEORIES OF CHEMICAL BONDING

7 Bondonic Picture of Chemical Bond

CONTENTS

1.1 INTRODUCTION

By employing the combined Bohmian quantum formalism with the U(1) and SU(2) gauge transformations of the non-relativistic wave-function and the relativistic spinor, within the Schrödinger and Dirac quantum pictures of electron motions, the existence of the chemical field is revealed along the associate bondon particle $\mathcal{B}$ characterized by its mass ($m_{\mathcal{B}}$), velocity ($v_{\mathcal{B}}$), charge ($e_{\mathcal{B}}$), and life-time ($t_{\mathcal{B}}$). This is quantized either in ground or excited states of the chemical bond in terms of reduced Planck constant $\hbar$, the bond energy E_{bond} and length X_{bond}, respectively. The mass-velocity-charge-time properties of bondons' particles were used in discussing various paradigmatic types of chemical bond towards assessing their covalent, multiple bonding, metallic and ionic features. Finally, its role in establishing the virtual states in Raman scattering was also established (Putz, 2010b).

7.2 HISTORICAL BACKGROUND FOR THE CHEMICAL BONDING QUANTUM PARTICLE

One of the first attempts to systematically use the electron structure as the basis of the chemical bond is due to the discoverer of the electron itself, J. J. Thomson, who published in 1921 an interesting model for describing one of the most puzzling molecules of chemistry, the benzene, by the aid of C–C portioned bonds, each with three electrons (Thomson, 1921) that were further separated into 2(s) + 1(p) lower and

---

Putz M. V. (2010). The bondons: The quantum particles of the chemical bond. *Int. J. Mol. Sci.*, **11**, 4227–4256.

higher energy electrons, respectively, in the light of Hückel s-p and of subsequent quantum theories (Doering and Detert, 1951; Hückel, 1931a, 1931b). On the other side, the electronic theory of the valence developed by Lewis (1916) and expanded by Langmuir (1919) had mainly treated the electronic behavior like a point-particle that nevertheless embodies considerable chemical information, due to the semi-classical behavior of the electrons on the valence shells of atoms and molecules. Nevertheless, the consistent quantum theory of the chemical bond was advocated and implemented by the works of Pauling (1931a, 1931b, 1931c) and Heitler and London (1927), which gave rise to the wave-function characterization of bonding through the fashioned molecular wave-functions (orbitals)–mainly coming from the superposition principle applied on the atomic wave-functions involved. The success of this approach, especially reported by spectroscopic studies, encouraged further generalization toward treating more and more complex chemical systems by the self-consistent wave-function algorithms developed by Slater (1928, 1929), Hartree (1957), Lowdin (1955a, 1955b, 1955c), Roothann (1951), Pariser and Parr (1953a, 1953b) and Pople (1953) in PPP theory, until the turn towards the density functional theory developed by Hohenberg and Kohn (1964), Kohn and Sham (1965) and by Pople group, see for instance (Pople et al., 1978; Head–Gordon et al., 1989) in the second half of the 20th century, which marked the subtle feed-back to the earlier electronic point-like view by means of the electronic density functionals and localization functions, see Chapter 3. The compromised picture of the chemical bond may be widely comprised by the emerging Bader's atoms-in-molecule theory (Bader, 1990; Bader, 1998; Bader, 1994), the fuzzy theory of Mezey's group (Mezey, 1993; Maggiora and Mezey, 1999; Szekeres et al., 2005), along with the chemical reactivity principles (Parr and Yang, 1989) as originating in the Sanderson's electronegativity (1988) and Pearson's chemical hardness (1973, 1990) concepts, and of their functionality in modern quantum chemistry (Chattaraj and Maiti, 2003; Chattaraj and Schleyer, 1994; Chattaraj et al., 1991; Parr et al., 1978; Mortier et al., 1985; Sen and Jørgenson, 1987).

Within this modern quantum chemistry picture, it seems that the Dirac dream (Dirac, 1929; Putz, 2010a) in characterizing the chemical bond (in particular) and the chemistry (in general) by means of the chemical field related with the Schrödinger wave-function (Schrödinger, 1926) or the Dirac spinor (Dirac, 1928) was somehow avoided by collapsing the undulatory quantum concepts into the (observable) electronic density.

Here is the paradoxical point: the dispersion of the wave function was replaced by the delocalization of density and the chemical bonding information is still beyond a decisive quantum clarification. Moreover, the quantum theory itself was challenged as to its reliability by the Einstein–Podolski–Rosen (-Bohr) entanglement formulation of quantum phenomena (Einstein et al., 1935; Bohr, 1935), qualitatively explained by the Bohm reformulation (Bohm, 1952a, 1952b) of the de Broglie wave packet through the combined de Broglie–Bohm wave-function (Bohm and Vigier, 1954; de Broglie and Vigier, 1953)

$$\Psi_0(t,x) = R(t,x)\exp\left(i\frac{S(t,x)}{\hbar}\right) \tag{7.1}$$

with the *R*-amplitude and *S*-phase action factors given, respectively, as

$$R(t,x) = \sqrt{\Psi_0(t,x)^2} = \rho^{1/2}(x) \tag{7.2}$$

$$S(t,x) = px - Et \tag{7.3}$$

in terms of electronic density ρ, momentum *p*, total energy *E*, and time-space (*t*, *x*) coordinates, without spin.

On the other side, although many of the relativistic effects were explored by considering them in the self-consistent equation of atomic and molecular structure computation (Lohr and Pyykkö, 1979; Pyykkö, 1978, 2000; Pyykkö and Zhao, 2003; Snijders and Pyykkö, 1980), the recent reloaded thesis of Einstein's special relativity (Einstein, 1905a, 1905b) into the algebraic formulation of chemistry (Whitney, 2008a, 2008b, 2009, 2010), widely asks for a further reformation of the chemical bonding quantum-relativistic vision (Boeyens, 2005, 2010).

In this respect, the present work advocates making these required steps toward assessing the quantum particle of the chemical bond as based on the derived chemical field released at its turn by the fundamental electronic equations of motion either within Bohmian non-relativistic (Schrödinger) or relativistic (Dirac) pictures and to explore the first consequences. If successful, the present endeavor will contribute to celebrate the dream in unifying the quantum and relativistic features of electron at the chemical level, while unveiling the particle-wave nature of the chemical bond.

7.3 BONDONIC IDENTIFICATION

The search for the bondons follows the algorithm (Putz, 2010b):

i. Considering the de Broglie–Bohm electronic wave-function/spinor Ψ_0 formulation of the associated quantum Schrödinger/Dirac equation of motion.
ii. Checking for recovering the charge current conservation law

$$\frac{\partial \rho}{\partial t} + \nabla \vec{j} = 0 \tag{7.4}$$

that assures for the circulation nature of the electronic fields under study.
iii. Recognizing the quantum potential V_{qua} and its equation, if it eventually appears.
iv. Reloading the electronic wave-function/spinor under the augmented U(1) or SU(2) group form

$$\Psi_G(t,x) = \Psi_0(t,x)\exp\left(\frac{i}{\hbar}\frac{e}{c}\aleph(t,x)\right) \tag{7.5}$$

with the standard abbreviation $e = e_0^2 / 4\pi\varepsilon_0$ in terms of the chemical field $\aleph$ considered as the inverse of the fine-structure order:

$$\aleph_0 = \frac{\hbar c}{e} \sim 137.03599976 \left[\frac{Joule \times meter}{Coulomb}\right] \tag{7.6}$$

since upper bounded, in principle, by the atomic number of the ultimate chemical stable element (Z = 137). Although apparently small enough to be neglected in the quantum range, the quantity (7.6) plays a crucial role for chemical bonding where the energies involved are around the order of 10^{-19} Joules (electron-volts)! Nevertheless, for establishing the physical significance of such chemical bonding quanta, one can proceed with the chain equivalences

$$\aleph_{\mathcal{B}} \sim \frac{\text{energy} \times \text{distance}}{\text{charge}} \sim \frac{\left(\text{charge} \times \begin{matrix}\text{potential}\\ \text{difference}\end{matrix}\right) \times \text{distance}}{\text{charge}} \sim \begin{pmatrix}\text{potential}\\ \text{difference}\end{pmatrix} \times \text{distance} \tag{7.7}$$

revealing that the chemical bonding field caries *bondons* with unit quanta $\hbar c / e$ along the distance of bonding within the potential gap of stability or by tunneling the potential barrier of encountered bonding attractors.

v. Rewriting the quantum wave-function/spinor equation with the group object Ψ_G, while separating the terms containing the real and imaginary ℵ chemical field contributions.
vi. Identifying the chemical field charge current and term within the actual group transformation context.
vii. Establishing the global/local gauge transformations that resemble the de Broglie–Bohm wave-function/spinor ansatz Ψ_0 of steps (i)–(iii).
viii. Imposing invariant conditions for Ψ_G wave function on pattern quantum equation respecting the Ψ_0 wave-function/spinor action of steps (i)–(iii).
ix. Establishing the chemical field ℵ specific equations.
x. Solving the system of chemical field ℵ equations.
xi. Assessing the stationary chemical field

$$\frac{\partial \aleph}{\partial t} \equiv \partial_t \aleph = 0 \tag{7.8}$$

that is the case in chemical bonds at equilibrium (ground state condition) to simplify the quest for the solution of chemical field ℵ.
xii. The manifested bondonic chemical field $\aleph_{bondon}$ is eventually identified along the bonding distance (or space).
xiii. Checking the eventual charge flux condition of Bader within the vanishing chemical bonding field (Bader, 1990)

$$\aleph_{\mathcal{B}} = 0 \Leftrightarrow \nabla \rho = 0 \tag{7.9}$$

xiv. Employing the Heisenberg time-energy relaxation-saturation relationship through the kinetic energy of electrons in bonding

$$v = \sqrt{\frac{2T}{m}} \sim \sqrt{\frac{2}{m}\frac{\hbar}{t}} \tag{7.10}$$

xv. Equate the bondonic chemical bond field with the chemical field quanta (7.6) to get the bondons' mass

$$\aleph_{B}\left(m_{B}\right)=\aleph_{0} \tag{7.11}$$

This algorithm will be next unfolded both for non-relativistic as well as for relativistic electronic motion to quest upon the bondonic existence, eventually emphasizing their difference in bondons' manifestations.

7.4 BONDONIC EXISTENCE

7.4.1 The Non-Relativistic Quantum Necessity

For the non-relativistic quantum motion, we will treat the above steps (i)–(iii) at once. As such, when considering the de Broglie–Bohm electronic wave-function into the Schrödinger equation

$$i\hbar\partial_t\Psi_0=-\frac{\hbar^2}{2m}\nabla^2\Psi_0+V\Psi_0 \tag{7.12}$$

it separates into the real and imaginary components as (Boeyens, 2005, 2010; Bohm, 1952a, 1952b; Putz, 2009)

$$\partial_t R^2+\nabla\left(\frac{R^2}{m}\nabla S\right)=0 \tag{7.13a}$$

$$\partial_t S-\frac{\hbar^2}{2m}\frac{1}{R}\nabla^2 R+\frac{1}{2m}(\nabla S)^2+V=0 \tag{7.13b}$$

While recognizing into the first equation (7.13a), the charge current conservation law with equation (7.2) along the identification

$$\vec{j}_S=\frac{R^2}{m}\nabla S \tag{7.14}$$

the second equation helps in detecting the quantum (or Bohm) potential

$$V_{qua}=-\frac{\hbar^2}{2m}\frac{\nabla^2 R}{R} \tag{7.15}$$

contributing to the total energy

$$E=T+V+V_{qua} \tag{7.16}$$

once the momentum-energy correspondences

$$\frac{1}{2m}(\nabla S)^2=\frac{p^2}{2m}=T \tag{7.17a}$$

$$\partial_t S=-E \tag{7.17b}$$

are engaged.

Next, when employing the associate U(1) gauge wave-function of equation (7.5) type, its partial derivative terms look like

$$\nabla\Psi_G = \left[\nabla R + \frac{i}{\hbar}R\left(\nabla S + \frac{e}{c}\nabla\aleph\right)\right]\exp\left[\frac{i}{\hbar}\left(S + \frac{e}{c}\aleph\right)\right] \tag{7.18a}$$

$$\nabla^2\Psi_G = \left\{\begin{aligned}&\nabla^2 R + 2\frac{i}{\hbar}\nabla R\left(\nabla S + \frac{e}{c}\nabla\aleph\right) + \frac{i}{\hbar}R\left(\nabla^2 S + \frac{e}{c}\nabla^2\aleph\right)\\ &-\frac{R}{\hbar^2}\left[(\nabla S)^2 + \left(\frac{e}{c}\nabla\aleph\right)^2\right] - 2\frac{e}{\hbar^2 c}R\nabla S\nabla\aleph\end{aligned}\right\}\exp\left[\frac{i}{\hbar}\left(S + \frac{e}{c}\aleph\right)\right] \tag{7.18b}$$

$$\partial_t\Psi_G = \left[\partial_t R + \frac{i}{\hbar}R\left(\partial_t S + \frac{e}{c}\partial_t\aleph\right)\right]\exp\left[\frac{i}{\hbar}\left(S + \frac{e}{c}\aleph\right)\right] \tag{7.18c}$$

Now, the Schrödinger equation (7.12) for Ψ_G in the form of (7.5) is decomposed into imaginary and real parts

$$-\partial_t R = \frac{1}{m}\left(\nabla R\cdot\nabla S + \frac{R}{2}\nabla^2 S\right) + \frac{e}{mc}\left(\nabla R\cdot\nabla\aleph + \frac{R}{2}\nabla^2\aleph\right) \tag{7.19}$$

$$-R\partial_t S - R\frac{e}{c}\partial_t\aleph = -\frac{\hbar^2}{2m}\nabla^2 R + \frac{R}{2m}\left[(\nabla S)^2 + \left(\frac{e}{c}\nabla\aleph\right)^2\right] + \frac{e}{mc}R\nabla S\cdot\nabla\aleph + VR \tag{7.20a}$$

that can be rearranged

$$-\left(\partial_t S + \frac{e}{c}\partial_t\aleph\right) = -\frac{\hbar^2}{2m}\frac{1}{R}\nabla^2 R + \frac{1}{2m}\left[(\nabla S)^2 + \left(\frac{e}{c}\nabla\aleph\right)^2\right] + \frac{e}{mc}\nabla S\cdot\nabla\aleph + V \tag{7.20b}$$

to reveal some interesting features of chemical bonding.

Firstly, through comparing the equation (7.19) with the charge conserved current equation form (7.4) from the general chemical field algorithm–the step (ii), the conserving charge current takes now the expanded expression:

$$\vec{j}_{U(1)} = \frac{R^2}{m}\left(\nabla S + \frac{e}{c}\nabla\aleph\right) = \vec{j}_S + \vec{j}_\aleph \tag{7.21}$$

suggesting that the additional current is responsible for the chemical field to be activated, namely

$$\vec{j}_\aleph = \frac{e}{mc}R^2\nabla\aleph \tag{7.22}$$

which vanishes when the *global gauge* condition is considered

$$\nabla\aleph = 0 \tag{7.23}$$

Therefore, in order that the chemical bonding is created, the *local gauge* transformation should be used that exists under the condition

$$\nabla\aleph \neq 0 \tag{7.24}$$

In this framework, the chemical field current $j_{\aleph}$ carries specific bonding particles that can be appropriately called *bondons*, closely related with electrons, in fact with those electrons involved in bonding, either as single, lone pair or delocalized, and having an oriented direction of movement, with an action depending on the chemical field itself $\aleph$.

Nevertheless, another important idea abstracted from the above results is that in the search for the chemical field $\aleph$ no global gauge condition is required. It is also worth noting that the presence of the chemical field does not change the Bohm quantum potential that is recovered untouched in (7.20b), thus preserving the entanglement character of interaction.

With these observations, it follows that in order for the de Broglie–Bohm–Schrödinger formalism to be invariant under the U(1) transformation (7.5), a couple of gauge conditions have to be fulfilled by the chemical field in equations (7.20a) and (7.20b), namely

$$\frac{e}{mc}\frac{\partial}{\partial x}\left(R^2\nabla\aleph\right)=0 \tag{7.25a}$$

$$\frac{e}{c}\partial_t\aleph+\frac{1}{2m}\left(\frac{e}{c}\nabla\aleph\right)^2+\frac{e}{mc}\nabla S\cdot\nabla\aleph=0 \tag{7.25b}$$

Next, the chemical field $\aleph$ is to be expressed through combining its spatial-temporal information contained in equations (7.25). From the first condition (7.25a) one finds that

$$\nabla\aleph=-\frac{R}{2}\frac{\nabla^2\aleph}{\nabla R\cdot\vec{\iota}}\vec{\iota} \tag{7.26}$$

where the vectorial feature of the chemical field gradient was emphasized on the direction of its associated charge current fixed by the versor $\vec{\iota}$ (i.e., by the unitary vector associate with the propagation direction, $\vec{\iota}^{\,2}=1$). We will apply such writing whenever necessary for avoiding scalar to vector ratios and preserving the physical sense of the whole construction as well. Replacing the gradient of the chemical field (7.26) into its temporal equation (7.25b) one gets the unified chemical field motion description

$$\frac{e}{8mc}\frac{R^2}{(\nabla R)^2}\left(\nabla^2\aleph\right)^2-\frac{R}{2m}\frac{\nabla S\cdot\nabla S}{\nabla R\cdot\nabla S}\left(\nabla^2\aleph\right)+\partial_t\aleph=0 \tag{7.27}$$

that can be further rewritten as

$$\frac{e}{2mc}\frac{\rho^2}{(\nabla\rho)^2}\left(\nabla^2\aleph\right)^2-\frac{\rho\vec{v}\cdot\vec{\iota}}{\nabla\rho\cdot\vec{\iota}}\left(\nabla^2\aleph\right)+\partial_t\aleph=0 \tag{7.28}$$

since calling the relations abstracted from equations (7.2) and (7.3)

$$R=\rho^{1/2};\nabla S=\vec{p}\Rightarrow\begin{cases}\nabla R=\dfrac{1}{2}\dfrac{\nabla\rho}{\rho^{1/2}};\ (\nabla R)^2=\dfrac{1}{4}\dfrac{(\nabla\rho)^2}{\rho}\\[2ex]\dfrac{\nabla S\cdot\nabla S}{\nabla R\cdot\nabla S}=\dfrac{2\rho^{1/2}\vec{p}\cdot\vec{\iota}}{\nabla\rho\cdot\vec{\iota}}\end{cases} \tag{7.29}$$

The (quadratic undulatory) chemical field equation (7.28) can be firstly solved for the Laplacian general solutions

$$\left(\nabla^2\aleph\right)_{1,2} = \frac{\frac{\rho\vec{v}\cdot\vec{\iota}}{\nabla\rho\cdot\vec{\iota}} \pm \sqrt{\frac{\rho^2 v^2}{\left(\nabla\rho\right)^2} - \frac{2e}{mc}\frac{\rho^2}{\left(\nabla\rho\right)^2}\partial_t\aleph}}{\frac{e}{mc}\frac{\rho^2}{\left(\nabla\rho\right)^2}} \tag{7.30}$$

that give special propagation equations for the chemical field since linking the spatial Laplacian with temporal evolution of the chemical field $(\partial_t\aleph)^{1/2}$; however, they may be considerably simplified when assuming the stationary chemical field condition (7.8), the step (xi) in the bondons' algorithm, providing the working equation for the stationary bondonic field

$$\nabla^2\aleph = 2\frac{mc}{e}\frac{\vec{v}\cdot\nabla\rho}{\rho} \tag{7.31}$$

Equation (7.31) may be further integrated between two bonding attractors, say X_A, X_B, to primarily give

$$\nabla\aleph = 2\frac{mc}{e}v\int_{X_A}^{X_B}\frac{\nabla\rho\cdot\vec{\iota}}{\rho}dx = \frac{mc}{e}v\left[\int_{X_A}^{X_B}\frac{\nabla\rho\cdot\vec{\iota}}{\rho}dx - \int_{X_B}^{X_A}\frac{\nabla\rho\cdot\vec{\iota}}{\rho}dx\right] \tag{7.32}$$

from where the generic bondonic chemical field is manifested with the form

$$\aleph_{\not{B}} = \frac{mc}{e}vX_{bond}\left(\int_{X_A}^{X_B}\frac{\nabla\rho\cdot\vec{\iota}}{\rho}dx\right) \tag{7.33}$$

The expression (7.33) has two important consequences. Firstly, it recovers the Bader zero flux condition for defining the *basins* of bonding (Bader, 1990) that in the present case is represented by the zero chemical boning fields, namely

$$\aleph_{\not{B}} = 0 \Leftrightarrow \nabla\rho\cdot\vec{\iota} = 0 \tag{7.34}$$

Secondly, it furnishes the bondonic (chemical field) analytical expression

$$\aleph_{\not{B}} = \frac{mc}{e}vX_{bond} \tag{7.35}$$

within the natural framework in which

$$X_B - X_A = X_{bond} \tag{7.36a}$$

$$\nabla\rho\cdot\vec{\iota} \rightarrow \frac{\rho}{X_{bond}} \tag{7.36b}$$

that is when one has

$$\int_{X_A}^{X_B} \frac{\nabla\rho \cdot \vec{\iota}}{\rho} dx = 1 \tag{7.36c}$$

The step (xiv) of the bondonic algorithm may be now immediately implemented through inserting the equation (7.10) into equation (7.35) yielding the simple chemical field form

$$\aleph_{\mathcal{B}} = \frac{c\hbar}{e}\sqrt{\frac{2m}{\hbar t}} X_{bond} \tag{7.37}$$

Finally, through applying the expression (7.11) of the bondonic algorithm–the step (xv) upon the result (7.37) with quanta (7.6) the *mass of bondons* carried by the chemical field on a given distance is obtained (Putz, 2010b)

$$m_{\mathcal{B}} = \frac{\hbar t}{2} \frac{1}{X_{bond}^2} \tag{7.38}$$

Note that the bondons' mass (7.38) directly depends on the time the chemical information "travels" from one bonding attractor to the other involved in bonding, while fast decreasing as the bonding distance increases.

This phenomenological behavior has to be in the sequel cross-checked by considering the generalized relativistic version of electronic motion by means of the Dirac equation.

Further quantitative consideration will be discussed afterwards.

7.4.2 The Relativistic Quantum Necessity

In treating the quantum relativistic electronic behavior, the consecrated starting point stays the Dirac equation for the scalar real valued potential w that can be seen as a general function of $(tc, \vec{x})$ dependency (Dirac, 1928; Putz, 2012)

$$i\hbar\partial_t \overleftrightarrow{\Psi}_0 = \left[-i\hbar c \sum_{k=1}^{3} \hat{\alpha}_k \partial_k + \hat{\beta} mc^2 + \hat{\beta} w \right] \overleftrightarrow{\Psi}_0 \tag{7.39}$$

with the spatial coordinate derivative notation $\partial_k \equiv \partial / \partial x_k$ and the special operators assuming the Dirac 4D representation

$$\hat{\alpha}_k = \begin{bmatrix} 0 & \hat{\sigma}_k \\ \hat{\sigma}_k & 0 \end{bmatrix}, \hat{\beta} = \begin{bmatrix} \hat{1} & 0 \\ 0 & -\hat{1} \end{bmatrix} \tag{7.40a}$$

in terms of bi-dimensional Pauli and unitary matrices

$$\hat{\sigma}_1 = \begin{bmatrix} 0 & 1 \\ 1 & 0 \end{bmatrix}, \hat{\sigma}_2 = \begin{bmatrix} 0 & -i \\ i & 0 \end{bmatrix}, \hat{\sigma}_3 = \begin{bmatrix} 1 & 0 \\ 0 & -1 \end{bmatrix}, \hat{1} \equiv \hat{\sigma}_0 = \begin{bmatrix} 1 & 0 \\ 0 & 1 \end{bmatrix} \tag{7.40b}$$

Written within the de Broglie–Bohm framework, the spinor solution of equation (7.39) looks like

$$\overleftrightarrow{\Psi}_0 = \frac{1}{\sqrt{2}} R(t,x) \begin{bmatrix} \varphi \\ \varphi \end{bmatrix} = \frac{1}{\sqrt{2}} R(t,x) \begin{bmatrix} \exp\left\{ \frac{i}{\hbar} [S(t,x)+s] \right\} \\ \exp\left\{ -\frac{i}{\hbar} [S(t,x)+s] \right\} \end{bmatrix}, s = \pm\frac{1}{2} \tag{7.41}$$

that from the beginning satisfies the necessary electronic density condition

$$\overleftrightarrow{\Psi}_0^* \overleftrightarrow{\Psi}_0 = R^* R = \rho \tag{7.42}$$

Going on, aiming for the separation of the Dirac equation (7.39) into its real/imaginary spinorial contributions, one firstly calculates the terms

$$\partial_t \overleftrightarrow{\Psi}_0 = \frac{1}{\sqrt{2}} \partial_t R \begin{bmatrix} \varphi \\ \varphi \end{bmatrix} + \frac{1}{\sqrt{2}} R \frac{i}{\hbar} \partial_t S \begin{bmatrix} \varphi \\ -\varphi \end{bmatrix} \tag{7.43a}$$

$$\partial_k \overleftrightarrow{\Psi}_0 = \frac{1}{\sqrt{2}} \partial_k R \begin{bmatrix} \kappa \\ \varphi \end{bmatrix} + \frac{1}{\sqrt{2}} R \frac{i}{\hbar} \partial_k S \begin{bmatrix} \varphi \\ -\varphi \end{bmatrix}$$

$$\sum_{k=1}^{3} \hat{a}_k \partial_k \overleftrightarrow{\Psi}_0 = \frac{1}{\sqrt{2}} \sum_{k=1}^{3} \partial_k R \begin{bmatrix} 0 & \hat{\sigma}_k \\ \hat{\sigma}_k & 0 \end{bmatrix} \begin{bmatrix} \varphi \\ \varphi \end{bmatrix} + \frac{1}{\sqrt{2}} R \frac{i}{\hbar} \sum_{k=1}^{3} \partial_k S \begin{bmatrix} 0 & \hat{\sigma}_k \\ \hat{\sigma}_k & 0 \end{bmatrix} \begin{bmatrix} \varphi \\ -\varphi \end{bmatrix} \tag{7.43b}$$

$$= \frac{1}{\sqrt{2}} \begin{bmatrix} \varphi \sum_k (\partial_k R) \hat{\sigma}_k \\ \varphi \sum_k (\partial_k R) \hat{\sigma}_k \end{bmatrix} + \frac{1}{\sqrt{2}} R \frac{i}{\hbar} \begin{bmatrix} -\varphi \sum_k (\partial_k S) \hat{\sigma}_k \\ \varphi \sum_k (\partial_k S) \hat{\sigma}_k \end{bmatrix} \tag{7.43c}$$

$$\hat{a} m c^2 \overleftrightarrow{\Psi}_0 = \frac{mc^2}{\sqrt{2}} R \begin{bmatrix} \hat{1} & 0 \\ 0 & -\hat{1} \end{bmatrix} \begin{bmatrix} \varphi \\ \varphi \end{bmatrix} = \frac{mc^2}{\sqrt{2}} R \begin{bmatrix} \varphi \\ -\varphi \end{bmatrix} \tag{7.43d}$$

$$\hat{a} w \overleftrightarrow{\Psi}_0 = \frac{w}{\sqrt{2}} R \begin{bmatrix} \varphi \\ -\varphi \end{bmatrix} \tag{7.43e}$$

to be then combined in (7.39) producing the actual de Broglie–Bohm–Dirac spinorial equation

$$\begin{bmatrix} i\hbar\varphi \partial_t R - R\phi \partial_t S \\ i\hbar\varphi \partial_t R + R\varphi \partial_t S \end{bmatrix} = \begin{bmatrix} -i\hbar c\varphi \sum_k (\partial_k R) \hat{\sigma}_k - Rc\varphi \sum_k (\partial_k S) \hat{\sigma}_k + (mc^2 + w) R\varphi \\ -i\hbar c\varphi \sum_k (\partial_k R) \hat{\sigma}_k + Rc\varphi \sum_k (\partial_k S) \hat{\sigma}_k - (mc^2 + w) R\varphi \end{bmatrix} \tag{7.44}$$

When equating the imaginary parts of (7.44) one yields the system

$$\begin{cases} \varphi \partial_t R + \varphi c \sum_k (\partial_k R) \hat{\sigma}_k = 0 \\ \varphi c \sum_k (\partial_k R) \hat{\sigma}_k + \varphi \partial_t R = 0 \end{cases} \tag{7.45}$$

that has non-trivial spinorial solutions only by canceling the associate determinant that is, by forming the equation

$$(\partial_t R)^2 = c^2 \left[\sum_k (\partial_k R)\hat{\sigma}_k\right]^2 \tag{7.46}$$

of which the minus sign of the squared root corresponds with the electronic conservation charge, while the positive sign is specific to the relativistic treatment of the positron motion. For proofing this, the specific relationship for the electronic charge conservation (7.4) may be unfolded by adapting it to the present Bohmian spinorial case by the chain equivalences

$$\begin{aligned}
0 &= \partial_t \rho + \nabla \vec{j} \\
&= \partial_t \left(R^2\right) + \sum_k \partial_k j_k \\
&= 2R\partial_t R + \sum_k \partial_k \left(c\vec{\Psi}_0^* \hat{\alpha}_k \vec{\Psi}_0\right) \\
&= 2R\partial_t R + \frac{c}{2}\sum_k \partial_k R^* R \begin{bmatrix} e^{-\frac{i}{\hbar}(S+s)} & e^{\frac{i}{\hbar}(S+s)} \end{bmatrix} \begin{bmatrix} 0 & \hat{\sigma}_k \\ \hat{\sigma}_k & 0 \end{bmatrix} \begin{bmatrix} e^{\frac{i}{\hbar}(S+s)} \\ e^{-\frac{i}{\hbar}(S+s)} \end{bmatrix} \\
&= 2R\partial_t R + \frac{c}{2}\sum_k \hat{\sigma}_k (\underbrace{\varphi^2}_{1} + \underbrace{\varphi^2}_{1})\partial_k R^2 \\
&= 2R\partial_t R + 2Rc\sum_k \hat{\sigma}_k (\partial_k R)
\end{aligned} \tag{7.47a}$$

The result

$$\partial_t R = -c\sum_k \hat{\sigma}_k (\partial_k R) \tag{7.47b}$$

indeed corresponds with the squaring root of (7.46) with the minus sign, certifying, therefore, the validity of the present approach that is, being in accordance with the step (ii) in bondonic algorithm of Section 4.2.

Next, let us see what information is conveyed by the real part of Bohmian decomposed spinors of Dirac equation (7.44); the system (7.48) is obtained

$$\begin{cases} \varphi\left(\partial_t S + mc^2 + w\right) - \varphi c \sum_k (\partial_k S)\hat{\sigma}_k = 0 \\ \varphi c \sum_k (\partial_k S)\hat{\sigma}_k - \left(\partial_t S + mc^2 + w\right)\varphi = 0 \end{cases} \tag{7.48}$$

that, as was previously the case with the imaginary counterpart (7.45), has no trivial spinors solutions only if the associate determinant vanishes, which gives the equation

$$c^2\left[\sum_k (\partial_k S)\hat{\sigma}_k\right]^2 = \left(\partial_t S + mc^2 + w\right)^2 \tag{7.49}$$

Now, considering the Bohmian momentum-energy (7.17) equivalences, the equation (7.49) further becomes

$$
\begin{aligned}
& c^2\left[\sum_k p_k\hat{\sigma}_k\right]^2 = \left(-E+mc^2+w\right)^2 \\
& \Leftrightarrow c^2\left(\vec{p}\cdot\hat{\vec{\sigma}}\right)^2 = \left(-E+mc^2+w\right)^2 \\
& \Leftrightarrow c^2 p^2 = \left(-E+mc^2+w\right)^2
\end{aligned}
\tag{7.50}
$$

from where, while retaining the minus sign through the square rooting (as prescribed above by the imaginary spinorial treatment in relation with charge conservation), one recovers the relativistic electronic energy-momentum conservation relationship

$$E = cp + mc^2 + w \tag{7.51}$$

thus confirming in full the reliability of the Bohmian approach over the relativistic spinors.

Moreover, the present Bohmian treatment of the relativistic motion is remarkable in that, except in the non-relativistic case, it does not produces the additional quantum (Bohm) potential (7.15)–responsible for entangled phenomena or hidden variables. This may be justified because within the Dirac treatment of the electron the entanglement phenomenology is somehow included throughout the Dirac Sea and the positron existence. Another important difference with respect to the Schrödinger picture is that the spinor equations that underlie the total charge and energy conservation do not mix the amplitude (7.2) with the phase (7.3) of the de Broglie–Bohm wave-function, whereas they govern now, in an independent manner, the flux and the energy of electronic motion. For these reasons, it seems that the relativistic Bohmian picture offers the natural environment in which the chemical field and associate bondons particles may be treated without involving additional physics.

Let us see, therefore, whether the Dirac–Bohmian framework will reveal (or not) new insight in the bondon (Schrödinger) reality. This will be done by reconsidering the working Bohmian spinor (7.41) as transformed by the internal gauge symmetry SU(2) driven by the chemical field ℵ related phase–in accordance with equation (7.5) of the step (iv) of bondonic algorithm

$$
\begin{aligned}
\overleftrightarrow{\Psi}_G(t,x) &= \overleftrightarrow{\Psi}_0(t,x)\exp\left(\frac{i}{\hbar}\frac{e}{c}\aleph(t,x)\right) \\
&= \frac{1}{\sqrt{2}}R(t,x)\begin{bmatrix}\varphi_G \\ \varphi_G\end{bmatrix} = \frac{1}{\sqrt{2}}R(t,x)\begin{bmatrix}\exp\left\{\frac{i}{\hbar}\left[S(t,x)+\frac{e}{c}\aleph(t,x)+s\right]\right\} \\ \exp\left\{-\frac{i}{\hbar}\left[S(t,x)+\frac{e}{c}\aleph(t,x)+s\right]\right\}\end{bmatrix}
\end{aligned}
\tag{7.52}
$$

Here it is immediate that expression (7.52) still preserves the electronic density formulation (7.2) as was previously the case with the gauge less field (7.41)

$$\overleftrightarrow{\Psi}_G^{*}\overleftrightarrow{\Psi}_G = R^{*}R = \rho \tag{7.53}$$

However, when employed for the Dirac equation terms, the field (7.52) modifies the previous expressions (7.43a)–(7.43c) as follows

$$\partial_t \vec{\Psi}_G = \frac{1}{\sqrt{2}}\partial_t R\begin{bmatrix}\varphi_G\\ \varphi_G\end{bmatrix} + \frac{1}{\sqrt{2}}R\frac{i}{\hbar}\left(\partial_t S + \frac{e}{c}\partial_t \aleph\right)\begin{bmatrix}\varphi_G\\ -\varphi_G\end{bmatrix} \tag{7.54a}$$

$$\partial_k \vec{\Psi}_0 = \frac{1}{\sqrt{2}}\partial_k R\begin{bmatrix}\varphi_G\\ \varphi_G\end{bmatrix} + \frac{1}{\sqrt{2}}R\frac{i}{\hbar}\left(\partial_k S + \frac{e}{c}\partial_k \aleph\right)\begin{bmatrix}\varphi_G\\ -\varphi_G\end{bmatrix} \tag{7.54b}$$

$$\sum_{k=1}^{3}\hat{a}_k \partial_k \vec{\Psi}_G = \frac{1}{\sqrt{2}}\sum\nolimits_k(\partial_k R)\hat{\sigma}_k\begin{bmatrix}\varphi_G\\ \varphi_G\end{bmatrix} + \frac{1}{\sqrt{2}}R\frac{i}{\hbar}\sum\nolimits_k\left(\partial_k S + \frac{e}{c}\partial_k \aleph\right)\hat{\sigma}_k\begin{bmatrix}-\varphi_G\\ \varphi_G\end{bmatrix} \tag{7.54c}$$

while producing the gauge spinorial equation

$$\begin{bmatrix} i\hbar\varphi_G\partial_t R - R\varphi_G\left(\partial_t S + \frac{e}{c}\partial_t\aleph\right)\\ i\hbar\varphi_G\partial_t R + R\varphi_G\left(\partial_t S + \frac{e}{c}\partial_t\aleph\right)\end{bmatrix} = \begin{bmatrix} -i\hbar c\varphi_G\sum\nolimits_k(\partial_k R)\hat{\sigma}_k - Rc\varphi_G\sum\nolimits_k\left(\partial_k S + \frac{e}{c}\partial_k\aleph\right)\hat{\sigma}_k + \left(mc^2+w\right)R\varphi_G\\ -i\hbar c\varphi_G\sum\nolimits_k(\partial_k R)\hat{\sigma}_k + Rc\varphi_G\sum\nolimits_k\left(\partial_k S + \frac{e}{c}\partial_k\aleph\right)\hat{\sigma}_k - \left(mc^2+w\right)R\varphi_G\end{bmatrix} \tag{7.55}$$

Now it is clear that since the imaginary part in (7.55) was not at all changed with respect to equation (7.44) by the chemical field presence, the total charge conservation (7.4) is naturally preserved; instead the real part is modified, respecting the case (7.44), in the presence of the chemical field (by internal gauge symmetry). Nevertheless, in order that chemical field rotation does not produce modification in the total energy conservation, it imposes that the gauge spinorial system of the chemical field must be as

$$\begin{cases}\varphi_G\partial_t\aleph - \varphi_G c\sum\nolimits_k(\partial_k\aleph)\hat{\sigma}_k = 0\\ \varphi_G c\sum\nolimits_k(\partial_k\aleph)\hat{\sigma}_k - \varphi_G\partial_t\aleph = 0\end{cases} \tag{7.56}$$

According to the already custom procedure, for the system (7.56) having no trivial gauge spinorial solution, the associated vanishing determinant is necessary, which brings to light the chemical field equation

$$c^2\left[\sum\nolimits_k(\partial_k\aleph)\hat{\sigma}_k\right]^2 = (\partial_t\aleph)^2 \tag{7.57a}$$

equivalently rewritten as

$$c^2\left[\nabla\aleph\cdot\hat{\vec{\sigma}}\right]^2 = (\partial_t\aleph)^2 \tag{7.57b}$$

that simply reduces to

$$c^2 (\nabla \aleph)^2 = (\partial_t \aleph)^2 \qquad (7.57c)$$

through considering the Pauling matrices (7.40b) unitary feature upon squaring.

At this point, one has to decide upon the sign of the square root of (7.57c); this was previously clarified to be minus for electronic and plus for positronic motions. Therefore, the electronic chemical bond is modeled by the resulting chemical field equation projected on the bonding length direction

$$\frac{\partial \aleph}{\partial X_{bond}} = -\frac{1}{c} \frac{\partial \aleph}{\partial t} \qquad (7.58)$$

The equation (7.58) is of undulatory kind with the chemical field solution having the general plane wave form

$$\aleph_{\not{B}} = \frac{\hbar c}{e} \exp\left[i\left(kX_{bond} - \omega t\right)\right] \qquad (7.59)$$

that agrees with both the general definition of the chemical field (7.6) as well as with the relativistic "traveling" of the bonding information. In fact, this is the paradox of the Dirac approach of the chemical bond: it aims to deal with electrons in bonding while they have to transmit the chemical bonding information—as waves—propagating with the light velocity between the bonding attractors. This is another argument for the need of bondons reality as a specific existence of electrons in chemical bond is compulsory so that such a paradox can be solved.

Note that within the Dirac approach, the Bader flux condition (7.9) is no more related to the chemical field, being included in the total conservation of charge; this is again natural, since in the relativistic case the chemical field is explicitly propagating with a percentage of light velocity (see the Discussion in Section 7.4 below) so that it cannot drive the (stationary) electronic frontiers of bonding.

Further on, when rewriting the chemical field of bonding (7.59) within the de Broglie and Planck consecrated corpuscular-undulatory quantifications

$$\aleph_{\not{B}}\left(t, X_{bond}\right) = \frac{\hbar c}{e} \exp\left[\frac{i}{\hbar}\left(pX_{bond} - Et\right)\right] \qquad (7.60)$$

it may be further combined with the unitary quanta form (7.6) in the equation (7.11) of the step (xv) in the bondonic algorithm to produce the phase condition

$$1 = \exp\left[\frac{i}{\hbar}\left(pX_{bond} - Et\right)\right] \qquad (7.61)$$

that implies the quantification

$$pX_{bond} - Et = 2\pi n\hbar, \, n \in \mathrm{N} \qquad (7.62)$$

By the subsequent employment of the Heisenberg time-energy saturated indeterminacy at the level of kinetic energy abstracted from the total energy (to focus on the motion of the bondonic plane waves)

$$E = \frac{\hbar}{t} \tag{7.63a}$$

$$p = mv = \sqrt{2mT} \rightarrow \sqrt{\frac{2m\hbar}{t}} \tag{7.63b}$$

the bondon equation (7.62) becomes

$$X_{bond}\sqrt{\frac{2m\hbar}{t}} = (2\pi n + 1)\hbar \tag{7.64}$$

that when solved for the bondonic mass yields the expression (Putz, 2010b)

$$m_{\text{\textit{B}}} = \frac{\hbar t}{2}\frac{1}{X_{bond}^2}(2\pi n + 1)^2 \quad , n = 0,1,2... \tag{7.65}$$

which appears to correct the previous non-relativistic expression (7.38) with the full quantification.

However, the Schrödinger bondon mass of equation (7.38) is recovered from the Dirac bondonic mass (7.65) in the ground state that is, by setting $n = 0$. Therefore, the Dirac picture assures the complete characterization of the chemical bond through revealing the bondonic existence by the internal chemical field symmetry with the quantification of mass either in ground or in excited states ($n \geq 0, n \in \mathbb{N}$).

Moreover, as always happens when dealing with the Dirac equation, the positronic bondonic mass may be immediately derived as well, for the case of the chemical bonding is considered also in the anti-particle world; it emerges from reloading the square root of the Dirac chemical field equation (7.57c) with a plus sign that will be propagated in all the subsequent considerations, for example, with the positronic incoming plane wave replacing the departed electronic one of (7.59), until delivering the positronic bondonic mass

$$\tilde{m}_{\text{\textit{B}}} = \frac{\hbar t}{2}\frac{1}{X_{bond}^2}(2\pi n - 1)^2, n = 0,1,2... \tag{7.66}$$

It nevertheless differs from the electronic bondonic mass (7.65) only in the excited spectrum, while both collapse in the non-relativistic bondonic mass (7.38) for the ground state of the chemical bond.

Remarkably, for both the electronic and positronic cases, the associated bondons in the excited states display heavier mass than those specific to the ground state, a behavior once more confirming that the bondons encompass all the bonding information that is, have the excitation energy converted in the mass-added-value in full agreement with the mass-energy relativistic custom Einstein equivalence (Einstein, 1905b).

7.5 CHEMICAL BOND CHARACTERIZATION BY BONDONS

Let us analyze the consequences of the bondon's existence, starting from its mass (7.38) formulation on the ground state of the chemical bond.

At one extreme (Putz, 2010c), when considering *atomic* parameters in bonding that is, when assuming the bonding distance of the Bohr radius size $a_0 = 0.52917 \cdot 10^{-10} [m]_{SI}$ the corresponding binding time would be given as $t \to t_0 = a_0 / v_0 = 2.41889 \cdot 10^{-17} [s]_{SI}$ while the involved bondonic mass will be half of the electronic one $m_0 / 2$, to assure fast bonding information. Of course, this is not a realistic binding situation; for that, let us check the hypothetical case in which the electronic m_0 mass is combined, within the bondonic formulation (7.38), into the bond distance $X_{bond} = \sqrt{\hbar t / 2m_0}$ resulting in it completing the binding phenomenon in the femtosecond time $t_{bonding} \sim 10^{-12} [s]_{SI}$ for the custom nanometric distance of bonding $X_{bond} \sim 10^{-9} [m]_{SI}$. Still, when both the femtosecond and nanometer time-space scale of bonding is assumed in (7.38), the bondonic mass is provided in the range of electronic mass $m_* \sim 10^{-31} [kg]_{SI}$ although not necessarily with the exact value for electron mass nor having the same value for each bonding case considered. Further insight into the time existence of the bondons will be reloaded for molecular systems below after discussing related specific properties as the bondonic velocity and charge.

For enlightenment on the last perspective, let us rewrite the bondonic mass (7.65) within the spatial-energetic frame of bonding that is, through replacing the time with the associated Heisenberg energy, $t_{bonding} \to \hbar / E_{bond}$, thus delivering another working expression for the bondonic mass

$$m_{\not{B}} = \frac{\hbar^2}{2} \frac{(2\pi n + 1)^2}{E_{bond} X_{bond}^2}, \; n = 0,1,2... \tag{7.67}$$

that is more practical than the traditional characterization of bonding types in terms of length and energy of bonding; it may further assume the numerical ground state ratio form

$$\varsigma_m = \frac{m_{\not{B}}}{m_0} = \frac{87.8603}{\left(E_{bond} [kcal \, / \, mol] \right) \left(X_{bond} [\overset{0}{A}] \right)^2} \tag{7.68}$$

when the available bonding energy and length are considered (as is the custom for chemical information) in kcal/mol and Angstrom, respectively. Note that having the bondon's mass in terms of bond energy implies the inclusion of the electronic pairing effect in the bondonic existence, without the constraint that the bonding pair may accumulate in the internuclear region (Berlin, 1951).

Moreover, since the bondonic mass general formulation (7.65) resulted within the relativistic treatment of electron, it is considering also the companion velocity of the bondonic mass that is reached in propagating the bonding information between the bonding attractors. As such, when the Einstein type relationship (Einstein, 1905c)

$$\frac{mv^2}{2} = h\upsilon \tag{7.69}$$

is employed for the relativistic bondonic velocity-mass relationship (Einstein, 1905a, 1905b)

$$m = \frac{m_{\text{B}}}{\sqrt{1 - \frac{v_{\text{B}}^2}{c^2}}} \tag{7.70}$$

and for the frequency of the associate bond wave

$$\upsilon = \frac{v_{\text{B}}}{X_{bond}} \tag{7.71}$$

it provides the quantified searched bondon to light velocity ratio

$$\frac{v_{\text{B}}}{c} = \frac{1}{\sqrt{1 + \frac{1}{64\pi^2} \frac{\hbar^2 c^2 (2\pi n + 1)^4}{E_{bond}^2 X_{bond}^2}}}, \ n = 0,1,2\ldots \tag{7.72}$$

or numerically in the bonding ground state as

$$\varsigma_v = \frac{v_{\text{B}}}{c} = \frac{100}{\sqrt{1 + \frac{3.27817 \times 10^6}{\left(E_{bond}[kcal/mol]\right)^2 \left(X_{bond}[\overset{0}{A}]\right)^2}}}[\%] \tag{7.73}$$

Next, dealing with a new matter particle, one will be interested also on its charge, respecting the benchmarking charge of an electron. To this end, one re-employs the step (xv) of bondonic algorithm, equation (7.11), in the form emphasizing the bondonic charge appearance, namely

$$\aleph_{\text{B}}(e_{\text{B}}) = \aleph_0 \tag{7.74}$$

Next, when considering for the left-hand side of (7.74), the form provided by equation (7.35), and for the right-hand side of (7.74), the fundamental hyperfine value of equation (7.6), one gets the working equation

$$c\frac{m_{\text{B}} v_{\text{B}}}{e_{\text{B}}} X_{bond} = 137.036 \left[\frac{Joule \times meter}{Coulomb}\right] \tag{7.75}$$

from where the bondonic charge appears immediately, once the associate expressions for mass and velocity are considered from equations (7.67) and (7.72), respectively, yielding the quantified form

$$e_{\text{B}} = \frac{4\pi\hbar c}{137.036} \frac{1}{\sqrt{1 + \frac{64\pi^2 E_{bond}^2 X_{bond}^2}{\hbar^2 c^2 (2\pi n + 1)^4}}}, \ n = 0,1,2\ldots \tag{7.76}$$

However, even for the ground state, and more so for the excited states, one may see that when forming the practical ratio respecting the unitary electric charge from (7.76), it actually approaches a referential value, namely

$$\varsigma_e = \frac{e_{B}}{e} = \frac{4\pi}{\sqrt{1+\frac{\left(E_{bond}[kcal/mol]\right)^2 \left(X_{bond}[\overset{0}{A}]\right)^2}{3.27817\times 10^6 (2\pi n+1)^4}}} \cong 4\pi \tag{7.77}$$

for, in principle, any common energy and length of chemical bonding. On the other side, for the bondons to have different masses and velocities (kinetic energy) as associated with specific bonding energy but an invariant (universal) charge seems a bit paradoxical. Moreover, it appears that with equation (7.77) the predicted charge of a bonding, even in small molecules such as $H_{2,}$ considerably surpasses the available charge in the system, although this may be eventually explained by the continuous matter-antimatter balance in the Dirac Sea to which the present approach belongs. However, to circumvent such problems, one may further use the result (7.77) and map it into the Poisson type charge field equation

$$e_{B} \cong 4\pi \times e \leftrightarrow \nabla^2 V \cong 4\pi \times \rho \tag{7.78}$$

from where the bondonic charge may be reshaped by appropriate dimensional scaling in terms of the bounding parameters (E_{bond} and X_{bond}) successively as

$$e_{B} \sim \frac{1}{4\pi}\left[\nabla_X^2 V\right]_{X=X_{bond}} \rightarrow \frac{1}{4}\frac{E_{bond}X_{bond}}{\aleph_0} \tag{7.79}$$

Now, equation (7.79) may be employed towards the working ratio between the bondonic and electronic charges in the ground state of bonding

$$\varsigma_e = \frac{e_{B}}{e} \sim \frac{1}{32\pi}\frac{\left(E_{bond}[kcal/mol]\right)\left(X_{bond}[\overset{0}{A}]\right)}{\sqrt{3.27817}\times 10^3} \tag{7.80}$$

with equation (7.80) the situation is reversed compared with the previous paradoxical situation, in the sense that now, for most chemical bonds (of Table 7.1, for instance), the resulted bondonic charge is small enough to be not yet observed or considered as belonging to the bonding wave spreading among the binding electrons.

Instead, aiming to explore the specific information of bonding reflected by the bondonic mass and velocity, the associated ratios of equations (7.68) and (7.73) for some typical chemical bonds (Findlay, 1955; Oelke, 1969) are computed in Table 7.1. They may be eventually accompanied by the predicted life-time of corresponding bondons, obtained from the bondonic mass and velocity working expressions (7.68) and (7.73), respectively, throughout the basic time-energy Heisenberg relationship—here restrained at the level of kinetic energy only for the bondonic particle; this way one yields the successive analytical forms

$$t_{B} = \frac{\hbar}{T_{B}} = \frac{2\hbar}{m_{B}v_{B}^2} = \frac{2\hbar}{(m_0\varsigma_m)(c\varsigma_v\cdot 10^{-2})^2} = \frac{\hbar}{m_0c^2}\frac{2\cdot 10^4}{\varsigma_m\varsigma_v^2} = \frac{0.0257618}{\varsigma_m\varsigma_v^2}\times 10^{-15}[s]_{SI} \tag{7.81}$$

and the specific values for various bonding types that are displayed in Table 7.1. Note that defining the bondonic life-time by equation (7.81) is the most adequate, since it involves the basic bondonic (particle!) information, mass and velocity; instead, when directly evaluating the bondonic life-time by only the bonding energy one deals with the working formula

$$t_{bond} = \frac{\hbar}{E_{bond}} = \frac{1.51787}{E_{bond}[kcal/mol]} \times 10^{-14}[s]_{SI} \tag{7.82}$$

that usually produces at least one order lower values than those reported in Table 7.1 upon employing the more complex equation (7.81).

This is nevertheless reasonable, because in the last case no particle information was considered, so that the equation (7.82) gives the time of the associate *wave* representation of bonding; this departs by the case when the time is computed by equation (7.81) where the information of bonding is contained within the *particle* (bondonic) mass and velocity, thus predicting longer life-times, and consequently a more susceptible timescale in allowing the bondonic observation. Therefore, as far as the chemical bonding is modeled by associate bondonic particle, the specific time of equation (7.81) rather than that of equation (7.82) should be considered.

While analyzing the values in Table 7.1, it is generally observed that as the bondonic mass is large as its velocity and the electric charge lower in their ratios, respecting the light velocity and electronic benchmark charge, respectively, however with some irregularities that allows further discrimination in the sub-bonding types. Yet, the life-time tendency records further irregularities, due to its complex and reversed bondonic mass-velocity dependency of equation (7.81), and will be given a special role in bondonic observation—see the Table 7.2 discussion below. Nevertheless, in all cases, the bondonic velocity is a considerable (non-negligible) percent of the photonic velocity, confirming therefore its combined quantum-relativistic nature. This explains why the bondonic reality appears even in the *non-relativistic* case of the Schrödinger equation when augmented with Bohmian entangled motion through the hidden quantum interaction.

TABLE 7.1 Ratios for the bondon-to-electronic mass and charge and for the bondon-to-light velocity, along the associated bondonic life-time for typical chemical bonds in terms of their basic characteristics such as the bond length and energy (Oelke, 1969; Findlay, 1955) through employing the basic formulas (7.68), (7.73), (7.80) and (7.81) for the ground states, respectively (Putz, 2010b).

Bond Type	X_{bond} **(Å)**	E_{bond} **(kcal/mol)**	$\varsigma_m = \frac{m_{\mathcal{B}}}{m_0}$	$\varsigma_v = \frac{v_{\mathcal{B}}}{c}[\%]$	$\varsigma_e = \frac{e_{\mathcal{B}}}{e}[\times 10^3]$	$t_{\mathcal{B}}[\times 10^{15}]$ **(seconds)**
H–H	0.60	104.2	2.34219	3.451	0.3435	9.236
C–C	1.54	81.2	0.45624	6.890	0.687	11.894
C–C[(a)]	1.54	170.9	0.21678	14.385	1.446	5.743
C=C	1.34	147	0.33286	10.816	1.082	6.616
C≡C	1.20	194	0.31451	12.753	1.279	5.037

TABLE 7.1 *(Continued)*

Bond Type	X_{bond} (Å)	E_{bond} (kcal/mol)	$\varsigma_m = \frac{m_{B}}{m_0}$	$\varsigma_v = \frac{v_{B}}{c}[\%]$	$\varsigma_e = \frac{e_{B}}{e}[\times 10^3]$	$t_{B}[\times 10^{15}]$ (seconds)
N≡N	1.10	225	0.32272	13.544	1.36	4.352
O=O	1.10	118.4	0.61327	7.175	0.716	8.160
F–F	1.28	37.6	1.42621	2.657	0.264	25.582
Cl–Cl	1.98	58	0.3864	6.330	0.631	16.639
I–I	2.66	36.1	0.3440	5.296	0.528	26.701
C–H	1.09	99.2	0.7455	5.961	0.594	9.724
N–H	1.02	93.4	0.9042	5.254	0.523	10.32
O–H	0.96	110.6	0.8620	5.854	0.583	8.721
C–O	1.42	82	0.5314	6.418	0.64	11.771
C=O(b)	1.21	166	0.3615	11.026	1.104	5.862
C=O(c)	1.15	191.6	0.3467	12.081	1.211	5.091
C–Cl	1.76	78	0.3636	7.560	0.754	12.394
C–Br	1.91	68	0.3542	7.155	0.714	14.208
C–I	2.10	51	0.3906	5.905	0.588	18.9131

(a) in diamond;
(b) in CH_2O;
(c) in O=C=O

Going now to particular cases of chemical bonding in Table 7.1, the hydrogen molecule maintains its special behavior through providing the bondonic mass as slightly more than double of the only two electrons contained in the whole system. This is not a paradox, but a confirmation of the fact the bondonic reality is not just the sum or partition of the available valence atomic electrons in molecular bonds, but a distinct (although related) existence that fully involves the undulatory nature of the electronic and nuclear motions in producing the chemical field. Remember the chemical field was associated either in Schrödinger as well in Dirac pictures with the internal rotations of the (Bohmian) wave function or spinors, being thus merely a phase property—thus inherently of undulatory nature. It is therefore natural that the risen bondons in bonding preserve the wave nature of the chemical field traveling the bond length distance with a significant percent of light.

Moreover, the bondonic mass value may determine the kind of chemical bond created, in this line the H_2 being the most covalent binding considered in Table 7.1 since it is most closely situated to the electronic pairing at the mass level. The excess in H_2 bond mass with respect to the two electrons in isolated H atoms comes from the nuclear motion energy converted (relativistic) and added to the two-sided electronic masses, while the heavier resulted mass of the bondon is responsible for the stabilization of the formed molecule respecting the separated atoms. The H_2 bondon seems to be also among the less circulated ones (along the bondon of the F_2 molecule) in bonding traveled information due to the low velocity and charge record—offering therefore another criterion of covalency that is, associated with better localization of the bonding space.

The same happens with the C–C bonding, which is predicted to be more *covalent* for its simple (single) bondon that moves with the *smallest velocity* ($\varsigma_v <<$) or fraction of the light velocity from all C–C types of bonding; in this case also the bondonic *highest mass* ($\varsigma_m >>$), *smallest charge* ($\varsigma_e <<$), and *highest (observed) life-time* ($t_{\mathcal{B}} >>$) criteria seem to work well. Other bonds with high covalent character, according with the bondonic velocity criterion only, are present in N≡N and the C=O bonding types and less in the O=O and C–O ones. Instead, one may establish the criteria for *multiple* (double and triple) *bonds* as having the series of current bondonic properties as: $\{\varsigma_m <, \varsigma_v >, \varsigma_e >, t_{\mathcal{B}} <\}$.

However, the diamond C–C bondon, although with the smallest recorded mass ($\varsigma_m <<$), is characterized by the highest velocity ($\varsigma_v >$) and charge ($\varsigma_e >$) in the CC series (and also among all cases of Table 7.1). This is an indication that the bond is very much delocalized, thus recognizing the solid state or *metallic* crystallized structure for this kind of bond in which the electronic pairings (the bondons) are distributed over all atomic centers in the unit cell. It is, therefore, a special case of bonding that widely informs us on the existence of conduction bands in a solid; therefore the metallic character generally associated with the bondonic series of properties $\{\varsigma_m <<, \varsigma_v >, \varsigma_e >, t_{\mathcal{B}} <\}$, thus having similar trends with the corresponding properties of multiple bonds, with the only particularity in the lower mass behavior displayed—due to the higher delocalization behavior for the associate bondons.

Very interestingly, the series of C–H, N–H, and O–H bonds behave similarly among them since displaying a shrink and medium range of mass (moderate high), velocity, charge and life-time (moderate high) variations for their bondons, $\{\varsigma_m \sim>, \varsigma_v \sim, \varsigma_e \sim, t_{\mathcal{B}} \sim>\}$; this may explain why these bonds are the most preferred ones in DNA and genomic construction of proteins, being however situated towards the *ionic character* of chemical bond by the lower bondonic velocities computed; they have also the most close bondonic mass to unity; this feature being due to the manifested polarizability and inter-molecular effects that allows the 3D proteomic and specific interactions taking place.

Instead, along the series of halogen molecules F_2, Cl_2, and I_2, only the life-time of bondons show high and somehow similar values, while from the point of view of velocity and charge realms only the last two bonding types display compatible properties, both with drastic difference for their bondonic mass respecting the F–F bond—probably due the most negative character of the fluorine atoms. Nevertheless, judging upon the higher life-time with respect to the other types of bonding, the classification may be decided in the favor of covalent behavior. At this point, one notes traces of covalent bonding nature also in the case of the rest of halogen-carbon binding (C–Cl, C–Br, and C–I in Table 7.1) from the bondonic life-time perspective, while displaying also the ionic manifestation through the velocity and charge criteria $\{\varsigma_v \sim, \varsigma_e\}$ and even a bit of metal character by the aid of small bondonic mass ($\varsigma_m <$). All these mixed features may be because of the joint existence of both inner electronic shells that participate by electronic induction in bonding as well as electronegativity difference potential.

Remarkably, the present results are in accordance with the recent signalized new binding class between the electronic pairs, somehow different from the ionic and covalent traditional ones in the sense that it is seen as a kind of resonance, as it appears in the molecular systems like F_2, O_2, N_2 (with impact in environmental chemistry) or in polar compounds like C–F (specific to ecotoxicology) or in the reactions that imply a competition between the exchange in the hydrogen or halogen (e.g., HF). The valence explanation relied on the possibility of higher orders of orbitals' existing when additional shells of atomic orbitals are involved such as <f> orbitals reaching this way the *charge-shift bonding* concept (Hiberty et al., 2006); the present bondonic treatment of chemical bonds overcomes the charge shift paradoxes by the relativistic nature of the bondon particles of bonding that have as inherent nature the time-space or the energy-space spanning towards electronic pairing stabilization between centers of bonding or atomic adducts in molecules.

However, we can also made predictions regarding the values of bonding energy and length required for a bondon to acquire either the unity of electronic charge or its mass (with the consequence in its velocity fraction from the light velocity) on the ground state, by setting equations (7.68) and (7.80) to unity, respectively. These predictions are summarized in Table 7.2.

From Table 7.2, one note is that the situation of the bondon having the same charge as the electron is quite improbable, at least for the common chemical bonds, since in such a case it will feature almost the light velocity (and almost no mass–that is, however, continuously decreasing as the bonding energy decreases and the bonding length increases). This is natural since a longer distance has to be spanned by lower binding energy yet carrying the same unit charge of electron while it is transmitted with the same relativistic velocity! Such behavior may be regarded as the present *zitterbewegung* (trembling in motion) phenomena, here at the bondonic level.

However, one records the systematic increasing of bondonic life-time towards being observable in the femtosecond regime for increasing bond length and decreasing the bonding energy–under the condition the chemical bonding itself still exists for certain $\{X_{bond}, E_{bond}\}$ combinations.

TABLE 7.2 Predicted basic values for bonding energy and length, along the associated bondonic life-time and velocity fraction from the light velocity for a system featuring unity ratios of bondonic mass and charge, respecting the electron values, through employing the basic formulas (7.81), (7.73), (7.68), and (7.80), respectively (Putz, 2010b).

X_{bond} $[\mathring{A}]$	E_{bond} $[kcal/mol]$	$t_{\mathcal{B}}[\times 10^{15}]$ **(seconds)**	$\varsigma_v = \frac{v_{\mathcal{B}}}{c}[\%]$	$\varsigma_m = \frac{m_{\mathcal{B}}}{m_0}$	$\varsigma_e = \frac{e_{\mathcal{B}}}{e}$
1	**87.86**	10.966	4.84691	1	0.4827×10^{-3}
1	**182019**	53.376	99.9951	4.82699×10^{-4}	1
10	**18201.9**	533.76	99.9951	4.82699×10^{-5}	1
100	**1820.19**	5337.56	99.9951	4.82699×10^{-6}	1

On the other side, the situation in which the bondon will weigh as much as one electron is a current one (see the Table 7.1); nevertheless, it is accompanied by quite reasonable chemical bonding length and energy information that it can carried at a low fraction of the light velocity, however with very low charge as well. Nevertheless, the unveiled bonding energy-length relationship from Table 7.2, based on equation (7.80), namely

$$E_{bond}[kcal/mol] \times X_{bond}[\overset{0}{A}] = 182019 \quad (7.83)$$

should be used in setting appropriate experimental conditions in which the bondon particle $\mathcal{B}$ may be observed as carrying the unit electronic charge yet with almost zero mass.

In this way, *the bondon is affirmed as a special particle of Nature, that when behaving like an electron in charge it is behaving like a photon in velocity and like neutrino in mass, while having an observable (at least as femtosecond) lifetime for nanosystems having chemical bonding in the range of hundred of Angstroms and thousands of kcal/mol*!

Such a peculiar nature of a bondon as the quantum particle of chemical bonding, the central theme of Chemistry, is not as surprising when noting that Chemistry seems to need both a particle view (such as offered by relativity) and a wave view (such as quantum mechanics offers), although nowadays these two physics theories are not yet fully compatible with each other, or even each fully coherent internally. Maybe the concept of "bondons" will help to improve the situation for all concerned by its further conceptual applications.

7.6 RAMAN SCATTERING BY BONDONS

Finally, just to give a conceptual glimpse of how the present bondonic approach may be employed, the scattering phenomena are considered within its Raman realization, viewed as a sort of generalized Compton scattering process that is, extracting the structural information from various systems (atoms, molecules, crystals, etc.) by modeling the inelastic interaction between an incident IR photon and a quantum system (here the bondons of chemical bonds in molecules), leaving a scattered wave with different frequency and the resulting system in its final state (Freeman, 1974). Quantitatively, one firstly considers the interaction Hamiltonian as being composed by two parts,

$$H^{(1)} = -\frac{e_{\mathcal{B}}}{m_{\mathcal{B}}}\sum_j \left[\vec{p}_{\mathcal{B}j} \cdot \vec{A}(\vec{r}_j, t)\right] \quad (7.84)$$

$$H^{(2)} = \frac{e_{\mathcal{B}}^2}{2m_{\mathcal{B}}}\sum_j \vec{A}^2(\vec{r}_j, t) \quad (7.85)$$

accounting for the linear and quadratic dependence of the light field potential vector $\vec{A}(\vec{r}_j, t)$ acting on the bondons "j", carrying the kinetic moment $p_{\mathcal{B}j} = m_{\mathcal{B}} v_{\mathcal{B}}$, charge $e_{\mathcal{B}}$ and mass $m_{\mathcal{B}}$.

Then, noting that, while considering the quantified incident $(\vec{q}_0, \upsilon_0)$ and scattered $(\vec{q}, \upsilon)$ light beams, the interactions driven by $H^{(1)}$ and $H^{(2)}$ model the changing in one- and two- occupation numbers of photonic trains, respectively. In this context, the transition probability between the initial $|B_i\rangle$ and final $|B_f\rangle$ bondonic states writes by squaring the sum of all scattering quantum probabilities that include absorption (A, with n_A number of photons) and emission (E, with n_E number of photons) of scattered light on bondons, see Figure 7.1.

Analytically, one has the *initial-to-final* total transition probability (Heitler, 1954) dependence here given as

$$\begin{aligned}
d^2\Pi_{fi} &\sim \frac{1}{\hbar}\left|\pi_{fi}\right|^2 \delta\left(E_{|B_i\rangle} + h\upsilon_0 - E_{|B_f\rangle} - h\upsilon\right)\upsilon^2 d\upsilon d\Omega \\
&= \frac{1}{\hbar}\Bigg|\langle f; n_A - 1, n_E + 1| H^{(2)} |n_A, n_E; i\rangle \\
&+ \sum_{B_v} \frac{\langle B_f; n_A - 1, n_E + 1| H^{(1)} |n_A - 1, n_E; B_v\rangle \langle B_v; n_A - 1, n_E | H^{(1)} | n_A, n_E; B_i\rangle}{E_{|B_i\rangle} - E_{|B_v\rangle} + h\upsilon_0} \\
&+ \sum_{B_v} \frac{\langle B_f; n_A - 1, n_E + 1| H^{(1)} |n_A, n_E + 1; B_v\rangle \langle B_v; n_A, n_E + 1| H^{(1)} | n_A, n_E; B_i\rangle}{E_{|B_i\rangle} - E_{|B_v\rangle} - h\upsilon}\Bigg|^2 \\
&\times \delta\left(E_{|B_i\rangle} + h\upsilon_0 - E_{|B_f\rangle} - h\upsilon\right)\upsilon^2 d\upsilon d\Omega
\end{aligned} \tag{7.86}$$

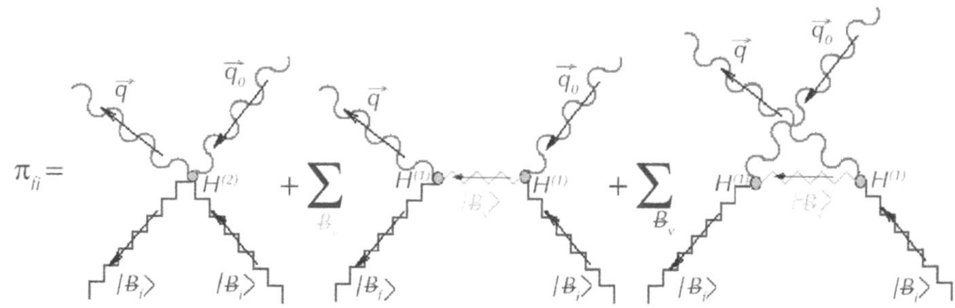

FIGURE 7.1 The Feynman diagrammatical sum of interactions entering the Raman effect by connecting the single and double photonic particles' events in absorption (incident wave light $\vec{q}_0, \upsilon_0$) and emission (scattered wave light $\vec{q}, \upsilon$) induced by the quantum first $H^{(1)}$ and second $H^{(2)}$ order interaction Hamiltonians of equations (7.84) and (7.85) through the initial $|B_i\rangle$, final $|B_f\rangle$, and virtual $|B_v\rangle$ bondonic states. The first term accounts for absorption (A)-emission (E) at once, the second term sums over the virtual states connecting the absorption followed by emission, while the third terms sums over virtual states connecting the absorption following the emission events (Putz, 2010b).

At this point, the conceptual challenge appears to explore the existence of the Raman process itself from the bondonic description of the chemical bond that turns the incoming IR photon into the (induced, stimulated, or spontaneous) structural frequencies

$$\upsilon_{v\leftarrow i} = \frac{E_{|\mathcal{B}_i\rangle} - E_{|\mathcal{B}_v\rangle}}{h} \tag{7.87}$$

As such, the problem may be reshaped in expressing the virtual state energy $E_{|\mathcal{B}_v\rangle}$ in terms of bonding energy associated with the initial state

$$E_{|\mathcal{B}_i\rangle} = E_{bond} \tag{7.88}$$

that can be eventually measured or computationally predicted by other means. However, this further implies the necessity of expressing the incident IR photon with the aid of bondonic quantification; to this end the Einstein relation (7.69) is appropriately reloaded in the form

$$h\upsilon_{v\leftarrow i} = \frac{m_{\mathcal{B}} v_{\mathcal{B}}^2}{2} = \frac{1}{4}\frac{v_{\mathcal{B}}^2 \hbar^2}{E_{bond} X_{bond}^2}(2\pi n_v + 1)^2 \tag{7.89}$$

where the bondonic mass (7.67) was firstly implemented. Next, in terms of representing the turn of the incoming IR photon into the structural wave-frequency related with the bonding energy of initial state, see equation (7.88); the time of wave-bond (7.82) is here considered to further transform equation (7.89) to the yield

$$h\upsilon_{v\leftarrow i} = \frac{1}{4}\frac{v_{\mathcal{B}}^2 E_{bond}^2 t_{bond}^2}{E_{bond} X_{bond}^2}(2\pi n_v + 1)^2 = \frac{1}{4}E_{bond}\frac{v_{\mathcal{B}}^2}{v_{bond}^2}(2\pi n_v + 1)^2 \tag{7.90}$$

where also the corresponding wave-bond velocity was introduced

$$v_{bond} = \frac{X_{bond}}{t_{bond}} = \frac{1}{\hbar}E_{bond} X_{bond} \tag{7.91}$$

It is worth noting that, as previously was the case with the dichotomy between bonding and bondonic times, sees equations (7.81) *vs*. (7.82), respectively the bonding velocity of equation (7.91) clearly differs by the bondonic velocity of equation (7.72) since the actual working expression

$$\frac{v_{bond}}{c} = \left(E_{bond}[kcal/mol]\right)\left(X_{bond}[\overset{0}{A}]\right)2.19758\times10^{-3}[\%] \tag{7.92}$$

provides considerably lower values than those listed in Table 7.1—again, due to missing the inclusion of the particle mass' information, unlike is the case for the bondonic velocity.

Returning to the bondonic description of the Raman scattering, one replaces the virtual photonic frequency of equation (7.90) together with equation (7.88) back in the Bohr-type equation (7.87) to yield the searched quantified form of virtual bondonic energies in equation (7.86) and Figure 7.1, analytically (Putz, 2010b)

$$E_{|\not{B}_v\rangle} = E_{bond} \left[1 - \frac{1}{4} \frac{v_{\not{B}}^2}{v_{bond}^2} \left(2\pi n_v + 1 \right)^2 \right]$$

$$= E_{bond} \left[1 - 16\pi^2 \frac{\left(2\pi n_v + 1 \right)^2}{64\pi^2 \dfrac{E_{bond}^2 X_{bond}^2}{\hbar^2 c^2} + \left(2\pi n_v + 1 \right)^4} \right] \tag{7.93}$$

or numerically

$$E_{|\not{B}_v\rangle} = E_{bond} \left[1 - \frac{16\pi^2 \left(2\pi n_v + 1 \right)^2}{0.305048 \times 10^{-6} \times \left(E_{bond}[kcal/mol] \right)^2 \times \left(X_{bond}[\overset{0}{A}] \right)^2 + \left(2\pi n_v + 1 \right)^4} \right], \ n_v = 0,1,2\ldots \tag{7.94}$$

Remarkably, the bondonic quantification (7.94) of the virtual states of Raman scattering varies from negative to positive energies as one moves from the ground state to more and more excited states of initial bonding state approached by the incident IR towards virtual ones, as may be easily verified by considering particular bonding data of Table 7.1. In this way, more space is given for future considerations upon the inverse or stimulated Raman processes, proving therefore the direct involvement of the bondonic reality in combined scattering of light on chemical structures. Overall, the bondonic characterization of the chemical bond is fully justified by quantum and relativistic considerations, to be advanced as a useful tool in characterizing chemical reactivity, times of reactions that is, when tunneling or entangled effects may be rationalized in an analytical manner. Note that further correction of this bondonic model may be realized when the present point-like approximation of nuclear systems is abolished and replaced by the bare-nuclear assumption in which additional dependence on the bonding distance is involved. This is left for future explorations.

7.7 TOWARDS CHEMICAL BONDING CLASSIFICATION BY BONDONS' PROPERTIES

The chemical bond, perhaps the greatest challenge in theoretical chemistry, has generated many inspiring theses over the years, although none definitive. Few of the most preeminent regard the orbitalic based explanation of electronic pairing, in valence shells of atoms and molecules, rooted in the hybridization concept (Pauling, 1931c) then extended to the valence-shell electron-pair repulsion (VSEPR) (Gillespie, 1970). Alternatively, when electronic density is considered, the atoms-in-molecule paradigms were formulated through the geometrical partition of forces by Berlin (1951), or in terms of core, bonding, and lone-pair lodges by Daudel (1983), or by the zero local flux in the gradient field of the density $\nabla\rho$ by Bader (1990), until the most recent employment of the chemical action functional in bonding, see Chapter 8.

Yet, all these approaches do not depart significantly from the undulatory nature of electronic motion in bonding, either by direct wave-function consideration or through its probability information in electronic density manifestation (for that is still considered as a condensed—observable version—of the undulatory manifestation of electron).

In other words, while passing from the Lewis point-like ansatz to the undulatory modeling of electrons in bonding, the reverse passage was still missing in an analytical formulation. Only recently the first attempt was formulated, based on the broken-symmetry approach of the Schrödinger Lagrangean with the electronegativity-chemical hardness parabolic energy dependency, showing that a systematical quest for the creation of particles from the chemical bonding fields is possible (Putz, 2008d).

Following this line, the present work makes a step forward and considers the gauge transformation of the electronic wave-function and spinor over the de Broglie–Bohm augmented non-relativistic and relativistic quantum pictures of the Schrödinger and Dirac electronic (chemical) fields, respectively. As a consequence, the reality of the chemical field in bonding was proved in either framework, while providing the corresponding bondonic particle with the associate mass and velocity in a full quantization form, see equations (7.67) and (7.72). In fact, the Dirac bondon (7.65) was found to be a natural generalization of the Schrödinger one (7.38), while supplementing it with its anti-bondon particle (7.66) for the positron existence in the Dirac Sea.

The bondon is the quantum particle corresponding to the superimposed electronic pairing effects or distribution in chemical bond; accordingly, through the values of its mass and velocity it may be possible to indicate the type of bonding (in particular) and the characterization of electronic behavior in bonding (in general).

However, one of the most important consequences of bondonic existence is that the chemical bonding may be described in a more complex manner than relaying only on the electrons, but eventually employing the fermionic (electronic)-bosonic (bondonic) mixture: the first preeminent application is currently on progress, that is, exploring the effect that the Bose–Einstein condensation has on chemical bonding modeling, see Chapter 9.

Yet, such possibility arises due to the fact that whether the Pauli principle is an independent axiom of quantum mechanics or whether it depends on other quantum description of matter is still under question (Kaplan, 2002), as is the actual case of involving hidden variables and the entanglement or non-localization phenomenology that may be eventually mapped onto the delocalization and fractional charge provided by quantum chemistry over and on atomic centers of a molecular complex/chemical bond, respectively.

As an illustration of the bondonic concept and of its properties such as the mass, velocity, charge, and life-time, the fundamental Raman scattering process was described by analytically deriving the involved virtual energy states of scattering sample (chemical bond) in terms of the bondonic properties above—proving its necessary existence and, consequently, of the associate Raman effect itself, while leaving space for further applied analysis based on spectroscopic data on hand.

On the other side, the mass, velocity, charge, and life-time properties of the bondons were employed for analyzing some typical chemical bonds (see Table 7.1), this way revealing a sort of fuzzy classification of chemical bonding types in terms of the bondonic-to-electronic mass and charge ratios ς_m and ς_e, and of the bondonic-to-light velocity percent ratio ς_v, along the bondonic observable life-time, $t_{\mathcal{B}}$ respectively–here summarized in Table 7.3.

TABLE 7.3 Phenomenological classification of the chemical bonding types by bondonic (mass, velocity, charge and life-time) properties abstracted from Table 7.1; the used symbols are: > and >> for "high" and "very high" values; < and << for "low" and "very low" values; ~ and ~> for "moderate" and "moderate high and almost equal" values in their class of bonding (Putz, 2010b).

Property Chemical bond	ς_m	ς_v	ς_e	$t_{\mathcal{B}}$
Covalence	>>	<<	<<	>>
Multiple bonds	<	>	>	<
Metallic	<<	>	>	<
Ionic	~>	~	~	~>

These rules are expected to be further refined through considering the new paradigms of special relativity in computing the bondons' velocities, especially within the modern algebraic chemistry (Whitney, 2007). Yet, since the bondonic masses of chemical bonding ground states seem untouched by the Dirac relativistic considerations over the Schrödinger picture, it is expected that their analytical values may make a difference among the various types of compounds, while their experimental detection is hoped to be some day completed.

KEYWORDS

- **Bondonic algorithm**
- **Bondonic mass**
- **Chemical Bond**
- **Non-relativistic case**
- **Wave-function**

8 Chemical Action Picture of Chemical Bond

CONTENTS

8.1 INTRODUCTION

New chemical bonding paradigm in terms of chemical action functional and of its reformulations by means of electronegativity, linear response and density softness kernels is advanced; it makes no use of traditional molecular orbital (MO) bonding analysis while providing reliability in identifying the bonding regions through appropriate specialization of the chemical action variational (conservation) principle along the bond length (Putz, 2009b, 2010a).

8.2 ON THE CHEMICAL ACTION CONCEPT

Be an N-electronic system with mono-electronic orbitals φ_i, $i=\overline{1,N}$ and the total density (1.5) (Hohenberg and Kohn, 1964). In these conditions, through the concerned system is engaging in exchanging of electrons that is display chemical reactivity, the variational energy principle (1.17), rewrites as leadings with the so called Kohn–Sham (KS) equations of the Schrödinger type (Kohn and Sham, 1965):

$$\int \frac{\delta\left(E[\rho]-\mu_i N[\rho]\right)}{\delta\varphi_i^*}\delta\varphi_i^* d\mathrm{r}=0 \tag{8.1}$$

$$\left[-\frac{1}{2}\nabla^2+V_{eff}\right]\varphi_i(\mathrm{r})=\mu_i\varphi_i(\mathrm{r}),\ i=\overline{1,N} \tag{8.2}$$

in terms of effective potential

$$V_{eff}(\mathrm{r}) = V(\mathrm{r}) + \int \frac{\rho(\mathrm{r'})}{|\mathrm{r}-\mathrm{r'}|}d\mathrm{r'} + V_{XC}(\mathrm{r}) \tag{8.3}$$

and the orbital eigen-values as the orbital chemical potential μ_i, that can, at any moment (Parr et al., 1978), be seen as the negative of the orbital electronegativities based on equation (1.19)

$$\mu_i = -\chi_i, \tag{8.4}$$

and where the exchange-correlation potential is written as the density functional derivative of the exchange-correlation energy

$$V_{XC}(\mathrm{r}) = \left(\frac{\delta E_{XC}[\rho]}{\delta\rho(\mathrm{r})}\right)_{V(\mathrm{r})} \tag{8.5}$$

Note that the exchange energy is here considered as the unknown density functional term of the total un-optimized total energy density functional:

$$E[\rho] = \sum_i^N \int \varphi_i^*(\mathrm{r})\left[-\frac{1}{2}\nabla^2\right]\varphi_i(\mathrm{r})d\mathrm{r} + \frac{1}{2}\iint \frac{\rho(\mathrm{r})\rho(\mathrm{r'})}{|\mathrm{r}-\mathrm{r'}|}d\mathrm{r}d\mathrm{r'} + E_{XC}[\rho] + \int V(\mathrm{r})\rho(\mathrm{r})d\mathrm{r} \tag{8.6}$$

Now, the optimized version of the total energy functional (8.6) is obtained when considering in it the above KS orbital equations (8.2); worth observing that taking the sum of the quantum average over orbitals in (8.2) one gets (Putz, 2008a):

$$-\sum_i^N \langle\chi_i\rangle_i = \sum_i^N \int \varphi_i^*(\mathrm{r})\left[-\frac{1}{2}\nabla^2\right]\varphi_i(\mathrm{r})d\mathrm{r} + \int V_{eff}(\mathrm{r})\rho(\mathrm{r})d\mathrm{r} \tag{8.7}$$

which combined with (8.6) provides the optimized total energy functional with the form:

$$E[\rho] = -\sum_i^N \langle\chi_i\rangle_i - \frac{1}{2}\iint \frac{\rho(\mathrm{r})\rho(\mathrm{r'})}{|\mathrm{r}-\mathrm{r'}|}d\mathrm{r}d\mathrm{r'} + E_{XC}[\rho] - \int V_{XC}(\mathrm{r})\rho(\mathrm{r})d\mathrm{r} \tag{8.8}$$

Further on, accounting for the electronegativity equalization principle (Mortier et al., 1985; Sanderson, 1988), here applied throughout all participant (valence) orbitals,

$$\langle\chi_1\rangle_1 = \ldots = \langle\chi_i\rangle_i = \ldots = \chi = -\mu = ct \tag{8.9}$$

equation (8.8) may be further rearranged as

$$E[\rho] \cong -N\chi + \int\left[\frac{E_{xc}[\rho]}{N} - V_{XC}(\mathrm{r}) - \frac{1}{2}\int\frac{\rho(\mathrm{r'})}{|\mathrm{r}-\mathrm{r'}|}d\mathrm{r'}\right]\rho(\mathrm{r})d\mathrm{r} \tag{8.10}$$

or even shorter as:

$$E[\rho] - N\mu \cong C_A^{SC} \tag{8.11}$$

where the electronegativity-chemical potential relationships (8.4) and (8.9) were implemented, while the second term in right side of (8.10) was denoted as the *semi-classical (SC) chemical action*:

$$C_A^{SC} = \int \left[\frac{E_{XC}[\rho]}{N} - V_{XC}(\mathrm{r}) - \frac{1}{2} \int \frac{\rho(\mathrm{r}')}{|\mathrm{r}-\mathrm{r}'|} d\mathrm{r}' \right] \rho(\mathrm{r}) d\mathrm{r} \tag{8.12}$$

seen as the convolution of the total density $\rho(\mathrm{r})$ with the SC potential:

$$V^{SC}(\mathrm{r}) = \frac{E_{XC}[\rho]}{N} - V_{XC}(\mathrm{r}) - \frac{1}{2} \int \frac{\rho(\mathrm{r}')}{|\mathrm{r}-\mathrm{r}'|} d\mathrm{r}' \tag{8.13}$$

which combines the quantum rest of exchange-correlation energy and potential contributions with the classical repulsion term. Note that when the first two exchange-correlation terms of (8.12) eventually cancel each other the SC chemical action reduces to the *classical (CS) chemical action*:

$$C_A^{CS} = \int \left[-\frac{1}{2} \int \frac{\rho(\mathrm{r}')}{|\mathrm{r}-\mathrm{r}'|} d\mathrm{r}' \right] \rho(\mathrm{r}) d\mathrm{r} \tag{8.14}$$

providing chemical action versatility in quantifying the Legendre transformation of type (8.11).

Further on, one can easily see that when the optimization of the density functional total energy takes place in the left side of (8.11) it is as well transferred to the associate chemical action functional in right side of (8.11) introducing therefore the first stage of the chemical action variational principle.

$$0 = \delta\left(E[\rho] - N\mu\right) \cong \delta C_A^{SC}[\rho] \tag{8.15}$$

On the other side, by assuming the total energy as dependent of electronic density and the external potential, $E = E[N, V(\mathrm{r})]$, its total differential variation may be written as:

$$dE[\rho] = \left(\frac{\delta E[\rho]}{\delta \rho} \right)_{V(\mathrm{r})} dN + \int \left(\frac{\delta E[\rho]}{\delta V(r)} \right)_N \delta V(\mathrm{r}) d\mathrm{r} = \mu dN + \int \rho(\mathrm{r}) \delta V(\mathrm{r}) d\mathrm{r} \tag{8.16}$$

If recalling that the chemical potential in (8.16) acts as a constant since its relation with equalized electronegativity (8.9) the energy (path) integration in (8.16) between infinity separations of electrons (with $E_\infty = 0; N_\infty = 0; V_\infty = 0$ in the bonding region) and the actual poly-electronic bonding system one yields with the chemical action CA introduced in (1.8), in accordance with the functional integration on (4.10).

$$E[\rho] - \mu N = C_A \tag{8.17}$$

With these, the chemical action variational principle is proved to hold at whatever potential level as a viable substitute for the stationary total energy. However, its advantage, and especially on the "simple" chemical action form (1.8) is that it constitutes the only completely known analytical term of the total electronic interaction in energy density functional (Kohn et al., 1996).

Finally, beyond the simple expression (1.8) the practical use of chemical action in relation with chemical reactivity may be accessed through its equivalence with electronegativity, rooting in the earlier March definition of electronegativity for the electrostatic potential of the electron cloud at the nucleus (March, 1993), see also (8.143):

$$\chi = \int \left[-\frac{1}{R} \right] \rho(\mathrm{r}) dr = C_A \tag{8.18}$$

Worth noting that the equality between chemical action and electronegativity may be considered in general as well since when performing it in (8.17) one finds that

$$C_A = \chi = -\frac{E[\rho_{N \to N-1}]}{N-1} \tag{8.19}$$

with the meaningfully interpretation that chemical action behaves as electronegativity manifested by any of concerned electrons (the valence ones) in the field of other N-1; alternatively, relation (8.19) tells us that chemical action equalizes electronegativity for the systems having *the distinctive electron* engaged in chemical reactivity and bonding.

However, observe that the chemical action concept encompasses the energetic convolution of the (valence state) electronic density with the bar potential acting on the electronic cloud while closely characterizing the mixed states in optimized total energy. The last assertion may be easily visualized by considering the equivalence of equations (8.11) and (8.17) under the appropriate form as:

$$E[\rho] = \sum_i^N \langle \mu_i \rangle_i + C_A \tag{8.20}$$

from where there is clear that the optimized total energy is unfolded as the (quantum) superposition of N-independent (virtually orthogonal) i-states while the remaining chemical action accounts for coupling or interaction between them.

Overall, chemical action under the form (1.8) looks as a fecund notion since it combines in an unique manner the first Hohenberg–Kohn theorem (assessing the bi-univocity of the applied potential with the density that optimizes the system), with the second Hohenberg–Kohn theorem (1964), consecrating the variational principle (8.5) both comprised by the actual chemical action (1.8) with its principle (8.15).

Not at last, worth remarking that chemical action (1.8), given that it corresponds with the quantum average of the applied potential (by nuclei $\langle V_{en} \rangle$, for instance), may be widely assumed as a suitable concept for treating (in principle any) chemical bonding: when seeing through the virial theorem (Preuss, 1969) leads at equilibrium with

$$C_A = \langle V_{en} \rangle = 2E - \langle W \rangle \tag{8.21}$$

thus necessarily accounting for the "accumulation" force acting in the base of the opposite averaged potential to that of inter-electronic repulsion $\langle W \rangle$ that prevents bonding.

Nevertheless, the chemical action principle under the form (8.15) remains to be in next employed towards density functional description of the chemical bond and bonding.

8.3 DENSITY KERNEL FUNCTIONALS OF CHEMICAL ACTION

In order to turn the chemical action functional into a current tool for bonding description its reformulation as related with reactivity softness and hardness concepts seems the appropriate endeavor. It starts with combining the unfolded variational principle equation (8.15)

$$0 = \int \delta\rho(\mathrm{r})V(\mathrm{r})d\mathrm{r} + \int \rho(\mathrm{r})\delta V(\mathrm{r})d\mathrm{r} \quad (8.22)$$

with the Hellmann–Feynman theorem (Feynman, 1939; Hellmann, 1937):

$$0 = \int \rho(\mathrm{r})\delta V(\mathrm{r})d\mathrm{r} \quad (8.23)$$

followed by the electronic density functional first order expansion in total number of electrons (eventually restricted to those participating in bonding) and in the applied potential upon them, $\rho = \rho[N, V(\mathrm{r})]$. The obtained equation

$$0 = \int \left(\frac{\delta\rho(\mathrm{r})}{\delta N}\right)_{V(\mathrm{r})} V(\mathrm{r})dNd\mathrm{r} + \int \left[\int \left(\frac{\delta\rho(\mathrm{r})}{\delta V(\mathrm{r}')}\right)_N \delta V(\mathrm{r}')d\mathrm{r}'\right] V(\mathrm{r})d\mathrm{r} \quad (8.24)$$

can be solved out for the electronic density with the form

$$\rho(\mathrm{r}) = -\int \kappa(\mathrm{r},\mathrm{r}')V(\mathrm{r}')d\mathrm{r}' \quad (8.25)$$

where the bilocal response function was introduced in (8.32) (Berkowitz and Parr, 1988; Garza and Robles, 1993). With electronic density (8.25) replaced in equation (1.8) the first kernel density functional of chemical action yields as

$$C_A = -\iint V(\mathrm{r})\kappa(\mathrm{r},\mathrm{r}')V(\mathrm{r}')d\mathrm{r}d\mathrm{r}' \quad (8.26)$$

Worth noted that the expression (8.26) gives the opportunity for understanding chemical action as an interaction quantity due to its close relationship with energy interaction $\varepsilon_{\mathrm{int}}$ relating polarizability α when the external potential $V(\mathrm{r}) = \mathrm{E}\cdot\mathrm{r}$, with the applied electric field amplitude E, is considered, viz. in polar coordinates

$$C_A = -\left[\iint \kappa(r,r')rr'\cos^2\theta\, drdr'\right]E^2 = \alpha E^2 = -\varepsilon_{\mathrm{int}} \quad (8.27)$$

Yet, the formulation (8.26) can be further refined since considering the chemical hardness and softness kernels, (8.18) and (8.21), respectively reconsidered here like (Parr and Gázquez, 1993; Parr and Yang, 1989)

$$\eta(\mathrm{r},\mathrm{r}') = -\frac{1}{2}\frac{\delta V(\mathrm{r})}{\delta \rho(\mathrm{r}')} \tag{8.28}$$

$$s(\mathrm{r},\mathrm{r}') = -\frac{\delta \rho(\mathrm{r})}{\delta V(\mathrm{r}')} \tag{8.29}$$

linked by the integration-differentiation chain rule of delta-Dirac bilocal function (8.24); now, making use of the fundamental density functional constraint (8.4) one gets from equation (8.24) the local hardness kernel softness density formulation (8.25) where the local chemical hardness was implemented as in equation (8.19). Finally, by comparison of densities (8.25) and (8.25) the softness-hardness response bilocal function can be immediately reached out

$$\kappa(\mathrm{r},\mathrm{r}') = -2N\frac{\eta(\mathrm{r}')s(\mathrm{r},\mathrm{r}')}{V(\mathrm{r}')} \tag{8.30}$$

leaving with the reactivity kernel expression of chemical action functional:

$$C_A = 2N\iint V(\mathrm{r})s(\mathrm{r},\mathrm{r}')\eta(\mathrm{r}')d\mathrm{r}d\mathrm{r}' \tag{8.31}$$

Note that the expression (8.31) enriches the chemical action foreground definition (1.8) with reactivity information compressed in softness kernel and chemical hardness whereas the bonding character is expressed by allowing the energetic double occupancy for the total (or bonding) number of electrons N, in close correspondence with the unrestricted Hartree–Fock orbital treatment (Hartree, 1957).

However, all double integrals involved in chemical action kernel formulations (8.26) and (8.31) are in close correspondence with the reactivity paths of the electronic pairs (r,r') of chemical bonding. Nevertheless, local and nonlocal consequences of the present chemical action functionals in bonding are therefore in next section explored.

8.5 LOCAL APPROXIMATION OF CHEMICAL ACTION

The first natural employed softness kernel specialization assumes the referential local approximation (Garza and Robles, 1993; Putz, 2009b):

$$s(\mathrm{r},\mathrm{r}')^{local} = -\frac{\delta\rho(\mathrm{r}')}{\delta V(\mathrm{r}')}\delta(\mathrm{r}-\mathrm{r}') \tag{8.32}$$

with the help of which the local version of the above response function (8.30) takes the successive appropriate forms

$$\begin{aligned}\kappa(\mathrm{r},\mathrm{r}')^{local} &= -\frac{1}{V(\mathrm{r}')}\left[\int\frac{\delta V(\mathrm{r}')}{\delta\rho(\mathrm{r})}\rho(\mathrm{r})d\mathrm{r}\right]\frac{\delta\rho(\mathrm{r}')}{\delta V(\mathrm{r}')}\delta(\mathrm{r}-\mathrm{r}')\\ &= -\frac{1}{V(\mathrm{r}')}\left[\int\frac{\delta\rho(\mathrm{r}')}{\delta\rho(\mathrm{r})}\rho(\mathrm{r})d\mathrm{r}\right]\delta(\mathrm{r}-\mathrm{r}') = -\frac{1}{V(\mathrm{r}')}\left[\int\frac{\rho(\mathrm{r})}{|\nabla_{r}\rho(\mathrm{r})|}\delta\rho(\mathrm{r}')\right]\delta(\mathrm{r}-\mathrm{r}')\\ &= -\frac{1}{V(\mathrm{r}')}\frac{\rho(\mathrm{r})\rho(\mathrm{r}')}{|\nabla_{r}\rho(\mathrm{r})|}\delta(\mathrm{r}-\mathrm{r}')\end{aligned} \tag{8.33}$$

It provides the analytical framework for the chemical action computation that according with equation (8.26) casts as the "local" realization

$$C_A^{local} = \int V(\mathrm{r})\rho(\mathrm{r})\frac{\rho(\mathrm{r})}{|\nabla_{r}\rho(\mathrm{r})|}d\mathrm{r} \tag{8.34}$$

while, when equating with the basic definition (1.8) it leads with the so called local equation of bonding:

$$\left|\nabla_r \rho(\mathrm{r})\right| = \rho(\mathrm{r}) \tag{8.35}$$

throughout fulfilling the Bader zero flux condition (Bader, 1990), see also equation (4.9) of Chapter 4

$$\rho(\mathrm{r}) = \nabla_r \rho \cdot \mathrm{n} = 0 \tag{8.36}$$

for the asymptotic densities

$$\rho(r)^{local} = \frac{N}{8\pi}\exp(-r) \tag{8.37}$$

as the solution of the equation (8.35) with the density functional radial constraint (8.4).

Worth noting that such electronic density expression is of the first importance in characterizing the exchange or Fermi holes in chemical structures (Becke, 1986) thus furnishing the backbone of the analytical chemical bonding analysis.

8.6 NON-LOCAL APPROXIMATION OF CHEMICAL ACTION

Going to the non-local level of the softness kernel one relevant choice should be shaped as (Garza and Robles, 1993; Putz, 2007)

$$s(\mathrm{r},\mathrm{r}')^{non-local} = s(\mathrm{r},\mathrm{r}')^{local} + \rho(\mathrm{r})\rho(\mathrm{r}') \tag{8.38}$$

producing the associate non-local response function (8.30) equivalencies

$$\begin{aligned}\kappa(\mathrm{r},\mathrm{r}')^{non-local} &= \kappa(\mathrm{r},\mathrm{r}')^{local} + \frac{\rho(\mathrm{r})\rho(\mathrm{r}')}{V(\mathrm{r}')}\int \frac{\delta V(\mathrm{r}')}{\delta\rho(\mathrm{r})}\rho(\mathrm{r})d\mathrm{r} \\ &= \kappa(\mathrm{r},\mathrm{r}')^{local} + \rho(\mathrm{r})\rho(\mathrm{r}')\int \frac{\delta V(\mathrm{r}')}{V(\mathrm{r}')}\frac{1}{\delta\rho(\mathrm{r})}\rho(\mathrm{r})d\mathrm{r} \\ &= \kappa(\mathrm{r},\mathrm{r}')^{local} - \rho(\mathrm{r})\rho(\mathrm{r}')\int \frac{\rho(\mathrm{r})}{\nabla\rho(\mathrm{r})\Delta\mathrm{r}}d\mathrm{r}\end{aligned} \tag{8.39}$$

where the Poisson finite difference or long range approximations (Putz, 2003; Putz et al., 2005):

$$\nabla V(\mathrm{r}') \cong -4\pi\rho(\mathrm{r}')\Delta\mathrm{r}' \tag{8.40}$$

$$V(\mathrm{r}') \cong 4\pi\rho(\mathrm{r}')\left[\Delta\mathrm{r}'\right]^2 \tag{8.41}$$

were involved for smearing out the explicit potential dependence. Still, the remaining displacement of the bonding localization (of the electronic pairs) can be finely tuned by employing once more the Hellmann–Feynman theorem under the form:

$$0 = \int \rho(\mathrm{r})\nabla V(\mathrm{r})d\mathrm{r} = \int \nabla\left[\rho(\mathrm{r})V(\mathrm{r})\right]d\mathrm{r} - \int V(\mathrm{r})\nabla\rho(\mathrm{r})d\mathrm{r} \tag{8.42}$$

from where the formal identity

$$\Delta[\rho(\mathrm{r})V(\mathrm{r})] \cong V(\mathrm{r})\nabla\rho(\mathrm{r})\Delta\mathrm{r} \tag{8.43}$$

can be used to rewrite the non-local response kernel function (8.39) until the chemical action dependency is achieved:

$$\begin{aligned}\kappa(\mathrm{r},\mathrm{r}')^{non-local} &\cong \kappa(\mathrm{r},\mathrm{r}')^{local} - \rho(\mathrm{r})\rho(\mathrm{r}')\int \frac{\rho(\mathrm{r})V(\mathrm{r})}{\Delta[\rho(\mathrm{r})V(\mathrm{r})]}d\mathrm{r}\\ &\cong \kappa(\mathrm{r},\mathrm{r}')^{local} - \rho(\mathrm{r})\rho(\mathrm{r}')\frac{C_A}{\rho(\mathrm{r})V(\mathrm{r})}\\ &= \kappa(\mathrm{r},\mathrm{r}')^{local} - \frac{\rho(\mathrm{r}')}{V(\mathrm{r})}C_A\end{aligned} \tag{8.44}$$

when the saddle point approximation (Hassani, 1991) was applied to the integral term becoming a local one.

In these conditions, as the relations (8.26) and (8.44) are combined the chemical action equation for non-local or delocalized bonding description is raised:

$$\begin{aligned}C_A^{non-local} &= \int V(\mathrm{r})\rho(\mathrm{r})\frac{\rho(\mathrm{r})}{|\nabla_r\rho(\mathrm{r})|}d\mathrm{r} + C_A\iint \rho(\mathrm{r}')V(\mathrm{r}')d\mathrm{r}d\mathrm{r}'\\ &= \int \frac{dC_A}{d\mathrm{r}}\frac{\rho(\mathrm{r})}{|\nabla_r\rho(\mathrm{r})|}d\mathrm{r} + C_A^2\Delta r = C_A\frac{\rho(\mathrm{r})}{|\nabla_r\rho(\mathrm{r})|} + C_A^2\Delta r\end{aligned} \tag{8.45}$$

Worth remarking that the explicit bonding displacement Δr in equation (8.45) modulates the chemical action amplitude; as Δr vanishes as the previous local case of bonding is recovered, see equation (8.35). Nevertheless, both local and non-local instants of chemical action produce the present chemical bonding picture in a complete non-orbital way. More details and discussions are in next addressed.

8.7 NON-RELATIVISTIC QUANTUM GEOMETRIZATION OF CHEMICAL BOND

Since, the non-local bonding equation (8.45) contains the local case (8.35) it may be rearranged to the convenient form recalling a sort of adapted Heisenberg relation for chemical bonding (Putz, 2009b):

$$C_A\Delta r = 1 - \frac{\rho(\mathrm{r})}{|\nabla_r\rho(\mathrm{r})|}\quad (in\ a.\ u.) \tag{8.46}$$

since $<\mathbf{energy}>\cdot<\mathbf{distance}>\sim Joule\cdot meter \sim h\cdot c$ with h the Planck constant and c the light velocity. Moreover, other arrangement of equation (8.46) provides the generalization of previous Bader relation for bonding flux of electronic density, see equations (8.35) and (8.36)

$$|\nabla_r\rho|(1-\Delta r C_A) = \rho \tag{8.47}$$

while it turns out that the chemical action appears as modulating the departures of electronic pairing from its localized version. Aiming to find the explicit localized-delocalized bonding density one has to integrate the equation (8.47) appropriately rewritten as

$$\frac{d\rho}{\rho} = \frac{1}{1-\Delta r C_A} dr \tag{8.48}$$

Now, we have two ways of integration depending on considering or not the delocalization as an integration variable; however we will treat both cases under the assumption that for initial condition of integration we have $r_0 = 0 \,\&\, \rho_0 = \rho^{local}$ of equation (8.37).

I. for a constant delocalization ($\Delta r \sim R$) we get:

$$\rho(r)^{bonding-I} = \rho_0 \exp\left(\frac{r}{1-RC_A}\right) \tag{8.49}$$

II. for a variable delocalization ($\Delta r \sim r$) we obtain:

$$\rho(r)^{bonding-II} = \rho_0 \exp\left(-\frac{1}{C_A}\ln\left(1-rC_A\right)\right) \tag{8.50}$$

Remarkably, either bonding densities (8.49) and (8.50) provide the same chemical action limits, viz.:

$$\lim_{C_A\to\infty} \rho(r)^{bonding-I} = \lim_{C_A\to\infty} \rho(r)^{bonding-II} = \rho_0 = \rho^{local} \tag{8.51}$$

$$\lim_{C_A\to 0} \rho(r)^{bonding-I} = \lim_{C_A\to 0} \rho(r)^{bonding-II} = \rho_0 \exp(r) = \frac{N}{8\pi} = ct. \tag{8.52}$$

$$\lim_{C_A\to 1/R} \rho(r)^{bonding-I} = \lim_{C_A\to 1/r} \rho(r)^{bonding-II} = \infty \tag{8.53}$$

Therefore, we may considerate the two bonding solutions (8.49) and (8.50) as expressing the same chemical interaction and equating them for a sort of universal chemical action-electronic pairing localization in bonding (atoms-in-molecules):

$$(1-RC_A)\ln(1-rC_A) = -rC_A \tag{8.54}$$

When *delocalization* towards *localization* that is $r \to R = \lambda$, equation (8.54) firstly becomes:

$$1-\lambda C_A = \exp\left(-\frac{\lambda C_A}{1-\lambda C_A}\right) \tag{8.55}$$

Now, taking the limit $C_A \to 0$ on the exponent of the right side of equation (8.55) is in fully accordance with the case (8.52) reflecting the fact that each of the $N/2$ pairing electrons is stabilized in spherical averaged sense ($1/4\pi$) with others electrons of bonding. In first order, the expression (8.55) rewrites as:

$$1-\lambda C_A = \exp\left(-\lambda C_A\right) \tag{8.56}$$

and have no explicit N-dependency thus confirming it holds for any pairing of electrons in bonding.

At the first level of comprehension, when equation (8.56) is seen only as a mathematical content, the resumation and expansion equalization of the chemical action may be regarded as a particular case of the chemical action principle (8.15), widely proved in Chapter 4, along the reactivity path. However, equation (8.56) should be regarded not just as a mathematical equivalence but as an expression that equate the expanded influences coming from two chemical interacting systems. That is, given two such systems A and B that may be atoms or molecules they chemically interact until the bond is established throughout equating left and right side terms of equation (8.56) coming from them in a mixed manner.

However, in order to be more explicitly in bonding description, let's firstly introduce from (8.56) the two working types of chemical binding functions:

$$f_{\alpha}(\lambda, C_A) = 1 - \Omega\lambda C_A = \begin{cases} 1, & \lambda \to 0 \\ -\infty, & \lambda \to \infty \end{cases} \tag{8.57}$$

$$f_{\beta}(\lambda, C_A) = \exp(-\Omega\lambda C_A) = \begin{cases} 1, & \lambda \to 0 \\ 0, & \lambda \to \infty \end{cases} \tag{8.58}$$

called as the *anti-bonding and bonding functions*, for the reason grounded on their asymptotical behavior, respectively; the introduced W-factor accounts for assumed dimensionless nature of functions (8.57) and (8.58), being adequately set as:

$$\Omega = \frac{1}{\hbar c} = 0.506773 \cdot 10^{-3} J^{-1} m^{-1} \tag{8.59}$$

in accordance with above adapted Heisenberg relationship, see equation (8.46), for the localization distance λ and chemical action C_A expressed in $\mathring{A}$ (Ångstrom) and eV (electron-volts), respectively.

With these, there is clear that while the bonding function (8.58) finitely localizes the electronic pairing on bonding asymptotic distance this is not the case for anti-bonding function (8.57), from where their different role in bonding. Now, considering the paradigmatic A–B chemically bonding system within a coordinate system centered in A, the binding functions (8.57) and (8.58) reciprocally combines, only in the way prescribed by equation (8.56), to provide the electronic pair-localization region within the bond length R_{AB} by means of the binding equations (see Figure 8.1),

$$(\mathrm{I}): f_{\alpha}^{A}\left(\lambda_{\mathrm{I}}, C_A^{A}\right) = f_{\beta}^{B}\left(R_{AB} - \lambda_{\mathrm{I}}, C_A^{B}\right) \tag{8.60}$$

$$(\mathrm{II}): f_{\alpha}^{B}\left(R_{AB} - \lambda_{\mathrm{II}}, C_A^{B}\right) = f_{\beta}^{A}\left(\lambda_{\mathrm{II}}, C_A^{A}\right) \tag{8.61}$$

as the interval $\lambda_{\mathrm{II}} - \lambda_{\mathrm{I}}$ or as the single point $\lambda_{\mathrm{II}} = \lambda_{\mathrm{I}}$ for the hetero- and homo- bonding systems that is having different or identical isolated chemical actions C_A^{A} and C_A^{B}, respectively. Note that in each of above (8.60) and (8.61) equations the binding "points" I and II appear as the informational crossing (transfer) between the anti-bonding and bonding functions of both A and B systems driven by their chemical actions; this way, the system (8.60) and (8.61) fixes the electronic pairing region on bonding as well as all types of involved bonding regions see Figure 8.1 (Putz, 2009b):

- The so called *sigma-bonding* region (σ-B in Figure 8.1) is delimited by the area under bonding functions f_β^A and f_β^B along the pairing interval $(\lambda_I, \lambda_{II})$ on the bond length; it is uniquely defined and has no "nodes" or discontinuities; it corresponds to the consecrated bonding obtained by the composed wave-function density $|\Psi_A + \Psi_B|^2$ in the conventional MO theory (Pauling, 1960).
- The so called *anti-bonding* region (⅂-B in Figure 8.1) also defined by the area under bonding functions f_β^A and f_β^B but outside of the interval $(\lambda_I, \lambda_{II})$ on the bond length; it is thus represented by the two parts spanning the space from the systems A and B until the sigma-bonding limit; it corresponds to the anti-bonding state density $|\Psi_A + \Psi_B|^2$ with separated parallel spin-electronic pair in MO theory.
- The so called *no-bonding* region (Ø-B in Figure 8.1) is formed by the area delimited by all the binding functions of (8.57) and (8.58) around the binding points I and II, outside of the interval $(\lambda_I, \lambda_{II})$ and not crossing the bond length; it is composed of two parts, one in each binding side respecting sigma-bonding.
- The so called *pi-bonding* region (π-B in Figure 8.1) is resulted by the area defined by all the binding functions of (8.57) and (8.58) around the binding points I and II, partially outside and partially inside (with a node) of the interval $(\lambda_I, \lambda_{II})$ while spanning the bond length entirely; such features make this region compatible with the consecrated pi-bond type of the MO theory.

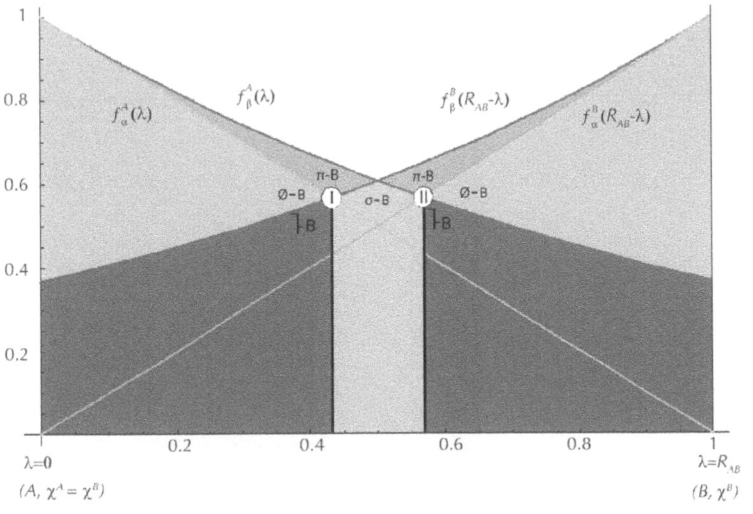

FIGURE 8.1 Phenomenological representation of chemical binding from chemical action influence of two systems A and B throughout equations (8.60) and (8.61) with binding functions (8.57) and (8.58). The binding points, solutions of the system (8.60) and (8.61), provide delimitation of the sigma-bonding (σ-B), anti-bonding (⅂-B), no-bonding (Ø-B), and pi-bonding (π-B) regions as describing the pair localization, spin pair separation, no-pairing and electronic delocalization along the bond length, respectively through constants and parametric settings as $\hbar = c = 1$, $\chi^A = \chi^B = 1$, $R_{AB} = 1$ (Putz, 2009b).

There is therefore obvious that the actual chemical action based treatment recovers the main bonding characteristics of the fashioned MO theory providing instead the geometrical and localization of the electronic pairs in bonding along the equilibrium bond lengths. However, in practice, worth making use of the chemical action equivalence with electronegativity, as in was in Section 8.2 revealed, see equation (8.18), providing therefore further chemical information in bonding picture. Yet, the "real" binding pictures may not have the geometrical resolution of Figure 8.1 so that all above identified bonding region to be clearly emphasized, although the (quantum) effects are present at the electronic level as above described. As well, note that actually, the binding regions of Figure 8.1 are formed with the bond length crossing the lowest anti-bonding function involved. This way, all above identified binding regions are defined within positive (0, 1) realm of binding functions (8.57) and (8.58) allowing the natural probabilistic interpretation for their inside.

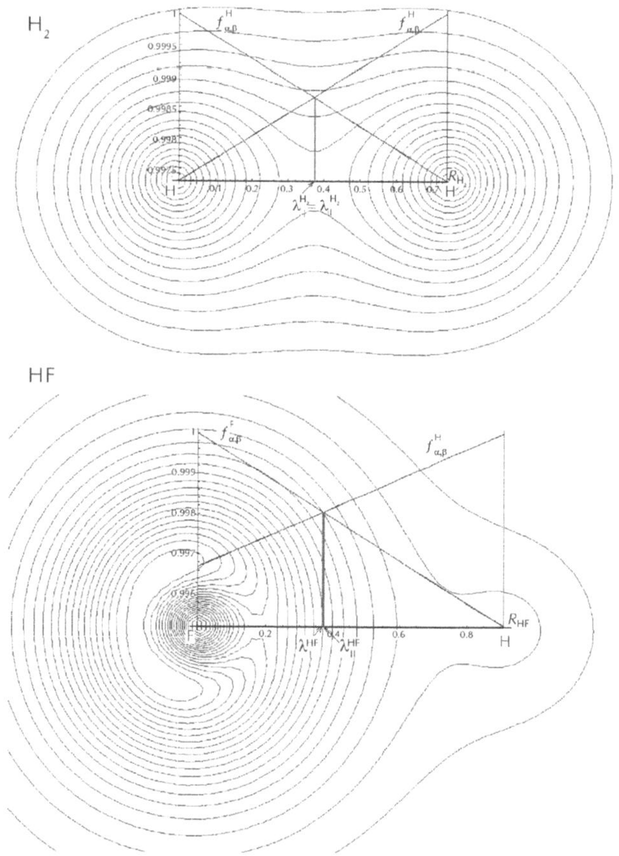

FIGURE 8.2 The density contour maps of H_2 (upper plot) and HF (bottom plot) with superimposed binding pictures specialized from Figure 8.1 through application of the binding equations (8.60) and (8.61) with binding functions (8.57) and (8.58) for the atomic chemical actions $C_A^{\mathrm{H}} = \chi_M^{\mathrm{H}} = 7.18\,eV$, $C_A^{\mathrm{F}} = \chi_M^{\mathrm{F}} = 10.4\,eV$ (Putz, 2006) and molecular bond lengths $R_{\mathrm{H_2}} = 0.741\,\mathring{A}$ (Pauling, 1960) and $R_{\mathrm{HF}} = 0.908378\,\mathring{A}$ (HyperChem, 2002), respectively (Putz, 2009b).

As an illustration, the Figure 8.2 presents the H_2 and HF molecular density contours with 1:1 superimposing the associate chemical bonding based chemical action (aka electronegativity) picture resulted as previously described through solving equations (8.60) and (8.61) for the atomic Mulliken electronegativity values $C_A^{\mathrm{H}} = \chi_M^{\mathrm{H}} = 7.18\ eV, C_A^{\mathrm{F}} = \chi_M^{\mathrm{F}} = 10.4 eV$ (Putz and Chiriac, 2008), and molecular equilibrium bond lengths $R_{\mathrm{H}_2} = 0.741\mathring{\mathrm{A}}$ (Pauling, 1960), $R_{\mathrm{HF}} = 0.908378\,\mathring{\mathrm{A}}$ (Hyperchem, 2002), respectively. As earlier anticipated, for the homo-atomic H_2 molecule only one bonding point is found, with $\lambda_{\mathrm{I}} = \lambda_{\mathrm{II}} = 0.3705\,\mathring{\mathrm{A}}$, while for the hetero-atomic HF molecule the bonding picture is non-symmetrical respecting the pairing range $(\lambda_{\mathrm{I}}, \lambda_{\mathrm{II}}) = (0.370784, 0.371213)\mathring{\mathrm{A}}$.

However, a part of the homo-atomic bond, there is apparent that the actual predicted localization for the electronic pairing in bonding is neither situated on the center of mass of the system nor on a single point but spanning rather a bonding interval as a direct consequence of quantum non-symmetrical effects rooting on bonding components (atoms-in-molecule).

Further studies are certainly needed in order to work out the present chemical bonding scheme on multi-center bonding, aromaticity, ionic and van der Waals interactions targeting on unitary theory of the chemical bonding based on chemical action and electronegativity.

8.8 RELATIVISTIC QUANTUM GEOMETRIZATION OF CHEMICAL BOND

There is a known fact that the basic Dirac equation unfolds as the temporal generalized operatorial form (Dirac, 1928; Harriman, 1978; Jauch and Rohrlich, 1976; Schweber, 1964)

$$i\hbar\partial_t[\Psi] = \hat{H}_{Dir}[\Psi] \tag{8.62}$$

in a very similar shape with the Schrödinger one, however with the Dirac Hamiltonian specialization:

$$\hat{H}_{Dir} = \hat{H}_{Dir}^{0} + \hat{v}(x) \tag{8.63}$$

with the free particle and applied potential components:

$$\hat{H}_{Dir}^{0} = -i\hbar c\hat{\vec{\alpha}}\cdot\hat{\vec{\nabla}} + mc^2\hat{\beta} \tag{8.64}$$

$$\hat{v}(x) = V(x)\hat{\beta} \tag{8.65}$$

respectively, with m - the particle mass, c - the light velocity, $\hbar$ - the Planck constant, while the introduced special operators $\hat{\alpha}, \hat{\beta}$ assume the Dirac 4D representation:

$$\hat{\alpha}_k = \begin{bmatrix} 0 & \hat{\sigma}_k \\ \hat{\sigma}_k & 0 \end{bmatrix}, k = 1,2,3 \tag{8.66}$$

$$\hat{\beta} = \begin{bmatrix} \hat{1} & 0 \\ 0 & -\hat{1} \end{bmatrix} \tag{8.67}$$

in terms of bi-dimensional Pauli and unitary matrices (operators)

$$\hat{\sigma}_0 = \hat{1} = \begin{bmatrix} 1 & 0 \\ 0 & 1 \end{bmatrix}, \hat{\sigma}_1 = \begin{bmatrix} 0 & 1 \\ 1 & 0 \end{bmatrix}, \hat{\sigma}_2 = \begin{bmatrix} 0 & -i \\ i & 0 \end{bmatrix}, \hat{\sigma}_3 = \begin{bmatrix} 1 & 0 \\ 0 & -1 \end{bmatrix} \tag{8.68}$$

and with the wave function featuring the so called spinorial (bi-dimensional) equivalent formulation

$$[\Psi] = \begin{bmatrix} \varphi \\ \varphi \end{bmatrix} e^{-\frac{i}{\hbar}E \cdot t} = \begin{cases} \begin{bmatrix} \varphi \\ 0 \end{bmatrix} e^{-\frac{i}{\hbar}|E| \cdot t} & , E > 0 \ \ anti-bonding\ states \\ \begin{bmatrix} 0 \\ \varphi \end{bmatrix} e^{+\frac{i}{\hbar}|E| \cdot t} & , E < 0 \ \ \ \ bonding\ states \end{cases} = \begin{bmatrix} \varphi e^{-\frac{i}{\hbar}|E| \cdot t} \\ \varphi e^{+\frac{i}{\hbar}|E| \cdot t} \end{bmatrix} \tag{8.69}$$

However, there also a rises the question whether the general Dirac equation (8.62) may be reduced or transformed so that to represent the eigen-equation for the electronic states for a given quantum system. For this, through closely analyzing the form of equation (8.62) with all its contribution, one may resume the free motion Dirac operator to the working form (Boeyens, 2005)

$$\hat{D} = -i\hbar\hat{A}\hat{\sigma}_0 \frac{\partial}{\partial t} + i\hbar\hat{\sigma}_1 \frac{\partial}{\partial x_1} + i\hbar\hat{\sigma}_2 \frac{\partial}{\partial x_2} + i\hbar\hat{\sigma}_3 \frac{\partial}{\partial x_3} + \hat{C} \tag{8.70}$$

and employing it to the stationary operatorial equation:

$$0 = \hat{D}[\Psi] = \hat{D} \begin{bmatrix} \Psi_1 e^{i(kx-\omega t)} \\ \Psi_2 e^{-i(kx-\omega t)} \end{bmatrix} = \hat{D} \begin{bmatrix} \zeta \\ \xi \end{bmatrix} \tag{8.71}$$

with the oscillatory phase written so that to be in accordance with the prescription of equation (8.69) for the Planck energy-frequency identification:

$$E = \hbar\omega \tag{8.72}$$

In these conditions, one notes that for the time and coordinate derivatives yields:

$$\frac{\partial}{\partial t}[\Psi] = i\omega \begin{bmatrix} -\zeta \\ \xi \end{bmatrix}, \tag{8.73}$$

$$\frac{\partial}{\partial x_k}[\Psi] = -ik_k \begin{bmatrix} -\zeta \\ \xi \end{bmatrix} \tag{8.74}$$

which reduce the above stationary condition (8.71) to the form

$$\left(\hbar\omega\hat{A}\hat{1} + \hbar k_k \hat{\sigma}_k \hat{1}\right) \begin{bmatrix} -\zeta \\ \xi \end{bmatrix} + \hat{C} \begin{bmatrix} \zeta \\ \xi \end{bmatrix} = 0 \tag{8.75}$$

Choosing now appropriately (that stands for the optimization procedure) the matrices the last form further rearranges as:

$$\hat{A} = \begin{bmatrix} 0 & 0 \\ 1 & 0 \end{bmatrix}, \hat{C} = \begin{bmatrix} 0 & 2m \\ 0 & 0 \end{bmatrix} \tag{8.76}$$

$$\begin{bmatrix} \hbar k_k \hat{\sigma}_k & 2m\hat{1} \\ \hbar\omega\hat{1} & \hbar k_k \hat{\sigma}_k \end{bmatrix} \begin{bmatrix} -\zeta \\ \xi \end{bmatrix} = 0 \tag{8.77}$$

leaving with the system:

$$\begin{cases} -\hbar k_k \hat{\sigma}_k \zeta + 2m\hat{1}\xi = 0 \\ -\hbar\omega\hat{1}\zeta + \hbar k_k \hat{\sigma}_k \xi = 0 \end{cases} \tag{8.78}$$

Now, since solving the first equation of the system (8.78) in one variable, say

$$\xi = \frac{\hbar k_k \hat{\sigma}_k \hat{1}}{2m}\zeta = \frac{p_k \hat{\sigma}_k \hat{1}}{2m}\zeta \tag{8.79}$$

and substituting into the second one, there is obtained:

$$E\hat{1}\zeta = \frac{(p_k \hat{\sigma}_k)(p_k \hat{\sigma}_k)}{2m}\zeta = \frac{p_k^2 \hat{\sigma}_k^2}{2m}\zeta = \frac{p^2}{2m}\hat{1}\zeta \tag{8.80}$$

where the above Planck relationship was supplemented by the companion de Broglie one for the momentum,

$$\hbar k_k = p_k \tag{8.81}$$

while the Pauli matrices basic property

$$\hat{\sigma}_k^2 = \hat{1} \tag{8.82}$$

applies, as may be immediately verified from their realization of equation (8.68). Nevertheless, the equation (8.79) represents in fact the eigen-equation for the free motion, supporting the latent generalization to the bounded state, either in anti-bonding or bonding existence

$$E\zeta = \varepsilon\zeta, \; E\xi = \varepsilon\xi \tag{8.83}$$

This is an interesting result because abolished many odd perception about Dirac equation and its meaning; actually, there follows that:

- Dirac equation is formally related with the temporal Schrödinger one, while producing the same eigen-problems, thus describing in essence the same nature of the electronic motion;
- The spin information modeled by the bi-dimensional spinors is not necessarily a relativistic effect (beside completing the 2 + 2 = 4 relativistic framework dimension of the Dirac equation) but merely a quantum one since fulfilling the eigen-value problems, each separately;
- The two spinors of the Dirac equation may be associated with the bonding (for negative energies) and anti-bonding (for positive energies) of a system, being thus suited for physically modeling of the chemical bond, beside the common

interpretation of negative/positive spectrum of free positronic/electronic energies in the Dirac Sea.

However, before effectively pursuit to the chemical bonding description based on Dirac equation one needs some background of the recent non-orbitalic quantum modeling of the chemical bond.

The above spinorial identification as bonding and anti-bonding, see equation (8.69), may be now combined with the introduced bonding and anti-bonding functions (8.57) and (8.58) so that to create the actual working *binding spinor* :

$$[\Psi]=\begin{bmatrix}\left(1-\frac{\lambda\chi}{\hbar c}\right)\exp\left(-\frac{i}{\hbar}|E|\cdot t\right)\\ \exp\left(-\frac{\lambda\chi}{\hbar c}\right)\exp\left(+\frac{i}{\hbar}|E|\cdot t\right)\end{bmatrix} \tag{8.84}$$

where the previous chemical action dependence was here reconsidered as the more generalized (density functional) electronegativity.

Next, we impose the condition the spinor (8.84) fulfilling the Dirac equation (8.62); for this we separately express the involved terms, while self-understanding the presence of the (bi-dimensional) unitary and other Dirac operators on both spinorial upper and down components so that the implicit total dimension of the wave-function to be completed to 4D space:

- the time derivative Dirac term is directly computed as:

$$i\hbar\partial_t[\Psi]=\begin{bmatrix}i\hbar\left(1-\frac{\lambda\chi}{\hbar c}\right)\left(-\frac{i}{\hbar}|E|\right)\exp\left(-\frac{i}{\hbar}|E|\cdot t\right)\\ i\hbar\frac{i}{\hbar}|E|\exp\left(-\frac{\lambda\chi}{\hbar c}\right)\exp\left(+\frac{i}{\hbar}|E|\cdot t\right)\end{bmatrix}=\begin{bmatrix}|E|\left(1-\frac{\lambda\chi}{\hbar c}\right)\exp\left(-\frac{i}{\hbar}|E|\cdot t\right)\\ -|E|\exp\left(-\frac{\lambda\chi}{\hbar c}\right)\exp\left(+\frac{i}{\hbar}|E|\cdot t\right)\end{bmatrix} \tag{8.85}$$

- the space coordinate Dirac derivative needs the pre-requisite of simple derivative

$$\partial_k\lambda=\partial_k\sqrt{\lambda_k\lambda_k}=\frac{\lambda_k}{\lambda} \tag{8.86}$$

providing the yield:

$$\begin{aligned}-i\hbar c\hat{\alpha}^k\partial_k[\Psi]&=\begin{bmatrix}-i\hbar c\hat{\alpha}^k\left(-\frac{\lambda_k\chi}{\hbar c\lambda}\right)\exp\left(-\frac{i}{\hbar}|E|\cdot t\right)\\ -i\hbar c\hat{\alpha}^k\left(-\frac{\lambda_k\chi}{\hbar c\lambda}\right)\exp\left(-\frac{\lambda\chi}{\hbar c}\right)\exp\left(+\frac{i}{\hbar}|E|\cdot t\right)\end{bmatrix}\\ &=\begin{bmatrix}0&\hat{\sigma}_k\\ \hat{\sigma}_k&0\end{bmatrix}\begin{bmatrix}i\frac{\chi}{\lambda}\lambda_k\exp\left(-\frac{i}{\hbar}|E|\cdot t\right)\\ i\frac{\chi}{\lambda}\lambda_k\exp\left(-\frac{\lambda\chi}{\hbar c}\right)\exp\left(+\frac{i}{\hbar}|E|\cdot t\right)\end{bmatrix}\\ &=\begin{bmatrix}i\frac{\chi}{\lambda}\hat{\sigma}_k\lambda_k\exp\left(-\frac{\lambda\chi}{\hbar c}\right)\exp\left(+\frac{i}{\hbar}|E|\cdot t\right)\\ i\frac{\chi}{\lambda}\hat{\sigma}_k\lambda_k\exp\left(-\frac{i}{\hbar}|E|\cdot t\right)\end{bmatrix}\end{aligned} \tag{8.87}$$

- the free + potential moving term:

$$\left(mc^2+V(x)\right)\hat{\beta}[\Psi]=\begin{bmatrix}\left(mc^2+V(x)\right) & 0\\ 0 & -\left(mc^2+V(x)\right)\end{bmatrix}\begin{bmatrix}\left(1-\frac{\lambda\chi}{\hbar c}\right)\exp\left(-\frac{i}{\hbar}|E|\cdot t\right)\\ \exp\left(-\frac{\lambda\chi}{\hbar c}\right)\exp\left(+\frac{i}{\hbar}|E|\cdot t\right)\end{bmatrix}$$
$$=\begin{bmatrix}\left(mc^2+V(x)\right)\left(1-\frac{\lambda\chi}{\hbar c}\right)\exp\left(-\frac{i}{\hbar}|E|\cdot t\right)\\ -\left(mc^2+V(x)\right)\exp\left(-\frac{\lambda\chi}{\hbar c}\right)\exp\left(+\frac{i}{\hbar}|E|\cdot t\right)\end{bmatrix} \tag{8.88}$$

With these, the Dirac equation (8.62) now provides the system of equations:

$$\hat{1}|E|\left(1-\frac{\lambda\chi}{\hbar c}\right)e^{-\frac{i}{\hbar}|E|\cdot t}=i\frac{\chi}{\lambda}\hat{\sigma}_k\lambda_k e^{-\frac{\lambda\chi}{\hbar c}}e^{\frac{i}{\hbar}|E|\cdot t}+\hat{1}\left(mc^2+V(x)\right)\left(1-\frac{\lambda\chi}{\hbar c}\right)e^{-\frac{i}{\hbar}|E|\cdot t} \tag{8.89}$$

$$-\hat{1}|E|e^{-\frac{\lambda\chi}{\hbar c}}e^{\frac{i}{\hbar}|E|\cdot t}=i\frac{\chi}{\lambda}\hat{\sigma}_k\lambda_k e^{-\frac{i}{\hbar}|E|\cdot t}-\hat{1}\left(mc^2+V(x)\right)e^{-\frac{\lambda\chi}{\hbar c}}e^{\frac{i}{\hbar}|E|\cdot t} \tag{8.90}$$

Next, through getting out from the second equation (8.90) the term containing the covariant product:

$$i\frac{\chi}{\lambda}\hat{\sigma}_k\lambda_k=\hat{1}\left(mc^2+V(x)-|E|\right)e^{-\frac{\lambda\chi}{\hbar c}} \tag{8.91}$$

it is then replaced in the first equation (8.89) to obtain:

$$\left(mc^2+V(x)-|E|\right)\left(1-\frac{\lambda\chi}{\hbar c}+e^{-2\frac{\lambda\chi}{\hbar c}}\right)=0 \tag{8.92}$$

whose solutions expresses the energy conservation

$$|E|=mc^2+V(x) \tag{8.93}$$

and the Dirac adapted bonding equation

$$-e^{-2\frac{\lambda\chi}{\hbar c}}=1-\frac{\lambda\chi}{\hbar c} \tag{8.94}$$

Yet, due to the negative sign of the left hand side of equation (8.94) one may infer that it is just one solution of a quadratic equation, say

$$e^{-4\frac{\lambda\chi}{\hbar c}}\cong 1-2\frac{\lambda\chi}{\hbar c} \tag{8.95}$$

providing the second accompanied root,

$$e^{-2\frac{\lambda\chi}{\hbar c}}\cong 1-\frac{\lambda\chi}{\hbar c} \tag{8.96}$$

However, there is noted the formal difference between equations (8.95) and (8.96) not only because of sign but also due to the approximate nature of the second, coming from the form (8.95) in short range of binding distance regime $\lambda \cong 0$. Nevertheless, the appearance of two (±) forms of Dirac chemical bonding equation is in accordance with the manifestation of the Dirac positive/negative manifestation of energies respecting the electronic/positronic motions within the Dirac Sea, respectively. Still, for the chemical bond description the difference in sign allows for further mixing of the bonding equations for a paradigmatic AB molecule generating more bonding points so modeling in more detail the bonding with spins in bonding and anti-bonding states.

Therefore, the actual working *Dirac binding functions* are (Putz, 2010a):

- The Dirac anti-bonding function remains the same as given within density kernel approach by equation (8.57):

$$f_{\alpha}^{Dir}(\lambda,\chi)=1-\frac{\lambda\chi}{\hbar c} \tag{8.97}$$

- The Dirac bonding function is modified respecting the previous one given by equation (8.58) while being two-folded:

$$f_{\beta(+)}^{Dir}(\lambda,\chi)=\exp\left(-2\frac{\lambda\chi}{\hbar c}\right) \tag{8.98}$$

$$f_{\beta(-)}^{Dir}(\lambda,\chi)=-\exp\left(-2\frac{\lambda\chi}{\hbar c}\right) \tag{8.99}$$

Now, the bonding geometric loci are determined, for the molecule AB, by the system of equations:

$$(I): f_{\alpha}^{A}\left(\lambda_{I},\chi^{A}\right)=f_{\beta(+)}^{Dir\text{-}B}\left(R_{AB}-\lambda_{I},\chi^{B}\right) \tag{8.100}$$

$$(\mathrm{II}): f_{\alpha}^{\mathrm{B}}\left(R_{\mathrm{AB}}-\lambda_{\mathrm{II}},\chi^{\mathrm{B}}\right)=f_{\beta(+)}^{Dir-\mathrm{A}}\left(\lambda_{\mathrm{II}},\chi^{\mathrm{A}}\right) \tag{8.101}$$

$$(III): f_{\alpha}^{A}\left(\lambda_{III},\chi^{A}\right)=f_{\beta(-)}^{Dir-A}\left(\lambda_{III},\chi^{A}\right) \tag{8.102}$$

$$(IV): f_{\alpha}^{B}\left(R_{AB}-\lambda_{IV},\chi^{B}\right)=f_{\beta(-)}^{Dir\text{-}B}\left(R_{AB}-\lambda_{IV},\chi^{B}\right) \tag{8.103}$$

which is regarded as Dirac generalization of the previous one of equations (8.60) and (8.61) by means of the last two equations which quantifies the "interference" effect of the anti-bonding and the negative bonding functions belonging to the same atom to be transferred towards a virtual bonding partner.

The representations of Figures 8.3–8.7 show how the Dirac binding functions and equations (8.97)–(8.103) provides more insight in modeling of chemical bonding respecting the previous density functional ones of Figures 8.1 and 8.2. The differences comes from two basic facts: the bonding function (8.58) takes through Dirac equation two forms, that is it degenerates into one positive and other negative, see equations (8.98) and (8.99), respectively, while having also the modified argumentum.

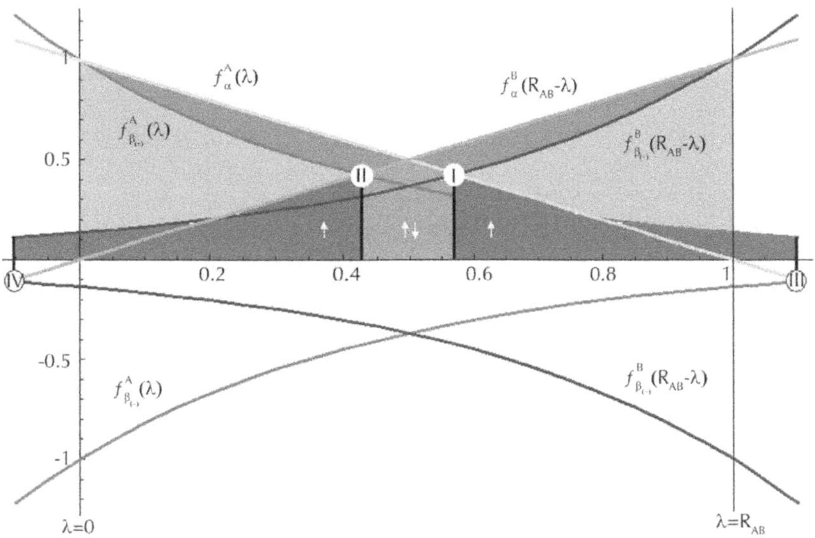

FIGURE 8.3 Geometric loci of the bonding regions as in Figure 8.1 for chemical binding from equal electronegativity influences of two systems A and B throughout equations (8.100)–(8.103) with Dirac binding functions (8.97)–(8.99) through constants and parametric settings as $\hbar = c = 1$, $\chi^A = \chi^B = 1$, $R_{AB} = 1$ (Putz, 2010a, Putz, 2011c).

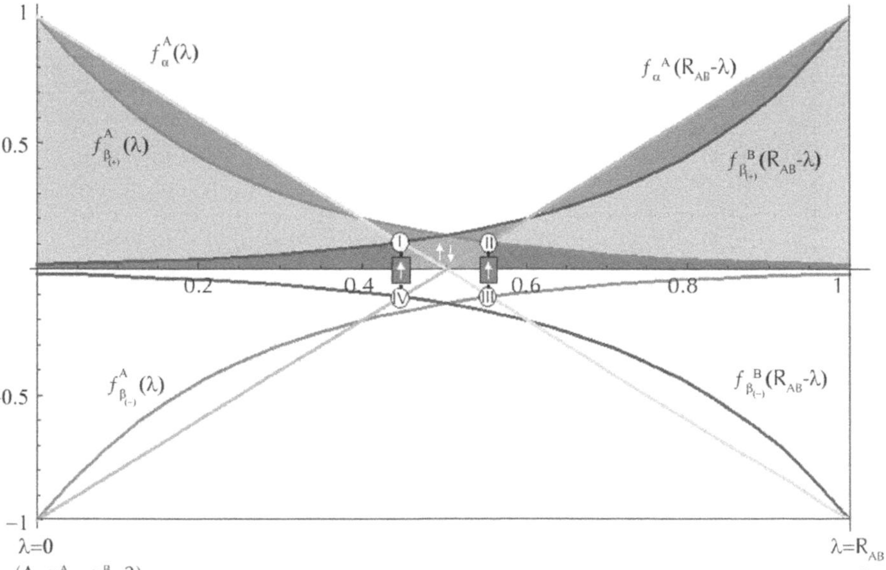

FIGURE 8.4 Geometric loci of the bonding regions as in Figure 8.1 for chemical binding from equal electronegativity influences of two systems A and B throughout equations (8.100)–(8.103) with Dirac binding functions (8.97)–(8.99) through constants and parametric settings as $\hbar = c = 1$, $\chi^A = \chi^B = 2$, $R_{AB} = 1$ (Putz, 2010a).

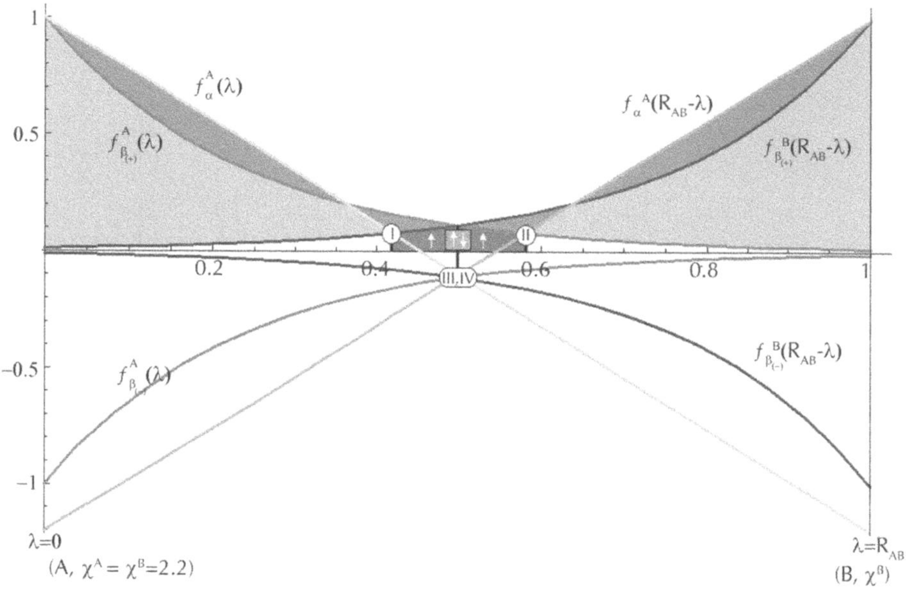

FIGURE 8.5 Geometric loci of the bonding regions as in Figure 8.1 for chemical binding from equal electronegativity influences of two systems A and B throughout equations (8.100)–(8.103) with Dirac binding functions (8.97)–(8.99) through constants and parametric settings as $\hbar = c = 1$, $\chi^A = \chi^B = 2.2$, $R_{AB} = 1$ (Putz, 2010a).

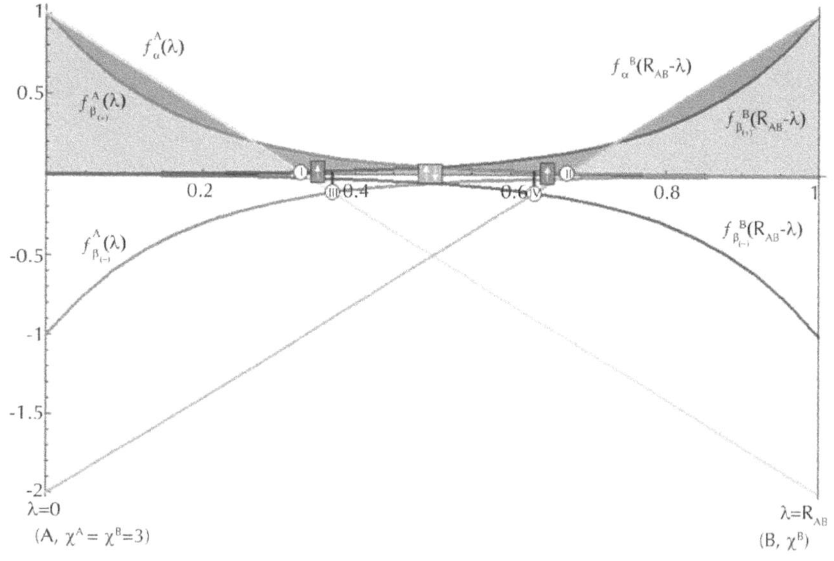

FIGURE 8.6 Geometric loci of the bonding regions as in Figure 8.1 for chemical binding from equal electronegativity influences of two systems A and B throughout equations (8.100)–(8.103) with Dirac binding functions (8.97)–(8.99) through constants and parametric settings as $\hbar = c = 1$, $\chi^A = \chi^B = 3$, $R_{AB} = 1$ (Putz, 2010a).

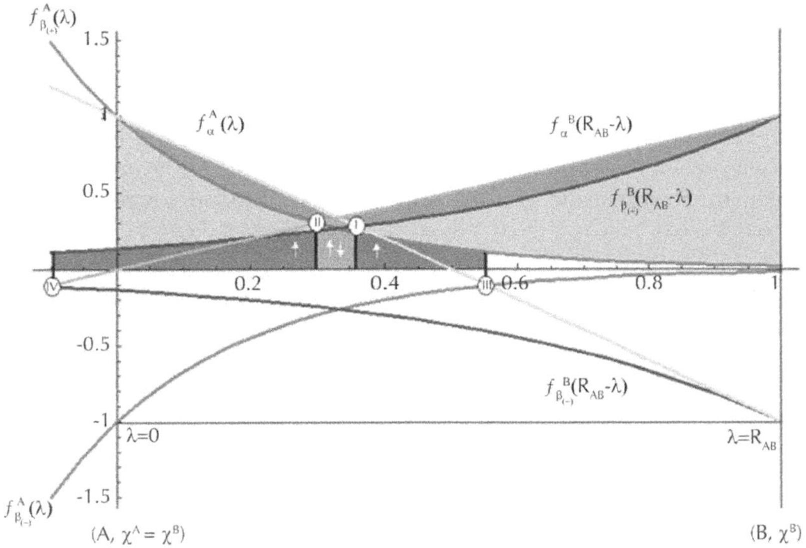

FIGURE 8.7 Geometric loci of the bonding regions as in Figure 8.2 for chemical binding from different electronegativity influences of two systems A and B throughout equations (8.100)–(8.103) with Dirac binding functions (8.97)–(8.99) through constants and parametric settings as $\hbar = c = 1$, $\chi^{A} = 2\chi^{B} = 2$, $R_{AB} = 1$ (Putz, 2010a).

Instead, the anti-bonding equation (8.58) is Dirac preserved either as in form and multiplicity. Due to this fact, depending on the electronegativity differences between the bonding partners the anti-bonding spin state may be located in various locations between the mixed (positive) bonding-antibonding crossing points I and II, equations (8.100) and (8.101), and the self (negative) bonding – antibonding crossing points III and IV, equations (8.102) and (8.103).

Even more, for equal electronegativity three types of parallel spin (antibonding) separation may arise as illustrated by the Figures 8.3, 8.4, and 8.6 that is as being delocalized outside, precisely localized at the edge and delocalized inside of the sigma-bonding region, respectively, while the Figure 8.5 illustrates the case when the bonding pairing is precisely localized on bond.

Actually, following the bonding points delivered by the system (8.100)–(8.103) one has, for the equal electronegativity cases of Figures 8.3–8.6, the following configurations respecting the electronegativity equal values, respectively:

$$\chi^{A} = \chi^{B} = 1:\ \underbrace{IV \le}_{\uparrow}\underbrace{II < I}_{\uparrow\downarrow}\underbrace{\le III}_{\uparrow} \tag{8.104}$$

providing the anti-bonding (parallel) spins delocalized outside of the region with delocalized anti-parallel pairing electrons;

$$\chi^{A} = \chi^{B} = 2:\ IV \underbrace{=}_{\uparrow} \underbrace{I < II}_{\uparrow\downarrow} \underbrace{=}_{\uparrow} III \tag{8.105}$$

providing the *precise localization of the anti-bonding (parallel) spins* at the margins of the region with delocalized anti-parallel pairing electrons;

$$\chi^A = \chi^B = 2.22: \underbrace{I < III}_{\uparrow} \underbrace{=}_{\uparrow\downarrow} \underbrace{IV < II}_{\uparrow} \tag{8.106}$$

providing the *precise localization of the anti-parallel pairing electrons* at the half of the bond length, being outside of it the delocalization of the anti-bonding (parallel) spins;

$$\chi^A = \chi^B = 3: \underbrace{I \leq}_{\uparrow} \underbrace{III < IV}_{\uparrow\downarrow} \underbrace{\leq II}_{\uparrow} \tag{8.107}$$

providing the limited delocalization of the anti-bonding (parallel) spins at the margins of regions with delocalized anti-parallel pairing electrons.

Worth observing that the case of the equal electronegativities presents the so called *critical electronegativity* fulfilling the equation (8.102), for instance, at the half of the bond length:

$$f_\alpha^{\mathrm{A}}\left(\frac{R_{\mathrm{AB}}}{2}, \chi\right) = f_{\beta(-)}^{Dir-\mathrm{A}}\left(\frac{R_{\mathrm{AB}}}{2}, \chi\right) \tag{8.108}$$

which give the information of precisely localization of pairing of anti-parallel electronic spins. As well, the precise localization of the anti-bonding parallel spins of equation (8.105), namely at the *critical distances*

$$\lambda_I = \lambda_{III} = 0.445571\,[R_{\mathrm{AB}}] \tag{8.109}$$

$$\lambda_{II} = \lambda_{IV} = 0.554429[R_{\mathrm{AB}}] \tag{8.110}$$

on the actual bond length scale, are candidate for universal application for the given (known) equal electronegativity as stipulated by the case (8.105), in adequate units.

However, for both the equation (8.108) and the set of equations (8.109) and (8.110), one providing localization of pairing anti-parallel spins at half distance ($R_{\mathrm{AB}}/2$) of bond length for critical electronegativity, and the other the critical localization of the parallel spins of the anti-bonding states for equal electronegativities equaling the twice of current units, respectively, furnishes important practical result when one likes to control the magnetic properties of quantum material composed by two aggregates.

The Figure 8.7 is nothing than the extension of the Figure 8.3 in ionic manner as was the case for Figure 8.2 different from Figure 8.1. Still, for the present case worth remarking that due to the Dirac treatment, that is by the involvement of the negative bonding function of (8.99) type, the actual model for anti-bonding state, although delocalized in Figures 8.3 and 8.7 is still finite respecting the infinite asymptotic space coverage in Figures 8.1 and 8.2.

The final remark regards the height of bonding probability that is more and more contracted for the increase equal electronegativity cases in Figures 8.3–8.6, while for different electronegativities the appreciable contraction of width of the sigma-bonding region is recorded, see Figures 8.2 and 8.7.

Overall, the inclusion of the binding functions of equations (8.97)–(8.99) into the Dirac spinor of equation (8.84) leaves with the binding equations (8.97)–(8.99) and the cases of bonding and anti-bonding spin states, while being able to identify the specific situation when either parallel or anti-parallel spin states are precisely localized along the bond lengths, with presumed higher importance in designing and controlling of atoms-in-molecules spin based reactivity, and of nano-composites.

KEYWORDS

- **Chemical action**
- **Chemical bonding**
- **Dirac binding functions**
- **Hellmann–Feynman theorem**
- **Hohenberg–Kohn theorem**
- **Kohn–Sham equations**
- **Molecular orbital theory**

9 Fermionic-Bosonic Picture of Chemical Bond

CONTENTS

9.1 INTRODUCTION

Within the Bose–Einstein condensates (BEC) – density functional Theory (DFT) context, the question of whether the quantum chemical bonding homopolar picture may allow for such special modeling extension is here responded in positive through employing the Gross–Pitaevsky equation for the condensed wave function of bosons as an extension to the consecrated Heitler–London quantum description when bonding and anti-bonding electrons are considered as pairing to form bosons with 0 and 1 magnetic moments (spins) respectively. The results show that the same form of the quantum chemical bonding and anti-bonding wave functions might characterize the bosonic condensed wave functions, respectively, however experiencing different energetic states (with the contribution of a $BEC_{ChemBond}$ term experiencing the gap of the so called bosonic-fermionic degeneracy of the chemical bonding) relying on the correcting the fermionic energetic spectrum by some fraction of it with the coupling related with the chemical hardness (from fermionic side) reduced by the bosonic strength (from the bosonic side). The present Heitler–London–Bose–Einstein (superfluid) picture may eventually lead with a reformulation and the control of the frontier reactivity as well as the more subtle quantum concepts as the delocalization, aromaticity or other collective manifestations in complex macromolecules and nanosystems assimilated with diluted systems of bosons of fermions (Putz, 2011d, 2011e, 2011f).

9.2 MOTIVATION FOR A BEC THEORY OF CHEMICAL BONDING

Traditionally, the chemical bond is quantum mechanically approached by the so called molecular orbital (MO) theory, according to which the pairing of atomic electrons in molecule results as an overlapping or superposition of mono-electronic atomic wave functions (orbitals or basis functions, usually of *hydrogen-like* atoms since these are known analytically that is *Slater-type orbitals* but other choices are possible like *Gaussian functions* from standard *basis sets*), through the molecular linear combinations of atomic orbitals

$$\psi_i = c_{1i}\varphi_1 + c_{2i}\varphi_2 + \ldots + c_{Ni}\varphi_N = \sum_{j=1}^{N} c_{ji}\varphi_j \tag{9.1}$$

as introduced Sir John Lennard–Jones (1929). The variational (or Hartree–Fock) procedure is used to determine the coefficients of the expansion and thus the MOs themselves (Corongiu, 2007; Glaesemann and Schmidt, 2010; Hartree, 1957; Pople and Nesbet, 1954; Slater, 1963). However, as the paradigmatic chemical bond of diatomic system is shown on Figure 9.1, depicting the bonding (lower) and anti-bonding (upper) states, in both spin and wave function symmetries, respectively; note that anti-bonding orbitals are symmetric in spin, asymmetric in spatial coordinates, being often labeled with an asterisk (*) on MO diagrams. However, the He–He interaction, that is too weak to produce chemical bonding, while resisting even from solidification under normal pressure even below of 1°K, opened the fruitful question of superfluid He, the prototype for the forthcoming BEC phenomena (London, 1938).

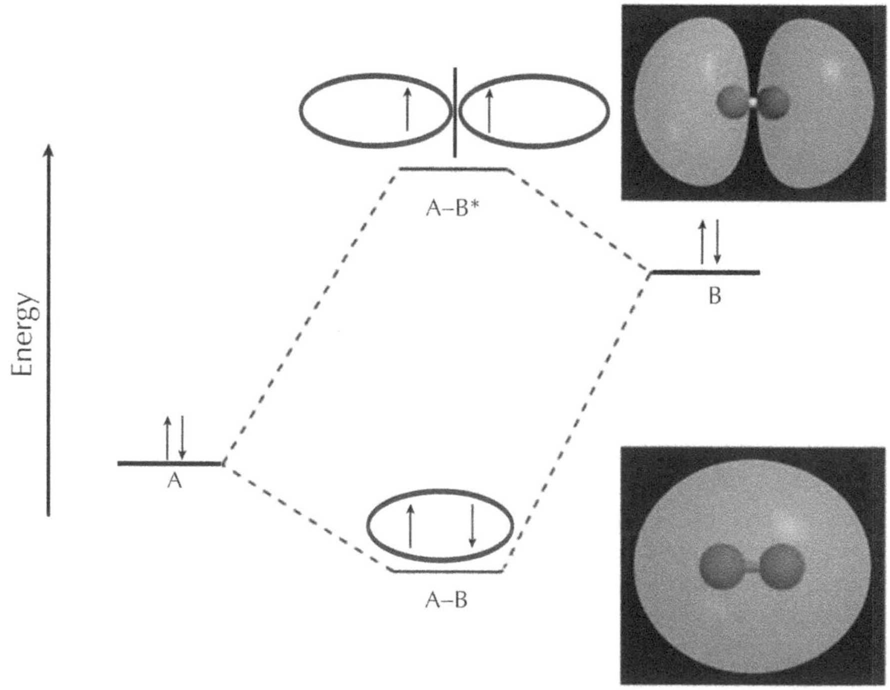

FIGURE 9.1 Typical bonding (lower) and anti-bonding (upper) MOs for an AB bonding between atoms A and B (Putz, 2011e).

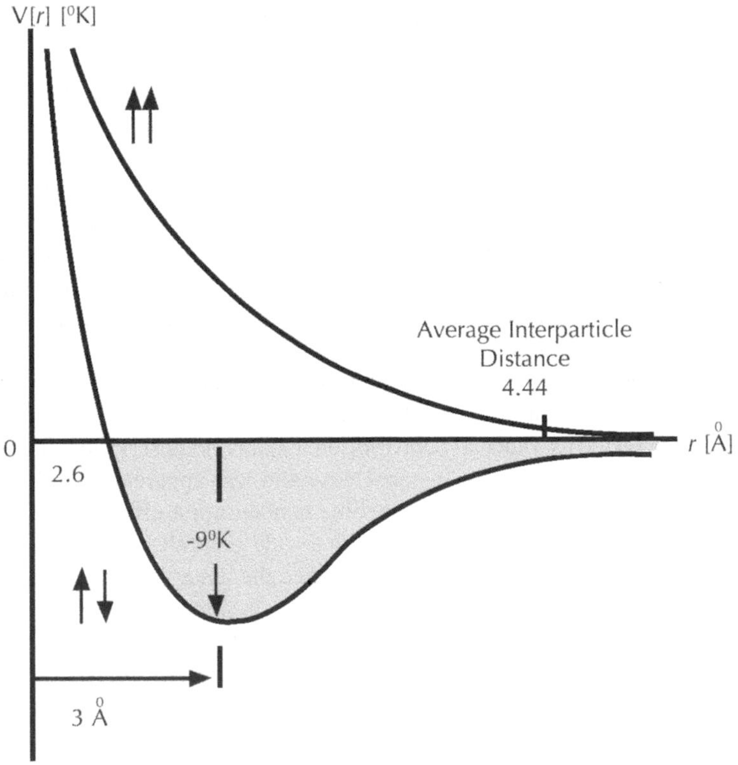

FIGURE 9.2 Potential energy representation (in kelvins) with homopolar He–He interaction along the inter-particle separation distance (Putz, 2011e).

The special behavior of He_2 is motivated by the inter-particle potential, see Figure 9.2, computed to have the analytical expression (Slater and Kirkwood, 1931),

$$V(r) = \left(5.67 \times 10^6\right) \exp\left[-21.5 \frac{r}{\sigma}\right] - 1.08 \left(\frac{\sigma}{r}\right)^6 \ldots [{}^\circ\mathrm{K}] \tag{9.2}$$

with the calculated average inter-particle distance $\sigma = 4.64\,\mathring{A}$, thus displaying a special kind of (Lennard–Jones) attractive potential; Nevertheless, on the other side, while investigating the quantum localization of He atoms (or, more precisely of their electrons) in bonding, for – say – a localization distance Δx relatively small compared with the potential range one has the energy uncertainty (in Boltzmann's constant units of temperature)

$$\Delta E \cong \frac{1}{2m}\left(\frac{\hbar}{\Delta x}\right) \xrightarrow{\Delta x \cong 0.5\,\mathring{A}} 10\,{}^\circ\mathrm{K} \tag{9.3}$$

that is comparable with the depths of the potential well, see the Figure 9.2; thus, the electronic localization in He_2 is impossible, in close agreement with the MO prediction of zero bonding order of chemical bond of He_2!

On the other side, BEC was primarily achieved in dilute gases of (weakly interacting) bosonic alkali atoms within external magnetic and/or optical traps (the external potential) with typical parameters as cc. 10^6 particles/condensate, the average density about 10^{12} cm^{-3} (i.e., the so called thermodynamic limit with practical infinite number of particle sin an infinite space $N \to \infty, V \to \infty$), with the transition temperature of order 10^{-7} K, however featuring smaller density (by ten orders magnitude) and temperature (by seven orders magnitude) (Anderson et al., 1995; Davis et al., 1995; Modugno et al., 2001). Such "extreme" behavior makes the BEC systems suitable for being treated by Thomas–Fermi approach, valid in systems with higher number of particles, as well as to its *N*-many body generalization as the celebrated DFT (Dreizler and Gross, 1995; Sholl and Steckel, 2009), especially for accurately treating the atoms and molecules in general and chemical bonding in special (Putz, 2008a).

Yet, the other way around, the resulted superfluid Landau theory of liquid Helium, and then incorporated in the more general BEC phenomenology (Bose, 1924; Einstein, 1925), should be reflected as a special contribution in quantum description of chemical bonding that eventually is dominant in some case and less apparent for others. This remark brings us in front of the quest of bosonic condensation effects in description of chemical bonding that will lead, for sure, with a wide generalization for the current quantum approaches of bonding and reactivity, since the assumed fermionic-bosonic mixture of electronic basins in molecules and complex systems.

Such approach is encouraged by the observation that since the chemical bonding is jointly represented by *bonding* and *anti-bonding* orbitals their electronic spin arrangements can be interpreted as states where electrons behave like *fermions* and *bosons*, respectively, coexisting as a mixture in the molecular space available.

Therefore, the present study unfolds the possible way of reconciling fermionic and bosonic behaviors in quantum description chemical bonding for the paradigmatic diatomic homopolar Heitler–London (1927) case.

To this end, the present chapter is organized as follows: in Section 9.3 the mean field theory of fermionic-bosonic mixed system is formulated so that proving such system may exists; in Section 9.4 the Gross–Pitaevsky (nonlinear) Schrödinger equation that appropriately describe bosonic diluted gas condensation is combined with the basic DFT concepts of Thomas–Fermi approximation to produce the first BEC–DFT connections; finally in the Section 9.5 the reformulation of the classical Heitler–London quantum description of chemical bond is undertaken with emphasis on the bosonic correction the replacement of the Schrödinger with Gross–Pitaevsky equation provides. The chapter ends with summary and perspective of the present BEC approach of chemical bonding.

9.3 PICTURES OF THE MEAN FIELD WITHIN FERMIONIC-BOSONIC MIXED SYSTEMS

Interesting picture of bosonic-fermionic mixture may be formulated when considered the above indicated functional integral formalism that it will now practice for illustrative purpose; for an in depth treatment (Rothel and Pelster, 2007). The starting point is to consider both the bosonic fields ψ_B & ψ_B^+ (commutative thus periodic on the imaginary time interval $\tau \in [0, \hbar\beta]$) and the fermionic fields ψ_F & ψ_F^+ (non-commutative so anti-periodic on the imaginary interval $\tau \in [0, \hbar\beta]$) as entering the Euclidean partition function while being weighted by the Boltzmann factor.

$$Z = \oint (D\psi_B)(D\psi_B^+)(D\psi_F)(D\psi_F^+)\exp\left\{-\frac{1}{\hbar}A\left[\psi_B,\psi_B^+,\psi_F,\psi_F^+\right]\right\} \tag{9.4}$$

Now, the total action of the bosonic-fermionic mixture may be decomposed as

$$A\left[\psi_B,\psi_B^+,\psi_F,\psi_F^+\right] = A_B\left[\psi_B,\psi_B^+\right] + A_F\left[\psi_F,\psi_F^+\right] + A_{BF}\left[\psi_B,\psi_B^+,\psi_F,\psi_F^+\right] \tag{9.5}$$

with the terms describing the bosonic, fermionic, and bosonic-fermionic contact interaction in the mixture, taking the forms, respectively

$$A_B\left[\psi_B,\psi_B^+\right] = \int_0^{\hbar\beta} d\tau \int d\mathrm{r}\psi_B^+(\mathrm{r},\tau)\left\{\hbar\frac{\partial}{\partial\tau} - \frac{\hbar^2}{2m_B}\Delta + \left[V_B(\mathrm{r}) - \mu_B\right] + \frac{g_{BB}}{2}\left|\psi_B(\mathrm{r},\tau)\right|^2\right\}\psi_B(\mathrm{r},\tau) \tag{9.6}$$

$$A_F\left[\psi_F,\psi_F^+\right] = \int_0^{\hbar\beta} d\tau \int d\mathrm{r}\psi_F^+(\mathrm{r},\tau)\left\{\hbar\frac{\partial}{\partial\tau} - \frac{\hbar^2}{2m_F}\Delta + \left[V_F(\mathrm{r}) - \mu_F\right]\right\}\psi_F(\mathrm{r},\tau) \tag{9.7}$$

$$A\left[\psi_B,\psi_B^+,\psi_F,\psi_F^+\right] = g_{BF}\int_0^{\hbar\beta} d\tau \int d\mathrm{r}\left|\psi_B(\mathrm{r},\tau)\right|^2\left|\psi_F(\mathrm{r},\tau)\right|^2 \tag{9.8}$$

as generalizing from the Gross–Pitaevsky time-dependent equation under the Wick rotation, $t = -i\hbar\beta = -i\hbar / k_B T$, see also (6.63). This happens when boson-boson and boson-fermion interactions are counted over the Fermi (chemical) potential influence (fermions), likely to happen in the "hidden" side of the chemical bonding too when bondons are staying for bosons formed by bindings of atoms-in-molecule with the coupling strengths

$$g_{BB} = \frac{4\pi a_{BB}\hbar^2}{m_B} \tag{9.9}$$

$$g_{BF} = 2\pi\hbar^2 a_{BF}\frac{m_{BB} + m_F}{m_{BB}m_F} \tag{9.10}$$

Next, the so called mean field theory is considered by assuming the condensate as composed by BEC + surrounding superfluid, paralleling the picture of ground + excited states of the condensate that can be further analytically represented by the so called background field Ψ (the total many-body BEC field in the ground state) and the fluctuation fields $\delta\psi$ that can be still associated with mono-bosonic fields that remains outside of BEC, yet as superfluid, in what is called *quantum depletion regime*. Note that the depletion is present even at 0K, so one cannot have 100% BEC like for the ideal gas! At T > 0K also the thermal effects add a thermal depletion to the already quantum depletion present, see Figure 9.3! Note that typical depletion is about 1% or less for alkali condensate, while being about 90% for liquid helium. However, one can rethink in quantum depletion terms about the chemical bonding, as a special (limiting) case of condensate (though pairing of electrons) in which for the mix gas of fermions (electrons of atoms-approaching-molecule) and bosons (electrons-in-bondons) if one has N_0 bosonic particle (bondons) in the condensate formed from the total of N valence contributing electrons, then the *bonding depletion* may be defined as

$$b_d = 1 - \frac{N_0(bondons)}{N(electrons)} \tag{9.11}$$

so resulting in the case of Hydrogen and Helium the interesting values

$$b_d = \begin{cases} 1 - \frac{1}{2} = 0.5 \ldots for \text{ H} \\ 1 - \frac{1}{4} = 0.75 \ldots for \text{ He} \end{cases} \tag{9.12}$$

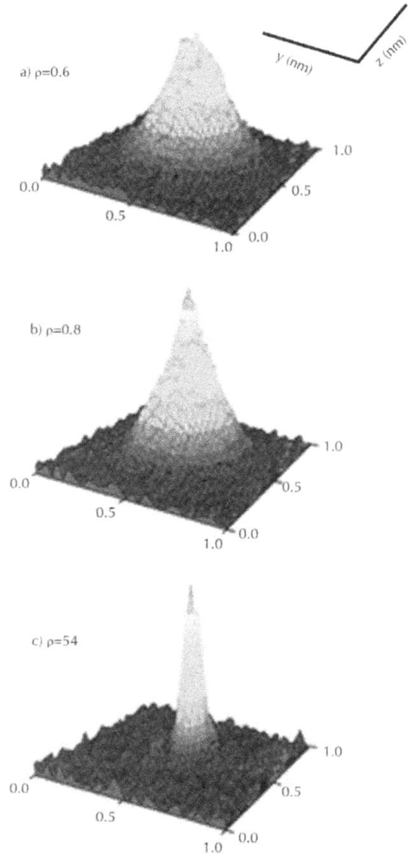

FIGURE 9.3 Bosonic condensates in terms of the BEC density of the background field takes as the order parameter $\rho(\mathbf{r}) \cong |\langle\Psi(\mathbf{r})\rangle|^2$, see Ψ of Figure 5.1(c), for ^{87}Rb realized by the experimental group of York University, Toronto: (a) Thermal cloud with N = 1.9 × 10^6 atoms at temperature T = 450 nK. (b) Mixed thermal atom cloud and BEC. where N = 1.8 × 10^6 and T = 400 nK. (c) Pure BE condensate where N = 4.2 × 10^5 atoms and T < 60 nK. (BEC-York Project, 2010).

These results offer a special vision on chemical bond (and its order) in the light of admitting the boson-fermion mixture as the electronic pairing arises and exists along the electronic valence of atoms-in-molecule. Namely, for Hydrogen the chemical bond is regarded as half bosonic condensation being accompanied by half fermionic depletion, while the Helium is predicted having more depletion than bonding condensation, in accordance with both facts that the He does not enter the ordinary MO-bonding picture, see Introduction, as well as behaving as a superfluid (i.e., even not as a BEC) with an depletion error departing only 15% by the recorded experimental result (see above) – that offers nevertheless fair qualitative description as far as only behavior is estimated by the introduced bonding depletion index. However, it is worth showing also by formal analytical arguments that in the fermionic-bosonic mixture (i.e., when the bosons are present) the BEC naturally appears. To this aim one considers the so called *background (or mean) field* framework (Jackiw, 1974), according which the bosonic condensation wave-function ψ is viewed as being driven by the *order parameter* $\langle\Psi\rangle$ (*or simply re-written as* Ψ) that is responsible for the BEC density $\rho(\mathbf{r}) \cong |\langle\Psi(\mathbf{r})\rangle|^2$ plus the fluctuating fields $\delta\psi$, namely written as

$$\psi_B(\mathbf{r},\tau) = \Psi_B(\mathbf{r},\tau) + \delta\psi_B(\mathbf{r},\tau) \tag{9.13}$$

$$\psi_B^+(\mathbf{r},\tau) = \Psi_B^+(\mathbf{r},\tau) + \delta\psi_B^+(\mathbf{r},\tau) \tag{9.14}$$

However, in the Gross–Pitaevsky picture the zero-order in fluctuation fields is considered that is the approach is restricted to the mean-field contribution – or the BEC itself, thus having the Euclidian action approximated as

$$\begin{aligned} &A\left[\psi_B, \psi_B^+, \psi_F, \psi_F^+\right] = A\left[\Psi_B + \delta\psi_B, \Psi_B^+ + \delta\psi_B^+, \psi_F, \psi_F^+\right] \cong A\left[\Psi_B, \Psi_B^+, \psi_F, \psi_F^+\right] \\ &= A_B\left[\Psi_B, \Psi_B^+\right] + A_F\left[\psi_F, \psi_F^+\right] + A_{BF}\left[\Psi_B, \Psi_B^+, \psi_F, \psi_F^+\right] \end{aligned} \tag{9.15}$$

So now, because of the order parametric (i.e., the average) nature of the mean fields, the functional integrations in above partition function for bosons simply become

$$\oint(D\psi_B) = \underbrace{\oint(D\langle\psi_B\rangle)}_{0} + \oint(D\delta\psi_B) = \oint(D\delta\psi_B) \tag{9.16}$$

$$\oint(D\psi_B^+) = \underbrace{\oint(D\langle\psi_B^+\rangle)}_{0} + \oint(D\delta\psi_B^+) = \oint(D\delta\psi_B^+) \tag{9.17}$$

with the consequence of being dropped out since we restrained to the zero-order of field fluctuations already.

In these conditions, one has the actual partition function rewritten as

$$\begin{aligned} \mathcal{Z}[\Psi_B, \Psi_B^+] = &\underbrace{\exp\left\{-\frac{1}{\hbar}A_B\left[\Psi_B, \Psi_B^+\right]\right\}}_{\mathcal{Z}_B[\Psi_B, \Psi_B^+]} \\ &\times \underbrace{\oint(D\psi_F)(D\psi_F^+)\exp\left\{-\frac{1}{\hbar}\left(A_F\left[\psi_F, \psi_F^+\right] + A_{BF}\left[\Psi_B, \Psi_B^+, \psi_F, \psi_F^+\right]\right)\right\}}_{\mathcal{Z}_{BF}[\Psi_B, \Psi_B^+]} \end{aligned} \tag{9.18}$$

This way, one yields the free energy with the form

$$\Omega[\Psi_B, \Psi_B^+] = -\frac{\ln \mathcal{Z}[\Psi_B, \Psi_B^+]}{\beta} = \frac{1}{\hbar\beta} A_B\left[\Psi_B, \Psi_B^+\right] - \frac{\ln \mathcal{Z}_{BF}[\Psi_B, \Psi_B^+]}{\beta} \tag{9.19}$$

having obtained therefore the conceptual confirmation that the BEC, whose basic index is the order parameter – here the mean or background fields Ψ_B, Ψ_B^+, appears indeed out of a bosonic-fermionic mixture. This picture allows therefore treating the chemical bonding as the particular bosonic condensation (with few bondons in a delocalizated bond, for instance) appeared out of the fermionic gas of valence electrons in atoms-encountering-into-molecule.

However, before exposing the chemical bonding formalism of BEC worth digressing a bit on the thermodynamic limit the condensation is required and of its influence and practical realization within the physical-chemical context, or more concrete within the Thomas–Fermi and DFT.

9.4 ANALYTICAL BEC-CHEMICAL HARDNESS CONNECTIONS

The practical implementation of BEC and of its mean field approximation usually makes use of thermodynamic limit constraint, namely for the systems with many-to-infinite number of particles ($N \to \infty$) the so called Thomas–Fermi approximation (Thomas, 1927; Fermi, 1927) may be used in such that the potential and the interaction energies are larger than the kinetic energy – that can be therefore neglected in the stationary Gross–Pitaevsky equation, reducing it to the algebraic form

$$|\psi|^2 \cong \frac{1}{g}[\mu - V(\mathbf{r})] \tag{9.20}$$

Now the discussion is whether this approximation is suitable to:

- BEC with finite numbers of bosons' interaction, as is the practical case;
- Chemical Bonding with single bosonic (bondonic) appearance.

The answer is fortunately given by a series of pioneering papers in physical philosophically questioning upon the statistical difference in systems with more or infinite more components (Anderson, 1972; Kadanoff, 2009; Park and Kim, 2010); the conclusion is simply that:

- "more is different" (Anderson, 1972)

where as

- "infinite more is the same" (Kadanoff, 2009)

These may be seen nothing but the re-definition of Fermi and Bose statistics, respectively. In other words, when more than one but finite number of particles are present in a system the statistical behavior is different than the single particle systems, while when infinite particle came into play they all end to behave like a (the) single particle case. For bosonic systems, further computational support is given by the plots of Figure 9.4, from where appears that within a given extent the bosonic systems with $N = 1,2,...\infty$ may be treated virtually the same.

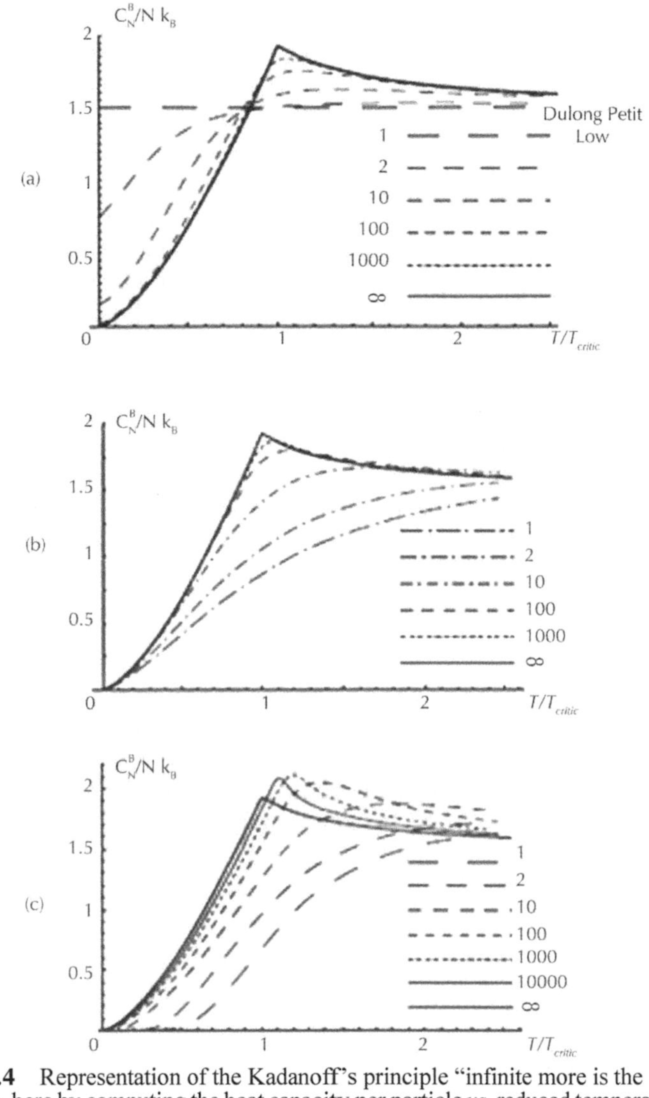

FIGURE 9.4 Representation of the Kadanoff's principle "infinite more is the same" (Park and Kim, 2010), here by computing the heat capacity per particle *vs.* reduced temperature of a bosonic sample with N = 1, 2, 10, 100, 1000, and ∞ size-dependent number of particles (i.e., for fixed particle density) for: (a) homogeneous non-interacting Bose gas; (b) homogeneous Bose gas with ground state (i.e., with separation of ground and excited modes); (c) Bose gas in a box potential (trapped potential); courtesy of Prof. Kleinert, from (Glaum et al., 2007; Kleinert, 2009).

Moreover, the Anderson assertion may be translated for chemical systems as

- "few electrons is different" in respecting their arrangement on atomic and MOs, while when they are regarded as bonding electrons in bondons the issue updates as
- "pairing electrons is the same" that consecrates the formed bondons like single-bosonic systems, formally mapped into the bosonic condensation of infinite bondons that behave the same as one!

Back to the Thomas–Fermi approximation, it is indeed in agreement with thermodynamic limit for bosonic condensation, while being usable also for the chemical systems from their chemical bond is present up to the limiting case of a single chemical bond systems (as the present paradigmatic homopolar bindings of molecular hydrogen and helium, H_2 and He_2).

However, the finite N-dependency may be restored from the Thomas–Fermi limit since the main equation of the DFT that minimizes the total energy density functional under the constrain of conservation of the total number of electrons in the system, (Chermette, 1999; Geerlings et al., 2003), being the equation (9.20) the striking *first connection between the custom BEC and DFT* formalisms of matter (we will specify latter what quantity is associate with fermions and what with bosons in the mixed bosonic-fermionic systems of chemical bonding). However, even more, once expanding the total energy functional according with the Hohenberg–Kohn–Sham rule (Hohenberg and Kohn, 1964; Kohn and Sham, 1965) in terms of the Hohenberg–Kohn functional $F_{HK}[\rho]$ and of the so called chemical action $C_A[\rho]$ - see Chapters 1 and 4 - the variational DFT principle release with the working relationship between the chemical potential, external potential and the density functional derivative of kinetic, Coulombic and exchange-correlation (XC) energies under equations (1.6) and (1.17). With these prescriptions the Thomas–Fermi BEC equation (9.20) takes the new form

$$|\psi|^2 \cong \frac{1}{g}\left[\frac{\delta F_{HK}[\rho]}{\delta\rho(\mathrm{r})}\right]_{V(\mathrm{r})} \tag{9.21}$$

that represents *the second BEC-DFT connection*, so to speak, since representing in fact another level of bosonic-fermionic coupling as far the Hohenberg–Kohn functional is associated with electrons and the rest of quantities represents the bosonic properties. Of course one can formulate also the bosonic version of the Hohenberg–Kohn functional, see Chapter 6.

However, the third working BEC-DFT connection can be established by employing the inter-bosonic average that successively gives

$$\begin{aligned}
\int d\mathrm{r}|\psi(\mathrm{r})|^4 &\cong \frac{1}{g^2}\int d\mathrm{r}[\mu - V(\mathrm{r})]^2 = \frac{1}{g^2}\int d\mathrm{r}\left[\nabla_\rho F_{HK}\right]^2 \\
&= \frac{1}{g^2}\left\{\int d\mathrm{r}\nabla_\rho\left[\left(\nabla_\rho F_{HK}\right)F_{HK}\right] - \int d\mathrm{r}F_{HK}\left[\nabla_\rho^2 F_{HK}\right]\right\} \\
&= \frac{1}{g^2}\int d\mathrm{r}\underbrace{\frac{d}{d\mathrm{r}}\left(\frac{d\mathrm{r}}{d\rho}\right)}_{\sim\frac{V^2}{N}}\left[\left(\nabla_\rho F_{HK}\right)F_{HK}\right] - \frac{1}{g^2}\int d\mathrm{r}F_{HK}\left[\nabla_\rho^2 F_{HK}\right] \\
&= \underbrace{\frac{V^2}{Ng^2}\int_\infty d\left[\left(\nabla_\rho F_{HK}\right)F_{HK}\right]}_{\downarrow\ 0} - \frac{F_{HK}}{g^2}\underbrace{\int d\mathrm{r}\frac{\delta^2 F_{HK}}{\delta\rho^2(\mathrm{r})}}_{\sim\eta} \\
\int d\mathrm{r}|\psi(\mathrm{r})|^4 &\cong -\frac{F_{HK}\eta}{g^2}
\end{aligned} \tag{9.22}$$

Where in above relationships the chemical hardness (Parr and Pearson, 1983) was recognized through its local integrated version (Berkowitz et al., 1985)

$$\eta = \int d\mathrm{r}\eta(\mathrm{r}), \tag{9.23}$$

$$\eta(\mathrm{r}) = \frac{\delta\mu}{\delta\rho(\mathrm{r})} = \frac{\delta^2 F_{HK}[\rho]}{\delta\rho^2(\mathrm{r})} \tag{9.24}$$

here adapted from the more general kernel version – see equation (4.18) and (Baekelandt et al., 1995), while using the (ρ, V) variable as independent ones (Putz, 2003, 2008b, 2008c, 2011a) and unitary Fukui function $f(\mathrm{r}) \to 1$ (Putz, 2011a) corresponding to a radicalic attack or reaens as an average of the HOMO and LUMO density influences (Parr and Yang, 1989).

These BEC–DFT connections that again are a sort of fermionic-bosonic mixture nature will be of first use in emphasizing the difference BEC phenomena induce on the chemical bonding, and in which degree is relevant to it, as it will be revealed in the next section.

9.5 REVISITING HEITLER-LONDON THEORY OF CHEMICAL BONDING

Now we are in position to rethink in the BEC terms upon the chemical bonding in its paradigmatic homopolar (A_2) version, firstly developed by Heitler and London (1927), while then becoming the pillar of the modern quantum (physical) approach of the chemical bonding (Ruedenberg, 1962). The starting point is to consider the Gross–Pitaevsky stationary Hamiltonian that can be formally reconsidered for chemical bond as being a mixed FB system thus casting as

$$\hat{H}_{FB} = \hat{H}_F + \hat{H}_B \tag{9.25}$$

with

$$\hat{H}_F = -\frac{\hbar^2}{2m_0}\nabla^2 + V(\mathrm{r}) \tag{9.26}$$

standing for electrons (with their mass m_0) in bonding, and

$$\hat{H}_B = g|\psi|^2 \tag{9.27}$$

accounting for the bosonic formation (and eventually of their condensation) by the self interaction driven by the self-interaction coupling – here looked as characterizing the bondons with the mass given in equation (7.67) and Putz (2010b) with the $(E_{bond}, X_{bond} = R_{bond})$ standing for the bonding equilibrium energy and distance. In other terms, the bondonic bosonic strength looks practically as

$$g_B = \left.\frac{4\pi a_B \hbar^2}{m_B}\right|_{a_B \to R_{bond}} = 8\pi E_{bond} R_{bond}^3 \tag{9.28}$$

from where the bonding volume and energy are resulting as main factors in diving bondons in bonding strength. Even more, since the bondonic volume dependence the bosonic self-interaction coupling can be further expressed through regaining the order parameter $\langle\psi(\mathrm{r})\rangle$ dependency by considering the cubic volume as the inscribing locus of the bonding so that we firstly have

$$R_{bond}^{3}=V_{bond}=\frac{N_{bondons}}{\underset{(bosons)}{\rho_{bodnons}(\mathrm{r})}}\overset{\begin{pmatrix}simple\\ bonding:\\ one\,bondon\end{pmatrix}}{=}\frac{1}{\left|\langle\psi(\mathrm{r})\rangle\right|^{2}}\tag{9.29}$$

and consecutively for the bosonic coupling

$$g_{B}=\frac{8\pi E_{bond}}{\left|\langle\psi(\mathrm{r})\rangle\right|^{2}}\tag{9.30}$$

This formula represents the bonding level reformulation for the inverse relationship between $\left|\langle\psi(\mathrm{r})\rangle\right|^{2}$ and g_B as already noted from the above Thomas–Fermi approximation. However, for consistency it is worth observing that the order parameter refers to the atomic gas bosonic property in isolate condition that is when it is treated only as bosonic condensate alone; it acquires the chemical bonding relevance when combined with the bonding energy in g_B.

Next, the expectation value of this bondonic-fermionic Hamiltonian is formed over the bondonic (bosonic) molecular wave-function raised from the superposition of the fermionic atomic wave-functions (the so called *build-in-bondonic* BB superposition of fermionic wave functions coming from the two identical atoms yet labeled as *A* and *B* for better emphasizing on the combined electronic effects among them)

$$\underbrace{\psi_{BB}(\mathrm{r})}_{BONDONIC}=\underbrace{c_A\psi_A(\mathrm{r})+c_B\psi_B(\mathrm{r})}_{FERMIONIC\ \ SUPERPOSITION},\ c_A,c_B\in\Re\tag{9.31}$$

thus giving the fermionic and bosonic contributions

$$\begin{aligned}&\left\langle\hat{H}_{FB}\right\rangle=\left\langle\hat{H}_{F}\right\rangle_{FERMIONIC\ SIDE}+\left\langle\hat{H}_{B}\right\rangle_{BOSONIC\ SIDE}\\&=\frac{\int d\mathrm{r}\left(c_A\psi_A(\mathrm{r})+c_B\psi_B(\mathrm{r})\right)^{*}\hat{H}_F\left(c_A\psi_A(\mathrm{r})+c_B\psi_B(\mathrm{r})\right)+g\int d\mathrm{r}\left|\psi_{BB}(\mathrm{r})\right|^{4}}{\int d\mathrm{r}\left(c_A\psi_A(\mathrm{r})+c_B\psi_B(\mathrm{r})\right)^{*}\left(c_A\psi_A(\mathrm{r})+c_B\psi_B(\mathrm{r})\right)}\end{aligned}\tag{9.32}$$

The homo-polar framework of binding is nevertheless reflected in the shortcuts for:

- Intra-atomic

$$H_{11}=\int d\mathrm{r}\psi_A^{*}(\mathrm{r})\hat{H}_F\psi_A(\mathrm{r})=\int d\mathrm{r}\psi_B^{*}(\mathrm{r})\hat{H}_F\psi_B(\mathrm{r})\tag{9.33}$$

- Inter-atomic

$$H_{12}=\int d\mathrm{r}\psi_A^{*}(\mathrm{r})\hat{H}_F\psi_B(\mathrm{r})=\int d\mathrm{r}\psi_B^{*}(\mathrm{r})\hat{H}_F\psi_A(\mathrm{r})\tag{9.34}$$

and overlapping

$$S=\int d\mathrm{r}\psi_A^{*}(\mathrm{r})\psi_B(\mathrm{r})=\int d\mathrm{r}\psi_B^{*}(\mathrm{r})\psi_A(\mathrm{r})\tag{9.35}$$

terms. Yet, through considering the above BEC-DFT expression for ψ^4 integral we further get the expression

$$\left\langle \hat{H}_{FB} \right\rangle \cong \frac{1}{c_A^2 + c_B^2 + 2c_A c_B S} \left\{ \left(c_A^2 + c_B^2\right) H_{11} + 2c_A c_B H_{12} - \frac{F_{HK}\eta_{Molec}}{g_B} \right\} \tag{9.36}$$

where we remain with bonding (bondonic) chemical hardness that is now assimilated with the chemical hardness of the molecule itself, while for the Hohenberg–Kohn functional we can still employ the previous basic DFT equation – see Section 4.2, equation (4.1) - adapted here as

$$\begin{aligned} F_{HK} &= \left\langle \hat{H}_{FB} \right\rangle - C_A \\ &= \left\langle \hat{H}_{FB} \right\rangle - \int d\mathrm{r} V(\mathrm{r}) \left[\left(c_A \psi_A(\mathrm{r}) + c_B \psi_A(\mathrm{r})\right)^* \left(c_A \psi_A(\mathrm{r}) + c_B \psi_A(\mathrm{r})\right) \right] \\ &= \left\langle \hat{H}_{FB} \right\rangle - \left(c_A^2 + c_B^2\right) V_{11} - 2c_A c_B V_{12} \end{aligned} \tag{9.37}$$

where we have introduced the natural potential integral notations paralleling the Coulombic and exchange contributions for electrons in bonding, namely

$$V_{11} = \int d\mathrm{r} \psi_A^*(\mathrm{r}) V(\mathrm{r}) \psi_A(\mathrm{r}) = \int d\mathrm{r} \psi_B^*(\mathrm{r}) V(\mathrm{r}) \psi_B(\mathrm{r}) \tag{9.38}$$

$$V_{12} = \int d\mathrm{r} \psi_A^*(\mathrm{r}) V(\mathrm{r}) \psi_B(\mathrm{r}) = \int d\mathrm{r} \psi_B^*(\mathrm{r}) V(\mathrm{r}) \psi_A(\mathrm{r}) \tag{9.39}$$

With these we obtain an algebraic equation for FB Hamiltonian expectation

$$\left\langle \hat{H}_{FB} \right\rangle \cong \frac{\left(c_A^2 + c_B^2\right)\left(H_{11} + \frac{\eta_{Molec}}{g_B} V_{11}\right) + 2c_A c_B \left(H_{12} + \frac{\eta_{Molec}}{g_B} V_{12}\right) - \left\langle \hat{H}_{FB} \right\rangle \frac{\eta_{Molec}}{g_B}}{c_A^2 + c_B^2 + 2c_A c_B S} \tag{9.40}$$

that can be conveniently rearranged as

$$\left\langle \hat{H}_{FB} \right\rangle \cong \frac{\left(c_A^2 + c_B^2\right)\left(H_{11} + \frac{\eta_{Molec}}{g_B} V_{11}\right) + 2c_A c_B \left(H_{12} + \frac{\eta_{Molec}}{g_B} V_{12}\right)}{c_A^2 + c_B^2 + 2c_A c_B S + \frac{\eta_{Molec}}{g_B}} \tag{9.41}$$

so that, comparing with the standard MO theory one can say that the BEC influence is "manifested" over MO binding in both energetic (nominator) and normalization (denominator) sides. However by the actual BEC-DFT connections we succeed in distributing the BEC effect such as the fundamental non-linear Schrödinger effects in Gross–Pitaevsky equation are absorbed by the chemical hardness (as fermionic-electronic quantity) and of its ratio respecting the self-bosonic (bondonic) coupling. This way the non-linear original problem, otherwise little solvable for non-zero potentials, was reformulated in an elegant manner within the traditional MO-binding problem, however corrected with additional terms through it all. Next, unfolding the variational principle in respecting the MO-coefficients

$$\begin{cases} \frac{\partial}{\partial c_A}\left\langle \hat{H}_{FB} \right\rangle = 0 \\ \frac{\partial}{\partial c_B}\left\langle \hat{H}_{FB} \right\rangle = 0 \end{cases} \tag{9.42}$$

the symmetric system in (c_A, c_B) coefficients is founded (Putz, 2011e, 2011f)

$$\begin{cases} c_A\left(H_{11} + \frac{\eta_{Molec}}{g_B}V_{11} - \left\langle \hat{H}_{FB} \right\rangle\right) + c_B\left(H_{12} + \frac{\eta_{Molec}}{g_B}V_{12} - \left\langle \hat{H}_{FB} \right\rangle S\right) = 0 \\ c_A\left(H_{12} + \frac{\eta_{Molec}}{g_B}V_{12} - \left\langle \hat{H}_{FB} \right\rangle S\right) + c_B\left(H_{11} + \frac{\eta_{Molec}}{g_B}V_{11} - \left\langle \hat{H}_{FB} \right\rangle\right) = 0 \end{cases} \tag{9.43}$$

whose secular determinant

$$\begin{vmatrix} H_{11} + \frac{\eta_{Molec}}{g_B}V_{11} - \left\langle \hat{H}_{FB} \right\rangle & H_{12} + \frac{\eta_{Molec}}{g_B}V_{12} - \left\langle \hat{H}_{FB} \right\rangle S \\ H_{12} + \frac{\eta_{Molec}}{g_B}V_{12} - \left\langle \hat{H}_{FB} \right\rangle S & H_{11} + \frac{\eta_{Molec}}{g_B}V_{11} - \left\langle \hat{H}_{FB} \right\rangle \end{vmatrix} = 0 \tag{9.44}$$

leads to the double-solution equation

$$H_{11} + \frac{\eta_{Molec}}{g_B}V_{11} - \left\langle \hat{H}_{FB} \right\rangle = \pm\left(H_{12} + \frac{\eta_{Molec}}{g_B}V_{12} - \left\langle \hat{H}_{FB} \right\rangle S\right) \tag{9.45}$$

- The first choice corresponds with the (+) sign in equation (9.45) and provides the *anti-bonding* energy and molecular wave-function

$$\begin{cases} \left\langle \hat{H}_{FB} \right\rangle^{\beta} \equiv E^{-} = \frac{H_{11} - H_{12}}{1-S} + \frac{\eta_{Molec}}{g_B}\frac{V_{11} - V_{12}}{1-S} \\ \psi_{BB}^{\beta}(\mathrm{r}) = \frac{1}{\sqrt{2-2S}}\left[\psi_A(\mathrm{r}) - \psi_A(\mathrm{r})\right] \end{cases} \tag{9.46}$$

- While the second choice of the (–) sign in equation (9.45) produces the *bonding* energy and molecular wave-function

$$\begin{cases} \left\langle \hat{H}_{FB} \right\rangle^{\alpha} \equiv E^{+} = \frac{H_{11} + H_{12}}{1+S} + \frac{\eta_{Molec}}{g_B}\frac{V_{11} + V_{12}}{1+S} \\ \psi_{BB}^{\alpha}(\mathrm{r}) = \frac{1}{\sqrt{2+2S}}\left[\psi_A(\mathrm{r}) + \psi_A(\mathrm{r})\right] \end{cases} \tag{9.47}$$

under the same condition of conservation of the number of bondons in the system (the N-body bosonic condition), however here reduced at unity since the basic *single chemical bond* considered that is

$$1 = \int d\mathbf{r}[\psi_{BB}(\mathbf{r})]^2 \tag{9.48}$$

However, one would like to comment and interpret the present BEC-DFT results on chemical bonding basic quantum-mechanically (orbital-molecular) theory. In this respect it is worth noting that

- The bosonic(bondonic)-fermionic mixture is assumed in chemical binding that is viewing the MO as representing the bondon yet formed by the fermionic electronic orbitalic contribution coming from the binding atoms;
- The obtained observed bonding and anti-bonding energies from the bondonic-fermionic mixture of chemical bonding display the custom fermionic part + additional potential part modulated by the bondonic coupling and the fermionic chemical hardness of the bonding and molecule, respectively;
- The resulted bonding and anti-bonding molecular wave functions shape essentially not-modified forms from the classical MO theory, leading with the idea that for the same orbital there are the fermionic and the FB forms of existence, however separated by the gaps $(\eta_{Molec}/g_B)[(V_{11} \pm V_{12})/(1 \pm S)]$, respectively. On the other side, one may argue that according with the Hugenholtz–Pines theorem (Hugenholtz and Pines, 1959), an interacting Bose gas does not exhibit an energy gap in the case of repulsive interactions, with $g > 0$, however we are talking here about Fermi–Bose mixture, so that the gap under discussion will be between the pure fermionic and fermionic-bosonic systems; we will call it therefore as the *bosonic-fermionic degeneration* of the chemical bonding.
- Regarding the position of the newly predicted energetic bosonic states, one can observe that since following the same algebraic forms of the pure fermionic leading terms, the bonding bosonic state will be pushed towards the total energy becoming more negative, while for the anti-bonding bosonic state the total energy is departed to become more positive – while being really observable in the bosonic temperature range of BEC that is far below 1°[K] for nanokelvins in fact; however, conceptually, since the quantum mechanical calculations for chemical systems tacitly assume the 0°[K] conditions – both the fermionic-bosonic lower ground state as well as the higher valence (anti-bonding) state are relevant in rethinking the chemical reactivity by the aid of frontier controlled orbitals.
- Remarkably, it is worth pointing that the bosonic nature of the chemical bonding primarily depends on the degree with which the atoms involved in bonding display the bosonic condensation when considered as single gas so that having non-zero order parameter $\langle\psi(\mathbf{r})\rangle \neq 0$ since the BEC side of the chemical bonding imply terms like (Putz, 2011d,2011e, 2011f)

$$BEC_{\substack{CHEMICAL\\BONDING}} = |\langle\psi(\mathbf{r})\rangle|^2 \frac{\eta_{Molec}}{8\pi E_{bond}} \frac{V_{11} \pm V_{12}}{1 \pm S} \tag{9.49}$$

- For the $(+)$ bonding and $(-)$ anti-bonding states of chemical bonding phenomenon, respectively. However, there is also relevant that replacing the BEC term in bonding and anti-bonding energies of equations (9.46) and (9.47) one gets the working equations (Putz, 2011d, 2011e, 2011f)

$$E_{bond}^{\pm} = \frac{H_{11} \pm H_{12}}{1 \pm S} + \left|\langle\psi(\mathrm{r})\rangle\right|^2 \frac{\eta_{Molec}}{8\pi E_{bond}^{\pm}} \frac{V_{11} \pm V_{12}}{1 \pm S} \tag{9.50}$$

that has the quadratic form

$$x = \Xi + \frac{\Theta}{x} \tag{9.51}$$

with identifications

$$\Xi \equiv \frac{H_{11} \pm H_{12}}{1 \pm S} \tag{9.52}$$

$$\Theta \equiv \left|\langle\psi(\mathrm{r})\rangle\right|^2 \frac{\eta_{Molec}}{8\pi} \frac{V_{11} \pm V_{12}}{1 \pm S} \tag{9.53}$$

The basic solutions of equation (9.51) look like

$$x_{1,2} = \frac{1}{2}\left[\Xi \pm \left(\Xi^2 + 4\Theta\right)^{1/2}\right] \tag{9.54}$$

and provides in the limit of small bosonic interaction that is $\Theta \sim \langle\psi\rangle \to 0$, as is the case of bondons in chemical bonding, the obtained working expressions

$$x_1 = -\frac{\Theta}{\Xi} \tag{9.55}$$

$$x_2 = \Xi + \frac{\Theta}{\Xi} \tag{9.56}$$

accounts for the actual BEC-DFT 2-fold bonding and anti-bonding energies:

$$E_{bond-BEC-I}^{\pm} = -\left|\langle\psi(\mathrm{r})\rangle\right|^2 \frac{\eta_{Molec}}{8\pi} \frac{V_{11} \pm V_{12}}{H_{11} \pm H_{12}} \tag{9.57}$$

$$E_{bond-BEC-II}^{\pm} = \frac{H_{11} \pm H_{12}}{1 \pm S} + \left|\langle\psi(\mathrm{r})\rangle\right|^2 \frac{\eta_{Molec}}{8\pi} \frac{V_{11} \pm V_{12}}{H_{11} \pm H_{12}} \tag{9.58}$$

The results of equations (9.57) and (9.58) considerably enlarge (and enrich) the chemical bonding paradigm – easily recognized in the first term of equation (9.58) when BEC phenomenology is completely neglected - such that to include the condensation and ordering phenomena, especially those associated with ^{4}He or alkali gases (Leggett, 2001).

- Equally relevant is the presence of the molecular chemical hardness on the nominator of the $BEC_{\mathrm{Chem.Bond}}$ term above – that further supports the condensation phenomenology by its consecrated maximum hardness principle for a stable molecule, see Chapter 4 and Chattaraj et al. (1995), Ayers and Parr (2000) that prescribes higher chemical stability for higher its chemical hardness.

Yet, it is still apparent that when an atomic system has higher propensity to form a bosonic condensate it does not mean that it will not support homopolar chemical bonding, as MO theory prescribes, but its superfluid property enters in the general scheme of chemical bonding (as here proved) with the amendment that indeed, due to its merely condensate manifestation, the superfluidic nature may prevail (see Figure 9.3) and hidden the chemical bonding mechanism.

In this respect, in fact, it was shown that *the chemical bonding quantum mechanically based scheme of matter condensation* (i.e., when bonding is seen as a particular case of condensation as well) *is quite general and applicable to any atomic system such as to include the Bose–Einstein effects*, when applicable, *as its "hidden" natural manifestation.* Further applications and extensions of the present ideas should be developed in the years to come for double and triple chemical bonding as well as to the delocalization of chemical bond over complex molecular structures.

9.6 THE PROMISES OF BOSE–EINSTEIN CONDENSATION IN MANY ELECTRONIC SYSTEMS

Viewed as the current comprehensive quantum chemical method, DFT begun being very popular in solid state physics since 1970s due to its versatility in providing relative low cost computations in comparison with the traditional (complicated) many-body wave-function approaches that is Hartree–Fock and its successors. Its main features that it makes particularly interesting for the future endeavours relay in:

- It is a many (N)-body theory whose *main vehicle is the density itself* $\rho(\mathrm{r})$ fulfilling the N- normalization condition (1.4) in remarkable similar way the Gross–Pitaevsky condition Section 5.4 stands; note that while reversing the equation (1.4) and the related ones, for the ground state density for instance ρ_0, the unique functional dependency of the associate ground state wave functional is (in principle) obtainable $\Psi_0 = \Psi[\rho_0]$; as a consequence, now, the ground state expectation values of an observable $\hat{O}$ is also a functional of ρ_0, $\hat{O}[\rho_0] = \langle \Psi[\rho_0] | \hat{O} | \Psi[\rho_0] \rangle$, and in particular the ground-state energy $E_0 = E[\rho_0] = \langle \Psi[\rho_0] | \hat{T} + \hat{V} + \hat{W} | \Psi[\rho_0] \rangle$ where the contribution of the *external potential* can be explicitly written for the ground state density but also for general density dependency $V[\rho] = \int d\mathrm{r} V(\mathrm{r})\rho(\mathrm{r})$ while $T[\rho]$ and $\hat{W}[\rho]$ are the kinetic inter-particle energetic functionals, respectively.
- It *supports the variational principle* in minimizing the energy functional $E[\rho]$ by means of Lagrangean method in terms of chemical potential $\delta\left(E[\rho] - \mu \int d\mathrm{r}\rho(\mathrm{r})\right) = 0$ while advancing for the energy functional the (Kohn–Sham) form that does not depend explicitly on the inter-particle interaction that is for the so called *auxiliary "KS-system"*.

On the other side, in the context of BEC phenomenology, the current theoretical methods in treating either the gas-to-condensate as well as the condensation-to-Mott insulator quantum transition phases for ultra cold bosonic diluted or trapped in an optical lattice systems are reviewed, while advancing the basic ideas and prospects in treating them unitarily within the DFT; the proposed DFT approach although never considered so far for modeling BEC quantum transition phases appears as a promising framework due to its inherent features: density as main variable, self-consistent equation, exchange and correlation effects included, local density and beyond approximations of density exchange and correlation density functionals available; as such, the future developments may address the outline in which employment of the DFT Kohn–Sham Hamiltonian in treating Bose condensates in optical lattices and critical phases should be provided, by various quantum techniques, furnishing therefore a viable alternative for the fashioned Bose–Hubbard model in describing the actual (and eventually predicting) the allied experiments.

For instance, one may proceed with various methods in determining the phase diagrams for bosonic gas/condensate or bosonic condensate/Mott insulator transitions by employing the Kohn–Sham Hamiltonian expectation values of Chapter 6 for main Bose–Hubbard models of Section 5.5 or their extension in perturbation or in variational perturbation ways according with the synopsis algorithm (Bradlyn et al., 2009; dos Santos and Pelster, 2009):

(i) Calculate grand-canonical partition function

$$\mathcal{Z}_{DFT} = \mathrm{Tr}\left(e^{-\beta \hat{H}_{KS}\left(\Psi,\Psi^{+}\right)}\right) \tag{9.59}$$

(ii) Form the grand-canonical free energy

$$\Omega_{DFT}\left(\Psi,\Psi^{+}\right) = -\frac{1}{\beta}\ln \mathcal{Z}_{DFT} \tag{9.60}$$

(iii) Landau expanding grand canonical free energy in order parameter, since it is small near the transition phase boundary

$$\Omega_{DFT}\left(\Psi,\Psi^{+}\right) = M\left\{a_0^{DFT}(T) + a_2^{DFT}(T)\left|\Psi\right|^2 + a_4^{DFT}(T)\left|\Psi\right|^4 + \ldots\right\} \tag{9.61}$$

(iv) Obtaining the parameter space with the points of phase boundary by imposing the second-order phase transition condition

$$a_2^{DFT}(T) = 0 \tag{9.62}$$

as the specific realization of the BEC phase transition conditions.

Overall, it appears the DFT as a promising self-consistent framework to model the dilute or strong bosonic interactions and their phase transitions in different optical lattice environments. Its main advantage, respecting the current available methods, resides in a plethora of ways the exchange-correlation functional $F_{XC}[\rho(\mathrm{r})]$ may be considered through the *exchange* and *correlation* contributions, which in Bose–Hubbard model's terms are seen as new realizations for the *hopping* and *onsite* effects of cold atomic bosons in optical lattices, respectively; this way the proposed DFT-BEC treatment may eventually provide a fruitful approach to be used in conceptual and computational treatment of BEC by improving the Bose–Hubbard model and of its theoretical predictions to available and further experiments (Lewenstein et al., 2007).

KEYWORDS

- **Bose–Einstein condensates**
- **Bosonic-fermionic mixture**
- **Density functional theory**
- **Fermionic-bosonic systems**
- **Gross–Pitaevsky equation**
- **Hohenberg–Kohn functional**
- **Molecular orbital theory**
- **Thomas–Fermi approximation**

References

Albus, A. P.; Illuminati, F.; and Wilkens, M. (2003). Ground-state properties of trapped Bose-Fermi mixtures: Role of exchange correlation. *Phys. Rev. A*, **67**, 063606.

Alonso, J. A. and Girifalco, L. A. (1978). Nonlocal approximation to the exchange potential and kinetic energy of an inhomogeneous electron gas. *Phys. Rev. B*, **17**, 3735–3743.

Anderson, M. H.; Ensher, J. R.; Matthews, M. R.; Wieman, C. E. and Cornell, E. A. (1995). Observation of Bose-Einstein condensation in a dilute atomic vapor. *Science*, **269**, 198–201.

Anderson, P. (1972). More is different. *Sci. New Ser.*, **177**, 393–396.

Anderson, P. W. (1966). Considerations on the flow of superfluid Helium. *Rev. Mod. Phys.*, **38**, 298–301.

Argaman, N. and Band Y. B. (2011). Finite-temperature density-functional theory of Bose-Einstein condensates. *Phys. Rev. A*, **83**, 023612.

Ayers, P. W. and Parr, R. G. (2000). Variational principles for describing chemical reactions: the Fukui function and chemical hardness revisited. *J. Am. Chem. Soc.*, **122**, 2010–2018.

Ayers, P. W. and Parr, R. G. (2001). Variational principles for describing chemical reactions: reactivity indices based on the external potential. *J. Am. Chem. Soc.*, **123**, 2007–2017.

Bader, R. F. W. (1990). *Atoms in Molecules—A Quantum Theory*. Oxford University Press, Oxford.

Bader, R. F. W. (1994). Principle of stationary action and the definition of a proper open system. *Phys. Rev. B*, **49**, 13348–13356.

Bader, R. F. W. (1998). A bond path: A universal indicator of bonded interactions. *J. Phys. Chem. A*, **102**, 7314–7323.

Bader, R. F. W.; Gillespie, R. J.; and MacDougall, P. J. (1988). A physical basis for the VSEPR model of molecular geometry. *J. Am. Chem. Soc.*, **110**, 7329–7336.

Baekelandt, B. G.; Cedillo, A.; and Parr, R. G. (1995). Reactivity indices and fluctuation formulas in density functional theory: Isomorphic ensembles and a new measure of local hardness. *J. Chem. Phys.*, **103**, 8548–8556.

Bakr, W. S.; Peng, A.; Tai, M. E.; Ma, R.; Simon, J.; Gillen, J. I.; Fölling, S.; Pollet, L.; and Greiner, M. (2010). Probing the superfluid–to–Mott insulator transition at the single-atom level. *Science*, **329**, 547–550.

Bamzai, A. S. and Deb, B. M. (1981). The role of single-particle density in chemistry. *Rev. Mod. Phys.*, **53**, 95–126.

Bartolotti, L. J. (1981). Time-dependent extension of the Hohenberg-Kohn-Levy energy-density functional. *Phys. Rev. A*, **24**, 1661–1667.

Bartolotti, L. J. (1982). A new gradient expansion of the exchange energy to be used in density functional calculations on atoms. *J. Chem. Phys.*, **76**, 6057–6059.

Bartolotti, L. J. and Acharya, P. K. (in press). On the functional derivative of the kinetic energy density functional. *J. Chem. Phys.*, **77**, 4576–4585.

Becke, A. D. (in press). Density functional calculations of molecular bond energies. *J. Chem. Phys.*, **84**, 4524–4529.

Becke, A. D. (1988). Density-functional exchange-energy approximation with correct asymptotic behavior. *Phys. Rev. A*, **38**, 3098–3100.

Becke, A. D. and Edgecombe, K. E. (1990). A Simple measure of electron localization in atomic and molecular systems. *J. Chem. Phys.*, **92**, 5397–5403.

BEC-York Project, http://www.yorku.ca/wlaser/projects/projects BEC.htm (accessed July 2010).

Berkowitz, M. (1986). Exponential approximation for the density matrix and the Wigner's distribution. *Chem. Phys. Lett.*, **129**, 486–488.

Berkowitz, M. (1987). Density functional approach to frontier controlled reactions. *J. Am. Chem. Soc.*, **109**, 4823–4825.

Berkowitz, M.; Ghosh, S. K.; and Parr, R. G. (1985). On the concept of local hardness in chemistry. *J. Am. Chem. Soc.*, **107**, 6811–6814.

Berkowitz, M. and Parr, R. G. (1988). Molecular hardness and softness, local hardness and softness, hardness and softness kernels, and relations among these quantities. *J. Chem. Phys.*, **88**, 2554–2557.

Berlin, T. (1951). Binding regions in diatomic molecules. *J. Chem. Phys.*, **19**, 208–213.

Bernier, J. S.; Kollath, C.; Georges, A.; De Leo, L.; Gerbier, F.; Salomon, C.; and Köhl, M. (2009). Cooling fermionic atoms in optical lattices by shaping the confinement. *Phys. Rev. A*, **79**, 061601.

Bloch, I. (2004). Quantum gases in optical lattices. *Physics World*, **17**, 25.

Bloch, I. (2008). Quantum coherence and entanglement with ultracold atoms in optical lattices. *Nature*, **453**, 1016–1022.

Boeyens, J. C. A. (2005). *New Theories for Chemistry*. Elsevier, New York, USA.

Boeyens, J. C. A. (2010). Emergent properties in Bohmian chemistry. In *Quantum Frontiers of Atoms and Molecules*, M. V. Putz, (Ed.), NOVA Science Inc., New York.

Bogoliubov, N. N. (1947). On the theory of superfluidity. *J. Phys. USSR*, **11**, 23–32.

Bohm, D. (1952a). A suggested interpretation of the quantum theory in terms of "hidden" variables. I. *Phys. Rev.*, **85**, 166–179.

Bohm, D. (1952b). A suggested interpretation of the quantum theory in terms of "hidden" variables. II. *Phys. Rev.*, **85**, 180–193.

Bohm, D. and Vigier, J. P. (1954). Model of the causal interpretation of quantum theory in terms of a fluid with irregular fluctuations. *Phys. Rev.*, **96**, 208–216.

Bohr, N. (1935). Can quantum-mechanical description of physical reality be considered complete? *Phys. Rev.*, **48**, 696–702.

Bose, S. N. (1924). Plancks Gesetz und Lichtquantenhypothese. *Z. Phys.*, **26**, 178–181.

Bradley, C. C.; Sackett, C. A.; Tollett, J. J.; and Hulet, R. G. (1995). Evidence of Bose-Einstein condensation in an atomic gas with attractive interactions. *Phys. Rev. Lett.*, **75**, 1687–1690.

Bradlyn, B.; dos Santos, F. E. A.; and Pelster, A. (2009). Effective action approach for quantum phase transitions in bosonic lattices. *Phys. Rev. A*, **79**, 13615.

Brand, J. (2004). A density-functional approach to fermionization in the 1D Bose gas. *J. Phys. B: At. Mol. Opt. Phys.*, **37**, S287–S300.

Capelle, K. (2006). A bird's-eye view of density-functional theory. *Brazilian J. Phys*, **36**, 1318–1343. [arXiv:cond-mat/0211443v5]

Capitani, J. F.; Nelewajski, R. F.; and Parr, R. G. (1982). Non-born-oppenheimer density functional theory of molecular systems. *J. Chem. Phys.*, **76**, 568–573.

Cedillo, A.; Robles, J.; and Gazquez, J. L. (1988). New nonlocal exchange-energy functional from a kinetic-energy-density padé-approximant model. *Phys. Rev. A*, **38**, 1697–1701.

Chan, G. K. L. and Handy, N. C. (1999). Kinetic-energy systems, density scaling, and homogeneity relations in density functional theory. *Phys. Rev. A*, **59**, 2670–2679.

Chattaraj, P. K.; Lee, H.; and Parr, R. G. (1991). Principle of maximum hardness. *J. Am. Chem. Soc.*, **113**, 1854–1855.

Chattaraj, P. K.; Liu, G. H.; and Parr, R. G. (1995). The maximum hardness principle in the Gyftpoulos-Hatsopoulos three-level model for an atomic or molecular species and its positive and negative ions. *Chem. Phys. Lett.*, **237**, 171–176.

Chattaraj, P. K. and Maiti, B. (2003). HSAB principle applied to the time evolution of chemical reactions. *J. Am. Chem. Soc.*, **125**, 2705–2710.

Chattaraj, P. K. and Schleyer, P. V. R. (1994). An ab initio study resulting in a greater understanding of the HSAB principle. *J. Am. Chem. Soc.*, **116**, 1067–1071.

Chen, J. and Stott, M. J. (1991a). V-representability for systems of a few fermions. *Phys. Rev. A*, **44**, 2809–2814.

Chen, J. and Stott, M. J. (1991b). V-representability for systems with low degeneracy. *Phys. Rev. A*, **44**, 2816–2821.

Chermette, H. (1999). Chemical reactivity indexes in density functional theory. *J. Comput. Chem.*, **20**, 129–154.

Cioslowski, J. (2007). Non-nuclear attractors in the Li_2 molecule. *J. Phys. Chem.*, **94**, 5496–5498.

Corongiu, G. (2007). The Hartree-Fock-Heitler-London method, III: Correlated diatomic hydrides. *J. Phys. Chem. A*, **111**, 5333–5342.

Dalfovo, F.; Giorgini, S.; Pitaevskii, L. P.; and Stringari, S. (1999). Theory of Bose-Einstein condensation in trapped gases. *Rev. Mod. Phys.*, **71**, 463–512.

Daudel, R.; Leroy, G.; Peeters, D.; and Sana, M. (1983). *Quantum Chemistry*. Wiley, New York, USA.

Davis, K. B.; Mewes, M. O.; Andrews, M. R.; van Druten, N. J.; Durfee, D. S.; Kurn, D. M.; and Ketterle, W. (1995). Bose-Einstein condensation in a gas of Sodium atoms. *Phys. Rev. Lett.*, **75**, 3969–3973.

de Broglie, L. (1923). Ondes et quanta. *Compt. Rend. Acad. Sci. (Paris)*, **177**, 507–510.

de Broglie, L. (1925). Sur la fréquence propre de l'électron. *Compt. Rend. Acad. Sci. (Paris)*, **180**, 498–500.

de Broglie, L. and Vigier, M. J. P. (1953). *La Physique Quantique Restera-t-elle Indéterministe?* Gauthier-Villars: Paris, France.

De Proft, F.; Liu, S.; Parr, R. G. (1997). Chemical potential, hardness and softness kernel and local hardness in the isomorphic ensemble of density functional theory. *J. Chem. Phys.*, **107**, 3000–3006.

DeMarco, B. (2010). An atomic view of quantum phase transition. *Science*, **329**, 523–524.

DePristo, A. E. (1996). Hohenberg-Kohn density-functional-theory as an implicit poisson equation for density changes from summed fragment densities. *Phys. Rev. A*, **54**, 3863–3869.

DePristo, A. E. and Kress, J. D. (1987). Kinetic-energy functionals via Padé approximations. *Phys. Rev. A*, **35**, 438–441.

Dirac, P. A. M. (1928). The quantum theory of the electron. *Proc. Roy. Soc. (London)*, **A117**, 610–624.

Dirac, P. A. M. (1929). Quantum mechanics of many-electron systems. *Proc. Roy. Soc. (London)*, **A123**, 714–733.

Doering, W. V. and Detert, F. (1951). Cycloheptatrienylium oxide. *J. Am. Chem. Soc.*, **73**, 876–877.

dos Santos, F. E. A. and Pelster A. (2009). Quantum phase diagram of bosons in optical lattices. *Phys. Rev. A*, **79**, 013614.

Drago, R. S. and Kabler, R. A. (1972). Quantitative evaluation of the HSAB [hard-soft acid-base] concept. *Inorg. Chem.*, **11**, 3144–3145.

Dreizler, R. M. and Gross, E. K. U. (1990). *Density Functional Theory*. Springer Verlag, Heidelberg.

Dufek, P.; Blaha, P.; Sliwko, V.; and Schwarz, K. (1994). Generalized-gradient-approximation description of band splittings in transition-metal oxides and fluorides. *Phys. Rev. B*, **49**, 10170–10175.

Dunlap, B. I. and Andzelm, J. (1992). Second derivatives of the local-density-functional total energy when the local potential is fitted. *Phys. Rev. A*, **45**, 81–86.

Dunlap, B. I.; Andzelm, J.; and Mintmire, J. W. (1990). Local-density-functional total energy gradients in the linear combination of gaussian-type orbitals method. *Phys. Rev. A*, **42**, 6354–6358.

Einstein, A. (1905a). On the electrodynamics of moving bodies. *Ann. Physik (Leipzig)*, **17**, 891–921.

Einstein, A. (1905b). Does the inertia of a body depend upon its energy content? *Ann. Physik (Leipzig)*, **18**, 639–641.

Einstein, A. (1905c). On a Heuristic viewpoint concerning the production and transformation of light. *Ann. Physik (Leipzig)*, **17**, 132–148.

Einstein, A. (1924). Quantentheorie des einatomigen idealen Gases, *Sitzber. Kgl. Preuss. Akad. Wiss.*, 261–267.

ibidem (1925), 3–14.

Einstein, A.; Podolsky, B. and Rosen, N. (1935). Can quantum-mechanical description of physical reality be considered complete? *Phys. Rev.*, **47**, 777–780.

Ernzerhof, M. (1994). Density-functional theory as an example for the construction of stationarity principles. *Phys. Rev. A*, **49**, 76–79.

Fermi, E. (1927). Un metodo statistico per la determinazione di alcune priopriètà dell'atomo. *Rend. Accad. Naz. Lincei*, **6**, 602–607.

Fermi, E. (1936). Motion of neutrons in hydrogenous substances. *Ricerca Sci.*, **2**, 13–52.

Fetter, A. L. and Walecka, J. D. (2003). *Quantum Theory of Many-Particle Systems*; Dover: New York, **2003**.

Feynman, R. P. (1939). Forces in molecules. *Phys. Rev.*, **56**, 340–343.

Findlay, A. (1955). *Practical Physical Chemistry*. Longmans: London, UK.

Fiolhais, C.; Nogueira, F.; and Marques, M. (Eds.) (2003). *A Primer in Density Functional Theory*. Springer-Verlag: Berlin.

Fisher, M. P. A.; Weichman, P. B.; Grinstein, G.; and Fisher, D. S. (1989). Boson localization and the superfluid-insulator transition. *Phys. Rev. B*, **40**, 546–570.

Flores, J. A. and Keller, J. (1992). Differential equations for the square root of the electronic density in symmetry-constrained density-functional theory. *Phys. Rev. A*, **45**, 6259–6262.

Freeman, S. (1974). *Applications of Laser Raman Spectroscopy*. John Wiley and Sons, New York, USA.

Fried, D. G.; Killian, T. C.; Willmann, L.; Landhuis, D.; Moss, S. C.; Kleppner, D.; and Greytak, T. J. (1998). Bose-Einstein condensation of atomic hydrogen. *Phys. Rev. Lett.*, **81**, 3811–3814.

Garza, J. and Robles, J. (1993). Density functional theory softness kernel. *Phys. Rev. A*, **47**, 2680–2685.

Gaspar, R. and Nagy, A. (1987). Local-density-functional approximation for exchange-correlation potential. Application of the self-consistent and statistical exchange-correlation parameters to the calculation of the electron binding energies. *Theor. Chim Acta*, **72**, 393–401.

Gázquez, J. L; Galván, M; and Vela, A. (1990). Chemical reactivity in density functional theory: The N-differentiability problem. *J. Mol. Structure (Theochem)*, **210**, 29–38.

Geerlings, P. and De Proft, F. (2001). Conceptual and computational DFT in the study of aromaticity. *Chem. Rev.*, **101**, 1451–1464.

Geerlings, P.; De Proft, F.; and Langenaeker, W. (2003). Conceptual density functional theory. *Chem. Rev.*, **103**, 1793–1874.

Ghosh, D. C. (2005). A new scale of electronegativity based on absolute radii of atoms. *J. Theor. Comp. Chem.*, **4**, 21–23.

Ghosh, D. C. and Biswas, R. (2002). Theoretical calculation of absolute radii of atoms and ions. Part 1. The atomic radii. *Int. J. Mol. Sci.*, **3**, 87–113.

Ghosh, S. K. and Parr, R. G. (1986). Phase-space approach to the exchange energy functional of density-functional theory. *Phys. Rev. A*, **34**, 785–791.

Gillespie, R. J. (1970). The electron-pair repulsion model for molecular geometry. *J. Chem. Educ.*, **47**, 18–23.

Ginzburg, V. L. (2004). On superconductivity and superfluidity (what I have and have not managed to do), as well as on the "physical minimum" at the beginning of the 21st century. *Chemphyschem*, **5**, 930–945.

Glaesemann, K. R. and Schmidt, M. W. (2010). On the ordering of orbital energies in high-spin ROHF. *J. Phys. Chem. A*, **114**, 8772–8777.

Glaum, K.; Kleinert, H.; and Pelster, A. (2007). Condensation of ideal Bose gas confined in a box within a canonical ensemble. *Phys. Rev. A*, **76**, 063604.

Greiner, M.; Mandel, O.; Esslinger, T.; Hänsch, T. W.; Bloch, I. (2002). Quantum phase transition from a superfluid to a Mott insulator in a gas of ultracold atoms. *Nature*, **415**, 39–44.

Griesmaier, A.; Werner, J.; Hensler, S.; Stuhler, J.; and Pfau, T. (2005). Bose-Einstein condensation of Chromium. *Phys. Rev. Lett.*, **94**, 160401.

Gross, E. P. (1961). Structure of a quantized vortex in boson systems. *Nuovo Cimento*, **20**, 454–477.

Gross, E. K. U.; Oliveira, L. N.; and Kohn, W. (1988a). Density-functional theory for ensembles of fractionally occupied states. I. Basic formalism. *Phys. Rev. A*, **37**, 2809–2820.

Gross, E. K. U.; Oliveira, L. N.; Kohn, W. (1988b). Density-functional theory for ensembles of fractionally occupied states. II. Application to the He atom. *Phys. Rev. A*, **37**, 2821–2833.

Gunton, J. D. and Buckingham, M. J. (1968). Condensation of the ideal Bose gas as a cooperative transition. *Phys. Rev.*, **166**, 152–158.

Guo, Y. and Whitehead, M. A. (1991). Generalized local-spin-density-functional theory. *Phys. Rev. A*, **43**, 95–108.

Harriman, J. E. (1978). *Theoretical Foundations of Electronic Spin Resonance*. Academic Press, New York.

Harriman, J. E. (1978a). Geometry of density matrices. I. Definitions, *N* matrices and 1 matrices. *Phys. Rev. A*, **17**, 1249–1255.

Harriman, J. E. (1978b). *Geometry of density matrices. II. Reduced density matrices and N representability. Phys. Rev. A*, **17**, 1257–1268.

Harriman, J. E. (1983). *Geometry of density matrices. IV. The relationship between density matrices and densities. Phys. Rev. A*, **27**, 632–645.

Harriman, J. E. (1984). *Geometry of density matrices. V. Eigenstates. Phys. Rev. A*, **30**, 19–29.

Harriman, J. E. (1986). Densities, operators, and basis sets. *Phys. Rev. A*, **34**, 29–39.

Hartree, D. R. (1957). *The Calculation of Atomic Structures*. Wiley & Sons: New York, USA.

Hassani, S. (1991). *Foundation of Mathematical Physics*. Prentice-Hall International, Inc., Chapter 7.

Head-Gordon, M.; Pople, J. A.; and Frisch, M. J. (1989). Quadratically convergent simultaneous optimization of wavefunction and geometry. *Int. J. Quantum Chem.*, **36**, 291–303.

Heitler, W. (1954). *The Quantum Theory of Radiation*. 3rd ed.. Cambridge University Press, New York, USA.

Heitler, W. and London, F. (1927). Wechselwirkung neutraler Atome und homöopolare Bindung nach der Quantenmechanik. *Z. Phys.*, **44**, 455–472.

Hellmann, H. (1937). *Einfürung in die Quantum-chemie*. Deuticke, Leipzig.

Hiberty, P. C.; Megret, C.; Song, L.; Wu, W.; and Shaik, S. (2006). Barriers of hydrogen abstraction *vs* halogen exchange: An experimental manifestation of charge-shift bonding. *J. Am. Chem. Soc.*, **128**, 2836–2843.

Hohenberg, P. and Kohn, W. (1964). Inhomogeneous Electronic Gas. *Phys. Rev.*, **136**, 864–871.

Huang, K. (1987). *Statistical Mechanics*. 2nd ed. Wiley & Sons, New York.

Huang, K. (2001). *Introduction to Statistical Physics*. Taylor & Francis, London.

Huang, K. and Yang, C. N. (1957). Quantum-mechanical many-body problem with hard-sphere interaction. *Phys. Rev.*, **105**, 767–775.

Hückel, E. (1931a). Quantentheoretische Beiträge zum Benzolproblem. I. *Z. Physik*, **71**, 204–286.

Hückel, E. (1931b).Quantentheoretische Beiträge zum Benzolproblem. II, *Z. Physik*, **72**, 310–337.

Hugenholtz, N. M. and Pines, D. (1959). Ground-state energy and excitation spectrum of a system of interacting bosons. *Phys. Rev.*, **116**, 489–506.

HyperChem (2002). 7.01, Hypercube, Inc. [Program package, Semiempirical, AM1, Polak-Ribier optimization procedure].

Iczkowski, R. P. and Margrave, J. L. (1961). Electronegativity. *J. Am. Chem. Soc.*, **83**, 3547–3551.

Jackiw, R. (1974). Functional evaluation of the effective potential. *Phys. Rev. D*, **9**, 1686–1701.

Jaksch, D.; Bruder, C.; Cirac, J. I.; Gardiner, C. W.; and Zoller, P. (1998). Cold bosonic atoms in optical lattices. *Phys. Rev. Lett.*, **81**, 3108–3111.

Janak, J. F. (1978). Proof that $\partial E/\partial n_i = \varepsilon$ in density-functional theory. *Phys. Rev. B*, **18**, 7165–7168.

Janke, W.; Pelster, A.; Schmidt, H. J.; and Bachmann, M. (Eds.) (2001). *Fluctuating Paths and Fields—Dedicated to Hagen Kleinert on the Occasion of his 60th Birthday*. World Scientific, Singapore.

Jauch, J. M. and Rohrlich, F. (1976). *The Theory of Photons and Electrons*. 2nd ed. Springer, New York.

Jördens, R.; Strohmaier, N.; Günter, K.; Moritz, H.; and Esslinger, T. (2008). A Mott insulator of fermionic atoms in an optical lattice. *Nature*, **455**, 204–207.

Kadanoff, L. P. (2009). More is the same; mean field theory and phase transitions. *J. Stat. Phys.*, **137**, 777–797.

Kaplan, I. G. (2002). Is the Pauli exclusive principle an independent quantum mechanical postulate? *Int. J. Quantum Chem.*, **89**, 268–276.

Keller, J. (1986). On the Formulation of the Hohenberg-Kohn-Sham Theory. *Int. J. Quantum Chem.*, **20**, 767–768.

Ketterle, W. (2002). Nobel Lecture: When atoms behave as waves: Bose-Einstein condensation and the atom laser. *Rev. Mod. Phys.*, **74**, 1131–1151.

Kim, Y. E. and Zubarev, A. L. (2003). Density-functional theory of bosons in a trap. *Phys. Rev. A*, **67**, 015602.

Kleinert, H. (2003). *Path Integrals in Quantum Mechanics Statistics, Polymer Physics, and Financial Markets*. 3rd ed. World Scientific, Singapore.

Kleinert, H. (2009). *Path Integrals in Quantum Mechanics, Statistics, Polymer Physics, and Financial Markets*. 5th ed. World Scientific, Singapore.

Kleinert, H.; Pelster, A.; and Putz, M. V. (2002). Variational perturbation theory for markov processes. *Phys. Rev. E*, **65**, 066128.

Kleinert, H.; Schmidt, S.; and Pelster, A. (2004). Reentrant phenomenon in the quantum phase transitions of a gas of bosons trapped in an optical lattice. *Phys. Rev. Lett.*, **93**, 160402.

Kleinert, H.; Schmidt, S.; and Pelster, A. (2005). Quantum phase diagram for homogeneous Bose-Einstein condensate. *Ann. Phys. (Leipzig)*, **14**, 214–230.

Koch, W. and Holthausen, M. C. (2002). *A Chemist's Guide to Density Functional Theory*. 2nd ed. Wiley-VCH, Weinheim.

Kohanoff, J. (2006). *Electronic Structure Calculations for Solids and Molecules: Theory and Computational Methods*. Cambridge University Press.

Kohn, W.; Becke, A. D.; and Parr, R. G. (1996). Density Functional Theory of Electronic Structure. *J. Phys. Chem.*, **100**, 12974–12980.

Kohn, W. and Sham, L. J. (1965). Self-Consistent Equations Including Exchange and Correlation Effects. *Phys. Rev.*, **140**, 1133–1138.

Kohn's Nobel Prize (1998). http://nobelprize.org/nobel_prizes/chemistry/laureates/1998/index.html.

Kohout, M.; Pernal, K.; Wagner, F. R.; and Grin, Y. (2004). Electron Localizability Indicator for Correlated Wavefunctions. I. Parallel-Spin Pairs. *Theor. Chem. Acc.*, **112**, 453–459.

Koopmans, T. (1934). Uber die Zuordnung von Wellen Funktionen und Eigenwerter zu den Einzelnen Elektronen Eines Atom. *Physica*, **1**, 104–113.

Kryachko, E. S. and Ludena, E. V. (1991a). Formulation of N-and V-representable density-functional theory. I. Ground States. *Phys. Rev. A*, **43**, 2179–2192.

Kryachko, E. S. and Ludena, E. V. (1991b). Formulation of N-and V-representable density-functional theory. II. Spin-dependent systems. *Phys. Rev. A*, **43**, 2194–2198.

Lam, K. C.; Cruz, F. G.; and Burke, K. (1998). Viral exchange-correlation energy density in Hooke's atom. *Int. J. Quantum Chem.*, **69**, 533–540.

Landau, L. (1941). The theory of superfluidity of Helium II. *J. Phys. USSR*, **5**, 71–90.

Langmuir, I. (1919). The arrangement of electrons in atoms and molecules. *J. Am. Chem. Soc.*, **41**, 868–934.

Lee, C. and Parr, R. G. (1987). Gaussian and other approximations to the first-order density matrix of electronic system, and the derivation of various local-density-functional-theories. *Phys. Rev. A*, **35**, 2377–2383.

Lee, C. and Parr, R. G. (1990). Exchange-Correlation Functional for Atoms and Molecules. *Phys. Rev. A*, **42**, 193–199.

Lee, C.; Yang, W.; and Parr, R. G. (1988). Development of the Colle-Salvetti correlation-energy formula into a functional of the electron density. *Phys. Rev. B*, **37**, 785–789.

Lee, C. and Zhou, Z. (1991). Exchange-energy density functional: Reparametrization of Becke's formula and derivation of second-order gradient correction. *Phys. Rev. A*, **44**, 1536–1539.

Lee, H. and Bartolotti, L. J. (1991). Exchange and exchange-correlation functionals based on the gradient correction of the electron gas. *Phys. Rev. A*, **44**, 1540–1542.

Leggett, A. J. (2001). Bose-Einstein condensation in the alkali gases: Some fundamental concepts. *Rev. Mod. Phys.*, **73**, 307–356.

Lennard-Jones, J. E. (1929). The electronic structure of some diatomic molecules. *Trans. Faraday Soc.*, **25**, 668–686.

Levy, M. (1982). Electron densities in search of Hamiltonians. *Phys. Rev. A*, **26**, 1200–1208.

Levy, M. (1991). Density-functional exchange correlation through coordinate scaling in adiabatic connection and correlation hole. *Phys. Rev. A*, **43**, 4637–4645.

Levy, M.; Ernzerhof, M.; and Gorling, A. (1996). Exact local exchange potential from fock equations at vanishing coupling constant, and $\delta T_c/\delta n$ from wave-function calculations at full coupling constant. *Phys. Rev. A*, **53**, 3963–3973.

Levy, M. and Gorling, A. (1996). Density-functional exchange identity from coordinate scaling. *Phys. Rev. A*, **53**, 3140–3150.

Levy, M. and Perdew, J. (1985). The constrained search formulation of density functional theory. In *Density Functional Methods in Physics*, R. M. Dreizler and da J. Providencia (Eds.). NATO ASI Series B: Physics Vol. 123, Plenum Press, New York, pp. 11–31.

Levy, M. and Perdew J. P. (1993). Tight bound and convexity constraint on the exchange-correlation-energy functional in the low-density limit, and other formal tests of generalized-gradient approximations. *Phys. Rev. B*, **48**, 11638–11645.

Lewenstein, M.; Sanpera, A.; Ahufinger, V.; Damski, B.; Sen, A.; and Sen, U. (2007). Ultracold atomic gases in optical lattices: Mimicking condensed matter physics and beyond. *Adv. Phys.* , **56**, 243–379.

Lewis, G. N. The atom and the molecule. *J. Am. Chem. Soc.* **1916**, *38*, 762–785.

Liberman, D. A.; Albritton, J. R.; Wilson, B. G.; and Alley, W. E. (1994). Self-consistent-field calculations of atoms and ions using a modified local-density approximation. *Phys. Rev. A*, **50**, 171–176.

Lieb, E. H. (1976). The stability of matter. *Rev. Mod. Phys.*, **48**, 553–569.

Lieb, E. H. (1981). Thomas-Fermi and related theories of atoms and molecules. *Rev. Mod. Phys.*, **53**, 603–641.

Liu, S. (1996). Local-density approximation, hierarchy of equations, functional expansion, and adiabatic connection in current-density-functional theory. *Phys. Rev. A*, **54**, 1328–1336.

Liu, S.; Nagy, A.; and Parr, R. G. (1999). Expansion of the density-functional energy components e_c and t_c in terms of moments of the electron density. *Phys. Rev. A*, **59**, 1131–1134.

Lohr, L. L., Jr. and Pyykkö, P. (1979). Relativistically parameterized extended Hückel theory. *Chem. Phys. Lett.*, **62**, 333–338.

London, F. (1938). On the Bose-Einstein condensation. *Phys. Rev.*, **54**, 947–954.

Löwdin, P. O. (1995a). Quantum theory of many-particle systems. I. Physical interpretations by means of density matrices, natural spin-orbitals, and convergence problems in the method of configurational interaction. *Phys. Rev.*, **97**, 1474–1489.

Löwdin, P. O. (1955b). Quantum theory of many-particle systems. II. Study of the ordinary Hartree-Fock approximation. *Phys. Rev.*, **97**, 1474–1489.

Löwdin, P. O. (1955c). Quantum theory of many-particle systems. III. Extension of the Hartree-Fock scheme to include degenerate systems and correlation effects. *Phys. Rev.*, **97**, 1509–1520.

MacDonald, J. K. L. (1934). On the Modified Ritz Variation Method. *Phys. Rev.*, **46**, 828.

Maggiora, G. M. and Mezey, P. G. (1999). A fuzzy-set approach to functional. *Int. J. Quantum Chem.*, **74**, 503–514.

Manoli, S. D. and Whitehead, M. A. (1988). Generalized-exchange local-spin-density-functional theory: Calculation and results for non-self-interaction-corrected and self-interaction-corrected theories. *Phys. Rev. A*, **38**, 3187–3199.

March, N. H. (1991). *Electron Density Theory of Many-Electron Systems*. Academic Press, New York.

March, N. H. (1993). The ground-state energy of atomic and molecular ions and its variation with the number of elections. *Structure and Bonding*, **80**, 71–86.

Matito, E.; Silvi, B.; Duran, M.; and Solà, M. (2006). Electron localization function at the correlated level. *J. Chem. Phys.*, **125**, 024301.

Mermin, N. D. (1965). Thermal Properties of the Inhomogeneous Electron Gas. *Phys. Rev.*, **137**, A1441–A1443.

Mezey, P. G. (1993). *Shape in Chemistry: An Introduction to Molecular Shape and Topology*. VCH Publishers, New York, USA.

Mineva, T.; Sicilia, E.; and Russo, N. (1998). Density functional approach to hardness evaluation and its use in the study of the maximum hardness principle. *J. Am. Chem. Soc.*, **120**, 9053–9058.

Modugno, G.; Ferrari, G.; Roati, G.; Brecha, R.J.; Simoni, A.; and Inguscio, M. (2001). Bose-Einstein condensation of Potassium atoms by sympathetic cooling. *Science*, **294**, 1320–1322.

Mortier, W. J.; Genechten, K. V.; and Gasteiger, J. (1985). Electronegativity equalization: Application and parametrization. *J. Am. Chem. Soc.*, **107**, 829–835.

Murphy, D. R. (1981). Sixth-order term of the gradient expansion of the kinetic-energy density functional. *Phys. Rev. A*, **24**, 1682–1688.

Nagy, A. (1998). Kohn-Sham Equations for multiplets. *Phys. Rev. A*, **57**, 1672–1677.

Nagy, A.; Liu, S.; and Parr, R. G. (1999). Density-functional formulas for atomic electronic energy components in terms of moments of the electron density. *Phys. Rev. A*, **59**, 3349–3354.

Nalewajski, R. F. (1998). Kohn-Sham description of equilibria and charge transfer in reactive systems. *Int. J. Quantum Chem.*, **69**, 591–605.

Neal, H. L. (1998). Density functional theory of one-dimension two-particle systems. *Am. J. Phys.*, **66**, 512–516.

Nunes, G. S. (1999). Density functional theory of the inhomogeneous Bose–Einstein condensate. *J. Phys. B: At. Mol. Opt. Phys.*, **32**, 4293–4299.

Oelke, W. C. (1969). *Laboratory Physical Chemistry*. Van Nostrand Reinhold Company, New York, USA.

Ou-Yang, H. and Levy, M. (1990). Nonuniform coordinate scaling requirements in density-functional theory. *Phys. Rev. A*, **42**, 155–159.

Ou-Yang, H. and Levy, M. (1991). Theorem for functional derivatives in density-functional theory. *Phys. Rev. A*, **44**, 54–58.

Paiva, T.; Scalettar, R.; Randeria, M.; and Trivedi, N. (2010). Fermions in 2D optical lattices: Temperature and entropy scales for observing antiferromagnetism and superfluidity. *Phys. Rev. Lett.*, **104**, 066406.

Pariser, R. and Parr, R. (1953a). A semi-empirical theory of the electronic spectra and electronic structure of complex unsaturated molecules. I. *J. Chem. Phys.*, **21**, 466–471.

Pariser, R. and Parr, R. (1953b). A semi-empirical theory of the electronic spectra and electronic structure of complex unsaturated molecules. II. *J. Chem. Phys.*, **21**, 767–776.

Park, J. H. and Kim, S. W. (2010). Thermodynamic instability and first-order phase transition in an ideal Bose gas. *Phys Rev. A*, **81**, 063636.

Parr, R. G. (1980). Density functional theory of atoms and molecules. In *Horizons of Quantum Chemistry*, Fukui K., Pullman B., Eds., Reidel: Dordrecht-Boston-London, **1980**, 5-15.

Parr, R. G. (1983). Density functional. *Annu. Rev. Phys. Chem.*, **34**, 631–656.

Parr, R. G.; Donnelly, R. A.; Levy, M.; and Palke, W. E. (1978). Electronegativity: The density functional viewpoint. *J. Chem. Phys.*, **68**, 3801–3808.

Parr, R. G. and Gázquez, J. L. (1993). Hardness functional. *J. Phys. Chem.*, **97**, 3939–3940.

Parr, R. G. and Pearson, R. G. (1983). Absolute hardness: companion parameter to absolute electronegativity. *J. Am. Chem. Soc.*, **105**, 7512–7516.

Parr, R. G. and Yang, W. (1984). Density functional approach to the frontier electron theory of chemical reactivity. *J. Am. Chem. Soc.*, **106**, 4049–4050.

Parr, R. G. and Yang, W. (1989). *Density Functional Theory of Atoms and Molecules*. Oxford University Press, New York.

Pauling, L. (1931a). Quantum mechanics and the chemical bond. *Phys. Rev.*, **37**, 1185–1186.

Pauling, L. (1931b). The nature of the chemical bond. I. Application of results obtained from the quantum mechanics and from a theory of paramagnetic susceptibility to the structure of molecules. *J. Am. Chem. Soc.*, **53**, 1367–1400.

Pauling, L. (1931c). The nature of the chemical bond II. The one-electron bond and the three-electron bond. *J. Am. Chem. Soc.*, **53**, 3225–3237.

Pauling L. (1960). *The Nature of the Chemical Bond.* Cornell University Press, Ithaca, New York.

Pearson, R. G. (1972). [Quantitative evaluation of the HSAB (hard-soft acid-base) concept]. Reply to the paper by Drago and Kabler. *Inorg. Chem.*, **11**, 3146.

Pearson, R. G. (1973). *Hard and Soft Acids and Bases*. Dowden, Hutchinson & Ross, Stroudsberg (PA).

Pearson, R. G. (1985). Absolute electronegativity and absolute hardness of Lewis acids and bases. *J. Am. Chem. Soc.*, **107**, 6801–6806.

Pearson, R. G. (1988). Absolute electronegativity and hardness: Application to inorganic chemistry. *Inorg. Chem.*, **27**, 734–740.

Pearson, R. G. (1990). Hard and soft acids and bases—the evolution of a chemical concept. *Coord. Chem. Rev.*, **100**, 403–425.

Pearson, R. G. (1997). *Chemical Hardness*. Wiley-VCH, Weinheim.

Penrose, O. and Osanger, L. (1956). Bose-Einstein condensation and liquid Helium. *Phys. Rev.*, **104**, 576–584.

Perdew, J. P. (1986). Density-functional approximation for the correlation energy of the inhomogeneous electron gas. *Phys. Rev. B*, **33**, 8822–8824.

ibid Erratum, **34**, 7406.

Perdew, J. P.; Burke, K.; and Ernzerhof, M. (1996). Generalized gradient approximation made simple. *Phys. Rev. Lett.*, **77**, 3865–3868.

Perdew, J. P.; Chevary, J. A.; Vosko, S. H.; Jackson, K. A.; Pederson, M. R.; Singh, D. J.; and Fiolhais, C. (1992). Atoms, molecules, solids, and surfaces: Applications of the generalized gradient approximation for exchange and correlation. *Phys. Rev. B*, **46**, 6671–6687.

Perdew, J. P.; Ernzerhof, M.; Zupan, A.; and Burke, K. (1998). Nonlocality of the density functional for exchange and correlation: Physical origins and chemical consequences. *J. Chem. Phys.*, **108**, 1522–1531.

Perdew, J. P. and Yue, W. (1986). Accurate and simple density functional for the electronic exchange energy: Generalized gradient approximation. *Phys. Rev. B*, **33**, 8800–8802.

Perdew, J. P. and Zunger, A. (1981). Self-Interaction Correction to Density-Functional Approximations for Many-Electron System. *Phys. Rev. B*, **23**, 5048–5079.

Pethick, C. J. and Smith, H. (2002). *Bose-Einstein Condensation in Dilute Gases*. Cambridge, University Press.

Pitaevskii, L. and Stringari, S. (2003). *Bose-Einstein Condensation*, Clarendon Press, Oxford.

Pitaevskii, L. P. (1961). Vortex lines in an imperfect Bose gas. *Zh. Eksp. Teor. Fiz.*, **40**, 646–651 [*Sov. Phys. JETP*, **13**, 451–454].

Pople, J. A. (1953). Electron interaction in unsaturated hydrocarbons. *Trans. Faraday Soc.*, **49**, 1375–1385.

Pople, J. A.; Binkley, J. S. and Seeger, R. (1976). Theoretical models incorporating electron correlation. *Int. J. Quantum Chem.*, **10**, 1–19.

Pople, J. A. and Nesbet, R. K. (1954). Self-consistent orbitals for radicals. *J. Chem. Phys.*, **22**, 571–572.

Preuss, H. (1969). *Quantenchemie fuer Chemiker*. Verlag Chemie, Weinheim, Germany.

Putz, M. V. (2003). *Contributions within Density Functional Theory with Applications to Chemical Reactivity Theory and Electronegativity*. Dissertation Com., Parkland, Florida.

Putz, M. V. (2005). Markovian approach of the electron localization functions. *Int. J. Quantum Chem.*, **105**, 1–11.

Putz, M. V. (2006). Systematic Formulation for Electronegativity and Hardness and Their Atomic Scales within Density Functional Softness Theory. *Int. J. Quantum Chem.*, **106**, 361–386.

Putz, M. V. (2007a). Unifying Absolute and Chemical Electronegativity and Hardness Density Functional Formulations through the Chemical Action Concept. In *Progress in Quantum Chemistry Research*, Erik O. Hoffman (Ed.). Nova Science Publishers Inc., New York, pp. 59–121.

Putz, M. V. (2007b). Can quantum-mechanical description of chemical bond be considered complete? In *Quantum Chemistry Research Trends*, Mikas P. Kaisas (Ed.). Nova Science Publishers Inc., New York, Expert Commentary.

Putz, M. V. (2008a). Density functionals of chemical bonding. *Int. J. Mol. Sci.*, **9**, 1050–1095.

Putz, M. V. (2008b). Maximum hardness index of quantum acid-base bonding. *MATCH Commun. Math. Comput. Chem.*, **60**, 845–868.

Putz M. V. (2008c). *Absolute and Chemical Electronegativity and Hardness*. Nova Publishers Inc., New York, USA.

Putz, M. V. (2008d). The chemical bond: Spontaneous symmetry–breaking approach. *Symmetr. Cult. Sci.*, **19**, 249–262.

Putz, M. V. (2008e). Density Functional Theory (in Romanian). Chapter 4, In *Quantum Information in Multiparticle Systems* (in Romanian), D. Popov, M. V. Putz, I. Zaharia (Eds.). Polytechnic University Publishing House, Timişoara.

Putz, M. V. (2009a). Levels of a unified theory of chemical interaction. *Int. J. Chem. Model.*, **1**, 141–147.

Putz, M. V. (2009b). Chemical action and chemical bonding. *J. Mol. Structure: THEOCHEM*, **900**, 64–70.

Putz M. V. (2009c). Path integrals for electronic densities, reactivity indices, and localization functions in quantum systems. *Int. J. Mol. Sci.* **2009 (c)**, *10*, 4816–4940.

Putz M. V. (2010a). Fulfilling the Dirac's promises on quantum chemical bond; In *Quantum Frontiers of Atoms and Molecules*, M.V. Putz (Ed.), NOVA Science Inc., New York.

Putz M. V. (2010b). The bondons: The quantum particles of the chemical bond. *Int. J. Mol. Sci.*, **11**, 4227–4256.

Putz M. V. (2010c). Beyond quantum nonlocality: chemical bonding field. *Int. J. Environ. Sci.*, **1**, 25–31.

Putz M. V. (2011a). Electronegativity and Chemical Hardness: Different Patterns in Quantum Chemistry. *Curr. Phys. Chem.*, **1**, 111–139.

Putz M. V. (2011b). Quantum Parabolic Effects of Electronegativity and Chemical Hardness on Carbon p-Systems. In *Carbon Bonding and Structures: Advances in Physics and Chemistry*, Mihai V. Putz (Ed.), Springer Verlag, London.

Putz M. V. (2011c). Chemical action concept and principle. *MATCH Commun. Math. Comput. Chem.*, **66**, 35–63.

Putz M. V. (2011d). Conceptual Density Functional Theory: From Inhomogeneous Electronic Gas to Bose-Einstein Condensates. In *Chemical Information and Computational Challenges in 21st Century,* Mihai V. Putz (Ed.), *Series „Chemistry Research and Applications" & „Chemical Engineering Methods and Technology"*, NOVA Science Publishers, Inc., New York, USA.

Putz M. V. (2011e). Hidden Side of Chemical Bond: The Bosonic Condensate. In *Advances in Chemistry Research. Volume 10*, James C. Taylor (Ed.), *Series „Advances in Chemistry Research"*, NOVA Science Publishers, Inc., New York, USA, Chapter 8.

Putz M. V. (2011f). Developing density functional theory for Bose-Einstein condensates. The case of chemical bonding. *AIP Conference Proceeeding Series* (dedicated to the 8th International Conference of Computational Methods in Sciences and Engineering ICCMSE-2010, Kos, Greece, October 3–8), in press.

Putz M. V. (2012). *First Principles of Quantum Chemistry ~ Vol. I. ~ General Quantum Theory and Observability*. NOVA Science Publishers, Inc., New York, USA.

Putz, M. V. and Chiriac, A. (2008). Quantum perspectives on the nature of the chemical bond. In *Advances in Quantum Chemical Bonding Structures*, M. V. Putz (Ed.). Transworld Research Network, Kerala, India, pp. 1–43.

Putz, M. V.; Chiriac, A. and Mracec, M. (2001). Foundations for a theory of the chemical field. II. The chemical action. *Rev. Roum. Chimie*, **46**, 1175–1181.

Putz, M. V.; Russo, N.; and Sicilia, E. (2003). Atomic radii scale and related size properties from density functional electronegativity formulation. *J. Phys. Chem. A*, **107**, 5461–5465.

Putz, M. V.; Russo, N.; and Sicilia, E. (2004). On the application of the HSAB principle through the use of improved computational schemes for chemical hardness evaluation. *J. Comput. Chem.*, **25**, 994–1003.

Putz, M. V.; Russo, N.; and Sicilia, E. (2005). About the Mulliken electronegativity in DFT. *Theor. Chem. Acc.*, **114**, 38–45.

Pyykkö, P. (1978). Relativistic quantum chemistry. *Adv. Quantum Chem.*, **11**, 353–409.

Pyykkö, P. (2000). *Relativistic theory of atoms and molecules. III. A Bibliography 1993–1999*. Lecture Notes in Chemistry, Springer-Verlag: Berlin, Germany, Volume 76.

Pyykkö, P. and Zhao, L. B. (2003). Search for effective local model potentials for simulation of QED effects in relativistic calculations. *J. Phys. B*, **36**, 1469–1478.

Rasolt, M. and Geldart, D. J. W. (1986). Exchange and correlation energy in a nonuniform fermion fluid. *Phys. Rev. B*, **34**, 1325–1328.

Richard, M. M. (2004). *Electronic Structure: Basic Theory and Practical Methods*. Cambridge University Press, New York.

Romera, E. and Dehesa, J. S. (1994). Weizsäcker energy of many-electron systems. *Phys. Rev. A*, **50**, 256–266.

Roothaan, C. C. J. (1951). New developments in molecular orbital theory. *Rev. Mod. Phys.*, **23**, 69–89.

Rothel, S. and Pelster, A. (2007). Density and stability in ultracold dilute boson-fermion mixtures. *Eur. Phys. J. B*, **59**, 343–356.

Ruedenberg, K. (1962). The physical nature of the chemical bond. *Rev. Mod. Phys.*, **34**, 326–376.

Runge, E. and Gross, E. K. U. (1984). Density-functional theory for time-dependent systems. *Phys. Rev. Lett.*, **52**, 997.

Sanderson, R. T. (1988). Principles of electronegativity. *J. Chem. Educ.*, **65**, 112–119.

Santos, J. C.; Tiznado, W.; Contreras, R.; and Fuentealba, P. (2000). *Sigma-pi* separation of the electron localization function and aromaticy. *J. Chem. Phys.*, **120**, 1670–1673.

Savin, A.; Preuss, H.; and Stoll, H. (1987). Non-local effects on atomic and molecular correlation energies studies with a gradient-corrected density functional. In *Density Matrices and Density Functionals*. R. Erhahy and V. H. Smith (Eds.). Reidel Publishing Company, pp. 457–465.

Savin, A.; Stoll, H.; and Preuss, H. (1986). An application of correlation energy density functionals to atoms and molecules. *Theor Chim Acta*, **70**, 407–419.

Savin, A; Wedig, U.; Preuss, H.; and Stoll, H. (1984). Molecular correlation energies obtained with a nonlocal density functional. *Phys. Rev. Lett.*, **53**, 2087–2089.

Scemama, A.; Chaquin, P.; and Caffarel, M. (2000). Electron pair localization function: A practical tool to visualize electron localization in molecules from quantum Monte Carlo data. *J. Chem. Phys.*, **121**, 1725–1735.

Schneider, U.; Hackermüller, L.; Will, S.; Best, Th.; Bloch, I.; Costi, T. A.; Helmes, R. W.; Rasch, D.; and Rosch, A. (2008). Metallic and insulating phases of repulsively interacting fermions in a 3D optical lattice. *Science*, **322**, 1520–1525.

Schrödinger, E. (1926). An undulatory theory of the mechanics of atoms and molecules. *Phys. Rev.*, **28**, 1049–1070.

Schweber, S. S. (1964). *An Introduction to Relativistic Quantum Field Theory*. Harper and Row, New York.

Seidl, M.; Perdew, J. P.; and Levy, M. (1999). Strictly correlated electrons in density-functional theory. *Phys. Rev. A*, **59**, 51–54.

Sen, K. D. and Jørgensen, C. K. (Eds.) (1987). Electronegativity, *Structure and Bonding*, Vol. 66, Springer Verlag, Berlin.

Sen, K. D. and Mingos, D. M. P. (Eds.) (1993). Chemical Hardness, *Structure and Bonding*, Vol. 80, Springer Verlag, Berlin.

Senatore, G. and March, N. H. (1994). Recent progress in the field of electron correlation. *Rev. Mod. Phys.*, **66**, 445–479.

Senet, P. (1996). Nonlinear electronic responses, Fukui functions and hardnesses as functionals of the ground-state electronic density. *J. Chem. Phys.*, **105**, 6471–6490.

Senet, P. (1997). Kohn-Sham orbital formulation of the chemical electronic responses, including the hardness. *J. Chem. Phys.*, **107**, 2516–2525.

Sholl, D. and Steckel, J. A. (2009). *Density Functional Theory: A Practical Introduction*, Wiley-Interscience, Hoboken.

Silvi, B. (2003). The spin-pair compositions as local indicators of the nature of the bonding. *J. Phys. Chem.*, **107**, 3081–3085.

Silvi, B. and Gatti, C. (2000). Direct space representation of the metallic bond. *J. Phys. Chem. A*, **104**, 947–953.

Silvi, B. and Savin, A. (1994). Classification of the chemical bonds based on topological analysis of electron localization functions. *Nature*, **371**, 683–686.

Slater, J. C. (1928). The self consistent field and the structure of atoms. *Phys. Rev.*, **32**, 339–348.

Slater, J. C. (1929). The theory of complex spectra. *Phys. Rev.*, **34**, 1293–1322.

Slater, J. C. (1951). A Simplification of the Hartree-Fock Method, *Phys. Rev.*, **81**, 385–390.

Slater, J. C. (1963). *Theory of Molecules and Solids, Vol. 1, Electronic Structure of Molecules*. Mc-Graw-Hill, New York.

Slater, J. C. (1974). *Quantum Theory of Molecules and Solids, Vol. 4, The Self-Consistent Field for Molecules and Solids*. McGraw-Hill, New York.

Slater, J. C. and Johnson K. H. (1972). Self-consistent-field Xα cluster method for polyatomic molecules and solids. *Phys. Rev. B*, **5**, 844–853.

Slater, J. C. and Kirkwood, J. G. (1931). The van der Waals forces in gases. *Phys Rev.*, **37**, 682–697.

Snijders, J. G. and Pyykkö, P. (1980). Is the relativistic contraction of bond lengths an orbital contraction effect? *Chem. Phys. Lett.*, **75**, 5–8.

Soncini, A. and Lazzeretti, P. (2003). Nuclear spin-spin coupling density functions and the Fermi hole. *J. Chem. Phys.*, **119**, 1343–1349.

Szekeres, Z.; Exner, T.; and Mezey, P. G. (2005). Fuzzy fragment selection strategies, basis set dependence and HF–DFT comparisons in the applications of the ADMA method of macromolecular quantum chemistry. *Int. J. Quantum Chem.,* **104**, 847–860.

Tachibana, A. (1987). Density functional rationale of chemical reaction coordinate. *Int. J. Quantum Chem.,* **21**, 181–190.

Tachibana, A.; Nakamura, K.; Sakata, K.; and Morisaki, T. (1999). Application of the regional density functional theory: the chemical potential inequality in the HeH^+ system. *Int. J. Quantum Chem.,* **74**, 669–679.

Tachibana, A. and Parr, R. G. (1992). On the redistribution of electrons for chemical reaction systems. *Int. J. Quantum. Chem.,* **41**, 527–555.

Taut, M. (1996). Generalized gradient correction for exchange: deduction from the oscillator model. *Phys. Rev. A*, **53**, 3143–3150.

Thomas, L. H. (1927). The calculation of atomic fields. *Proc. Cambridge Phil. Soc.*, **23**, 542–548.

Thomson, J. J. (1921). On the structure of the molecule and chemical combination. *Philos. Mag.*, **41**, 510–538.

Torrent-Sucarrat, M. and Solà, M. (2006). Gas-phase structures, rotational barriers, and conformational properties of hydroxyl and mercapto derivatives of cyclohexa-2,5-dienone and cyclohexa-2,5-dienthione. *J. Phys. Chem. A*, **110**, 8901–8911.

Tozer, D. J. and Handy, N. C. (1998). The development of new exchange-correlation functionals. *J. Chem. Phys.*, **108**, 2545–2555.

Valone, S. M. and Capitani, J. F. (1981). Bound excited states in density-functional theory. *Phys. Rev. A,* **23**, 2127–2133.

van Oosten, D.; van der Straten, P.; and Stoof, H. T. C. (2001). Quantum phases in an optical lattice. *Phys. Rev. A*, **63**, 053601.

Vetter, A. (1997). *Density Functional Theory for BEC* (in German). Thesis, Institute of Theoretical Physics, Bayerische Julius-Maximilians University, Würzburg.

von Barth, U. and Hedin, L. (1972). A local exchange-correlation potential for the spin polarized case. I. *J. Phys. C: Solid State Phys.,* **5**, 1629–1642.

Vosko, S. J.; Wilk, L.; and Nusair, M. (1980). Accurate spin-dependent electron liquid correlation energies for local spin density calculations: a critical analysis. *Can. J. Phys.*, **58**, 1200–1211.

Wang, Y. and Perdew, J. P. (1989). Spin scaling of the electron-gas correlation energy in the high-density limit. *Phys. Rev. B*, **43**, 8911–8916.

Wang, Y.; Perdew, J. P.; Chevary, J. A.; Macdonald, L. D.; and Vosko, S. H. (1990). Exchange potentials in density-functional theory. *Phys. Rev. A*, **41**, 78–85.

Whitney, C. K. (2007). Relativistic dynamics in basic chemistry. *Found. Phys.*, **37**, 788–812.

Whitney, C. K. (2008a). Closing in on chemical bonds by opening up relativity theory. *Int. J. Mol. Sci.*, **9**, 272–298.

Whitney, C. K. (2008b). Single-electron state filling order across the elements. *Int. J. Chem. Model.*, **1**, 105–135.

Whitney, C. K. (2009). Visualizing electron populations in atoms. *Int. J. Chem. Model.*, **1**, 245–297.

Whitney, C. K. (2010). The algebraic chemistry of molecules and reactions. In *Quantum Frontiers of Atoms and Molecules*. M. V. Putz (Ed.). NOVA Science Inc., New York..

Wilson, L. C.; Levy, M. (1990). Nonlocal Wigner-like correlation-energy density functional through coordinate scaling. *Phys. Rev. B*, **41**, 12930–12932.

Yang, W. and Parr, R. G. (1985). Hardness, softness, and the Fukui function in the electronic theory of metals and catalysis. *Proc. Natl. Acad. Sci. U.S.A.*, **82**, 6723–6726.

Yang, W.; Parr, R. G.; and Pucci, R. (1984). Electron density, Kohn–Sham frontier orbitals, and Fukui functions. *J. Chem. Phys.*, **81**, 2862–2863.

Yukalov, V. I. (2004). Principal Problems in Bose–Einstein condensation of dilute gases. *Laser Phys. Lett.*, **1**, 435–461.

Zagrebnov V. A. and Bru, J. B. (2001). The Bogoliubov model of weakly imperfect Bose gas. *Phys. Reports*, **350**, 291–434.

Index

D

N

O

P

Q

R

Milton Keynes UK
Ingram Content Group UK Ltd.
UKHW031146141024
449569UK00024B/1023

9 781774 632031